Face Biometrics for Personal Identification
Multi-Sensory Multi-Modal Systems

Riad I. Hammoud • Besma R. Abidi •
Mongi A. Abidi (Eds.)

Face Biometrics for Personal Identification

Multi-Sensory Multi-Modal Systems

With 118 Figures, 76 in Color and 24 Tables

 Springer

Dr. Riad I. Hammoud
Delphi Electronics & Safety
World Head Quarters
P.O. Box, M/C: E110 9005
Kokomo, IN 46904-9005
USA
e-mail: riad.hammoud@delphi.com

Dr. Besma R. Abidi
The University of Tennessee
Electrical and Computer Engineering
College of Engineering
Ferris Hall 317
Knoxville, TN 37996-2100
USA
e-mail: besma@utk.edu

Mongi A. Abidi
The University of Tennessee
Electrical and Computer Engineering
College of Engineering
Ferris Hall 328
Knoxville, TN 37996-2100
USA
e-mail: abidi@utk.edu

Library of Congress Control Number: 2006936971

ISBN-10 3-540-49344-1 Springer Berlin Heidelberg New York
ISBN-13 978-3-540-49344-0 Springer Berlin Heidelberg New York

This work is subject to copyright. All rights are reserved, whether the whole or part of the material is concerned, specifically the rights of translation, reprinting, reuse of illustrations, recitation, broadcasting, reproduction on microfilm or in any other way, and storage in data banks. Duplication of this publication or parts thereof is permitted only under the provisions of the German Copyright Law of September 9, 1965, in its current version, and permission for use must always be obtained from Springer. Violations are liable to prosecution under the German Copyright Law.

Springer is a part of Springer Science+Business Media.

springer.com

© Springer-Verlag Berlin Heidelberg 2007

The use of general descriptive names, registered names, trademarks, etc. in this publication does not imply, even in the absence of a specific statement, that such names are exempt from the relevant protective laws and regulations and therefore free for general use.

Typesetting by editors and SPi using a Springer LaTeX macro package
Cover design: WMX-Design GmbH, Heidelberg

Printed on acid-free paper   SPIN 11580904   62/3100/SPi   5 4 3 2 1 0

To

Ja'far al-Sadiq,

Bahjat and his friends,

Samy and Ramzy.

Preface

From time immemorial, the security realm and personal identification task had shown progress by employing technological means like *secret knowledge* as passwords and Personal Identification Numbers, and by using *personal possessions* as identity cards and Radio Frequency Identification chips. As opposed to these means which are generally easy targets for circulation and fraud, biometric traits (or modalities) like facial geometry, iris, voice timbre, and biological DNA are universal, difficult to copy, and for most, consistent over time with no expiration date and comfortable to use. The purpose of this book is to provide an up-to-date ample coverage of theoretical and experimental state-of-the-art work as well as new trends and directions in the biometrics field.

The *biometrics field* – the science of measuring physical properties of human beings – has marked a substantial leap over the past two decades. It will continue to climb, as a result of a strong demand from both the private sector and governmental agencies, on an ever ramping curve until practical objectives are achieved in terms of high accuracy, ease of use, and low-cost. As the cost of biometric sensors – visible, multispectral and thermal imagers, microphones, capacitive, pressure and motion sensors – continues to sink due to higher demand, biometric systems have the tendency to employ more than a single sensor to capture and identify an individual upon as many as nonredundant biometric traits.

While this book covers a range of biometric traits including facial geometry, 3D ear form, fingerprints, vein structure, voice, and gait, its main emphasis is placed on *multisensory* and *multimodal* face biometrics algorithms and systems. By "multisensory" we refer to combining data from two or more biometric sensors, such as synchronized reflectance-based and temperature-based face images. Likewise, by "multimodal" biometrics, we refer to fusing two or more biometric modalities, like face images and voice timber. These two multisensory and multimodal aspects, as they pertain to face Biometrics, are covered in details in the course of this book. Reported experimental results support the idea that adequate fusion of complementary biometric data is a step in the right direction to remedy to the limitations of existing uni-sensor and uni-modal biometric systems.

This book contains four distinctive parts and a brief introduction chapter. The latter places the reader into the context of the book's terminologies, motivations, addressed problems, and summary of solutions portrayed in the remaining 12 chapters. The basic fact that a high performance multibiometrics system is hard to achieve without maximizing the performance of each sensor and each modality independently, led us to reserve all four chapters of the first part to address *new and emerging face biometrics*. Emphasis is placed on biometric systems where single sensor and single modality are employed in challenging imaging conditions, like high magnification, illumination changes, head pose variations, and facial complexity over age progression. Higher levels of illumination tolerance were achieved by merging correlation filters with linear subspace methods. Pose variations were addressed in a multisample approach, integrating motion, lighting and shape in video data. Space and time variations were addressed in two separate chapters. One chapter looked at the effects of long distance and large zoom on the recognition of faces with the design of quality measures and restoration algorithms to remedy to degradations caused by magnification blur. The other chapter focused on time lapses between gallery and probe face images using a Bayesian age difference classifier to determine the age difference between pairs of face images. The second part on *multisensory face biometrics* is composed of three chapters. It addresses the personal identification task in challenging variable illuminations and outdoor operating scenarios by employing visible and thermal sensors. Various fusion approaches to counter illumination challenges are addressed and their performances compared. By fusing thermal and visible face data in the presence of eyeglasses and large pose variations, higher recognition rates are achieved as compared to the use of single sensors. Visual imagery is even mixed with physiological traits from vein structure emanating from thermal imagery. Thermal Minutia Points on the vein structure are shown to be specific to each individual and used for recognition in combination with the visible facial data. The third part of the book focuses on *multimodal face biometrics* by integrating voice, ear, and gait modalities with facial data. It presents numerous novel methodologies to combine 3D and 2D face data with ear forms, face profile images with gait silhouette templates, and finally speaker voice with face images. The fusion technique of these last two modalities is presented in chapter nine. Experimental validation is performed in the framework of a user verification system on a handheld cell device. It showed that this bimodal biometrics system significantly outperformed, in terms of the Equal Error Rate, voice-based, and face-based biometrics systems. The last part presents two generic chapters on multibiometrics fusion methodologies and performance prediction techniques. An excellent rendering of the different classes and advantages of multibiometrics systems is given and five different levels of fusion are discussed. The computation of performance bounds of multibiometrics systems is also formulated by modeling the realizations of biometric signatures as those of a random process and using the fidelity and desired error rate of the system as guides.

This practical reference offers students and software engineers a thorough understanding of how some core low-level building blocks of a multibiometric system are implemented. It contains enough material to fill a two-semester upper-division or advanced graduate course in the field of biometrics, face recognition, data fusion and human–computer interaction in general. The University of Tennessee's Imaging Robotics and Intelligent Systems Laboratory is already planning to assign selected chapters for a graduate class on multisensory data fusion and biometrics systems. Scientist and teachers will find in-depth coverage of recent state-of-the-art face biometrics algorithms and experiments. Moreover, the book helps readers of all levels understand the motivations, activities, trends, and directions of researchers and engineers in the biometrics field in today's market, and offers them a view of the future of this rapidly evolving technological area. However, reading through the details of the algorithms in this book requires a basic knowledge in biometrics, pattern recognition, computer vision, signal processing, and statistics.

This effort could not have been achieved without the excellent scientific contributions made by a number of pioneering scientists and experts in the biometrics field. We are thankful for their participation and the support of their institutions: O. Arandjelovic and R. Cipolla with *University of Cambridge, UK*; V. Asari and S. Gundimada with *Old Dominion University*; B. Bhanu, A.R. Chowdhury, Y. Xu, and X. Zhou with *University of California, Riverside*; P. Buddharaju, I. Kakadiaris, N. Murtuza, G. Passalis, I. Pavlidis, T. Theoharis, and G. Toderici with *University of Houston*; R. Chellappa and N. Ramanathan with *University of Maryland*; J. Han with *Lawrence Berkeley National Laboratory*; T.J. Hazen and A. Park with *Massachusetts Institute of Technology*; B. Heisele with *Honda Research Institute*; A. K. Jain with *Michigan State University*; D.P. Khosla, V. Kumar, and M. Savvides with *Carnegie Mellon University*; J. Ming with *Queen's University, Belfast*; J.A. O'Sullivan with *Washington University*; A. Ross and N.A. Schmid with *West Virginia University*; E. Weinstein with *New York University*; L.B. Wolff with *Equinox Corporation*; and Y. Yao with *The University of Tennessee*. Their expertise, contributions, feedbacks, and reviewing added significant value to this groundbreaking resource.

We are also very grateful to Diego Socolinsky, Andrea Salgian, and Andreas Koschan for their valuable corrections and comments. We would like to extend thanks to all folks at Springer-Verlag, and in particular to Christoph Baumann and Dieter Merkle for their warm support.

August 14, 2006

Riad I. Hammoud
Delphi Electronics & Safety,

Besma R. Abidi
The University of Tennessee,

Mongi A. Abidi
The University of Tennessee

Contents

1 Introduction
Lawrence B. Wolff 1
1.1 Motivations, General Addressed Problems, Trends, Terminologies 1
1.2 Inside This Book 2
1.3 Evaluation of This Book 5

Part I Space/Time Emerging Face Biometrics

2 Pose and Illumination Invariant Face Recognition Using Video Sequences
Amit K. Roy-Chowdhury and Yilei Xu 9
2.1 Introduction 9
    2.1.1 Overview of the Approach 9
    2.1.2 Relation to Previous Work 10
    2.1.3 Organization of the Chapter 13
2.2 Integrating Illumination and Motion Models in Video 13
2.3 Learning Joint Illumination and Motion Models from Video 16
    2.3.1 Algorithm 17
    2.3.2 Handling Occlusions 17
2.4 Face Recognition From Video 18
2.5 Experimental Results 20
    2.5.1 Tracking and Synthesis Results 20
    2.5.2 Face Recognition Results 22
2.6 Conclusions 25

3 Recognizing Faces Across Age Progression
Narayanan Ramanathan and Rama Chellappa 27
3.1 Introduction 27
    3.1.1 Previous work on Age Progression 27
    3.1.2 Problem Statement 30
3.2 Age Difference Classifier 31
    3.2.1 Bayesian Framework 32
    3.2.2 Experiments and Results 35
3.3 Facial Similarity 36

3.4	Craniofacial Growth Model	38
	3.4.1 Model Computation: An Optimization Problem	39
3.5	Conclusions	42

4 Quality Assessment and Restoration of Face Images in Long Range/High Zoom Video
Yi Yao, Besma Abidi, and Mongi Abidi 43

4.1	Introduction	43
	4.1.1 Scope	43
	4.1.2 Related Work	44
	4.1.3 Chapter Organization	46
4.2	Database Acquisition	46
	4.2.1 Indoor Sequence Acquisition	47
	4.2.2 Outdoor Sequence Acquisition	49
4.3	Face Image Quality Assessment	49
	4.3.1 Face Recognition Rate vs. System Magnification	49
	4.3.2 Adaptive Sharpness Measures	50
	4.3.3 Image Sharpness and System Magnification	53
4.4	Face Image Enhancement	54
4.5	Result Validation	56
4.6	Conclusions	60

5 Core Faces: A Shift-Invariant Principal Component Analysis (PCA) Correlation Filter Bank for Illumination-Tolerant Face Recognition
Marios Savvides, B.V.K. Vijaya Kumar, and Pradeep K. Khosla 61

5.1	Introduction	61
	5.1.1 Advanced Correlation Filters	62
5.2	Eigenphases vs. Eigenfaces	64
5.3	CoreFaces	68
5.4	Discussion	71

Part II Multi-Sensory Face Biometrics

6 Towards Person Authentication by Fusing Visual and Thermal Face Biometrics
Ognjen Arandjelović, Riad Hammoud, and Roberto Cipolla 75

6.1	Introduction	75
	6.1.1 Mono-Sensor Based Techniques	75
	6.1.2 Multi-Sensor Based Techniques	77
6.2	Method Details	77
	6.2.1 Matching Image Sets	77
	6.2.2 Data Preprocessing and Feature Extraction	79
	6.2.3 Single Modality-Based Recognition	80
	6.2.4 Fusing Modalities	81
	6.2.5 Dealing with Glasses	83

6.3	Empirical Evaluation	84
	6.3.1 Results	85
6.4	Conclusion	90

7 Multispectral Face Recognition: Fusion of Visual Imagery with Physiological Information
Pradeep Buddharaju and Ioannis Pavlidis 91
7.1	Introduction	91
7.2	Physiological Feature Extraction from Thermal Images	92
	7.2.1 Face Segmentation	92
	7.2.2 Segmentation of Superficial Blood Vessels	96
	7.2.3 Extraction of TMPs	99
	7.2.4 Matching of TMPs	100
7.3	PCA-Based Feature Extraction from Visual Images	102
7.4	Experimental Results and Discussion	103
7.5	Conclusions	108

8 Feature Selection for Improved Face Recognition in Multisensor Images
Satyanadh Gundimada and Vijayan Asari 109
8.1	Introduction	109
	8.1.1 Sensors and Systems	109
	8.1.2 Related Work	109
	8.1.3 Proposed Methodologies	110
	8.1.4 Organization of the Chapter	111
8.2	Phase Congruency Features	111
8.3	Feature Selection	113
8.4	Image Fusion	114
	8.4.1 Data Level Fusion	115
	8.4.2 Decision Level Fusion	115
8.5	Experimental Results	115
8.6	Conclusion	120

Part III Multimodal Face Biometrics

9 Multimodal Face and Speaker Identification for Mobile Devices
Timothy J. Hazen, Eugene Weinstein, Bernd Heisele, Alex Park, and Ji Ming .. 123
9.1	Introduction	123
9.2	Person Identification Technologies	124
	9.2.1 Speaker Identification	124
	9.2.2 Face Identification	126
	9.2.3 Multimodal Fusion	128
9.3	Multimodal Person ID on a Handheld Device	128
	9.3.1 Overview	128
	9.3.2 Data Collection	128

	9.3.3 Training 130
	9.3.4 Face Detection Issues 130
	9.3.5 Experimental Results 130
9.4	The Use of Dynamic Lip-Motion Information 132
9.5	Noise Robust Speaker Identification 134
	9.5.1 The Posterior Union Model 134
	9.5.2 Universal Compensation 135
	9.5.3 Experimental Results 136
9.6	Summary 138

10 Quo Vadis: 3D Face and Ear Recognition?

I. Kakadiaris, G. Passalis, G. Toderici, N. Murtuza, and T. Theoharis 139

10.1	Introduction 139
10.2	Related Work 140
	10.2.1 Face Recognition 140
	10.2.2 Ear Recognition 141
10.3	Methods 142
	10.3.1 Generic 3D-Driven Recognition System 142
	10.3.2 Data Preprocessing 143
	10.3.3 Annotated Model 144
	10.3.4 Alignment 144
	10.3.5 Deformable Model Fitting 145
	10.3.6 Geometry Image Representation 146
	10.3.7 Distance Metrics 148
10.4	3D Face Recognition 150
	10.4.1 Databases 150
	10.4.2 Results 150
	10.4.3 Discussion 156
	10.4.4 3D Face Recognition Hardware Prototype 156
10.5	3D Ear Recognition 157
	10.5.1 Ear-Specific Issues 157
	10.5.2 Annotated Ear Model 158
	10.5.3 Ear-Specific Algorithm 159
	10.5.4 3D Ear Databases 160
	10.5.5 Results 161
	10.5.6 Discussion 163
10.6	Conclusion 164

11 Human Recognition at a Distance in Video by Integrating Face Profile and Gait

Xiaoli Zhou, Bir Bhanu, and Ju Han 165

11.1	Introduction 165
11.2	Technical Approach 166
	11.2.1 High-Resolution Image Construction for Face Profile 167
	11.2.2 Face Profile Recognition 170

	11.2.3 Gait Recognition 175
	11.2.4 Integrating Face Profile and Gait for Recognition at a Distance .. 177
11.3	Experimental Results 178
	11.3.1 Data 178
	11.3.2 Experiments.. 178
11.4	Conclusions 181

Part IV Generic Approaches to Multibiometric Systems

12 Fusion Techniques in Multibiometric Systems
Arun Ross and Anil K. Jain 185
- 12.1 Introduction 185
- 12.2 Multibiometric Systems 188
- 12.3 Taxonomy of Multibiometric Systems 190
- 12.4 Levels of Fusion 193
  - 12.4.1 Sensor-Level Fusion 193
  - 12.4.2 Feature-Level Fusion 196
  - 12.4.3 Score-Level Fusion 200
  - 12.4.4 Rank-Level Fusion 207
  - 12.4.5 Decision-Level Fusion 208
- 12.5 Summary 212

13 Performance Prediction Methodology for Multibiometric Systems
Natalia A. Schmid and Joseph A. O'Sullivan 213
- 13.1 Introduction 213
- 13.2 Stochastic Model for Multimodal Biometric Signatures 215
- 13.3 Performance of a Multimodal Biometric Recognition System with M Templates 216
  - 13.3.1 Exponential Error Rate Analysis 218
- 13.4 Recognition Capacity 221
- 13.5 Examples 222
  - 13.5.1 M-ary Gaussian Example 222
  - 13.5.2 Capacity of the Multimodal System Based on PCA Signatures of the Face and Iris 225
- 13.6 Summary 227

Part V Acknowledgments, Biographies, References and Index items

Acknowledgments 231

Biographies 233

References 247

Index 273

1 Introduction

Lawrence B. Wolff

1.1 Motivations, General Addressed Problems, Trends, Terminologies

Development of face recognition systems that will exhibit high performance under most real world circumstances is an extremely challenging endeavor. Amongst the phenomenology and effects that confound such systems are variations due to illumination, facial expression, pose, aging, partial occlusion, optical blurring, and noise degradation. And this does not even touch upon the more insidious problems inherent to recognition of evasive subjects. Fortunately researchers today have at their disposal a more diverse arsenal of higher quality sensors than their counterparts of 20 years ago. It is natural and perhaps even of compelling necessity to explore how multiple sources of information from different sensors and different biometrics can be brought to bear to more effectively tackle this demanding challenge. Such is the subject matter of this book comprised of an edited collection of chapters that reflect some of the latest trends in this area.

It is well known from other engineering application areas that the fusing of multiple sources of information does not guarantee a superior solution to the problem at hand, particularly if these sources of information are highly correlated or can be derived from one another. At the least they must separately provide complementary measurement data. In recent years there have been largely three categories of approaches to face recognition for obtaining complementary information that can be loosely termed *multisample*, *multisensor*, and *multimodal*. In the first category data is acquired with the same sensor measuring the same biometric but under different conditions such as varying facial expression, illumination, pose or over different points in time. Also frequently employed are image frames extracted from continuous video sequences of a dynamic subject. Technically speaking almost all face recognition systems use with varying degree some form of multiple sampling, for instance, in a gallery with two or more images of the same subject. The ability to exploit the diversity of images representing a subject in a gallery and/or in a probe set strongly influences the overall performance of the system. In the second category two or more different sensors are used which can include nonimaging sensors such as 3D range measurement. As thermal imaging measures emissive properties complementary to reflective properties measured by conventional video cameras, fused combinations of visible and thermal imaging for face recognition has become very popular over the last few years driven in part by thermal cameras becoming more prevalent with decreasing cost. More recently a couple of research groups have further combined this with 3D geometric data. The complementarity of visible imaging with near-infrared and SWIR imaging has also been used for face recognition. Not surprisingly the best multisensor face recognition systems are the ones that also exploit the use of multisample. The third category of approaches, multimodal refers to complementary information obtained by combining face with other biometrics such as fingerprint, iris, voice, gait, and ear.

Biometric systems are beginning to take root in mainstream parts of society. Fingerprint readers are becoming more common as password access to desktop and laptop computers. Physical access control systems using fingerprint and/or face recognition are also more common. Some airlines have instituted voluntary enrollment "trusted passenger" programs for expediting boarding and immigration processing for individuals with electronically readable identity card containing biometric data. Depending upon the program these identity cards contain some subset of various combinations of fingerprint, face, and iris biometric data, in a sense already applying rudimentary multimodal fusion. Some countries are in the process of incorporating RFID sensors containing similar biometric information into passports. And then there are national identification cards containing same that are under controversial debate. Related is biometric technology incorporated into future driver licenses to aid in law enforcement. However with all these technological advances just mentioned there still is one important capability lacking; passive (i.e., completely noninvasive) identification of a subject at a significant stand-off distance. It is this capability that face recognition biometrics currently offers a large hope of achieving, and it is information fusion that bolsters this potential.

1.2 Inside This Book

Part I of this book discusses new emerging technologies for face recognition, generally having a bearing on different aspects of multisampling over time and space and even image synthesis resulting from this type of fusion. Chapter 2 presents a novel technique for comparing the multisampling from images obtained from video sequences both real and synthesized using a motion and illumination model. Parameters are estimated for the illumination model under which a probe video sequence is taken of a subject. Pose of the face is also estimated for each image in the probe video sequence. The gallery consists of 3D face models that were derived from previously obtained video sequences. From these 3D face models, video sequences of each subject in the gallery can be *synthesized* under the same illumination modeling and pose conditions as for the probe video sequences. Thus real probe video sequences can be directly compared with synthesized gallery video sequences adapting to general illumination and motion conditions. This paper illustrates the way multisample fusion can be effectively utilized for face recognition at different levels. Implicit to this paper (but referenced and described in other papers) is derivation of 3D face models from fusion of 2D images in video sequences.

Chapters 3 and 4 do not explicitly discuss use of fusion but they represent important aspects of video face recognition just beginning to be explored that can be of large significance to multisample fusion systems. Chapter 3 presents a Bayesian age-difference classifier to recognize adult subjects imaged over protracted periods of time up to a decade. They also propose a craniofacial growth model to predict intrapersonal variations in children and younger adults for video images taken across long periods of time. Face recognition being such a young discipline has not yet had the opportunity to fully exploit conglomerate biometric information over long periods of time, but this is certainly an area that is receiving more and more attention. Chapter 4 discusses methods for compensating degraded optical effects from video imaging over long distances. Not much work has been done for face recognition of subjects at large distances over 100 m where atmospheric effects as well as reduction in MTF due to high lens magnification become significant issues. The authors use an adaptive sharpness measure with special metric to evaluate the level of how degraded a face image appears. They identify multiscale processing based on wavelet transforms for image enhancement to be the most effective for improving performance of the FaceIt algorithm. Although not discussed in this chapter one can speculate how multisampling can aid in increasing face recognition performance at such remote stand-off distances for which there are a number of practical applications.

Chapter 5, the last in Part I, develops a new filter method for face recognition that is robust to handling variations in appearance of multisample images in both the gallery and probe sets. The authors develop a theory based upon the observation that PCA on face images in the Fourier domain, when restricted to phase spectrums, both encodes key discrimination detail and is tolerant to changes in illumination and occlusion (i.e., partial faces). The component *eigenphases* are formulated into a filter bank to make them shift invariant, and therefore tolerant to face image registration shifts, resulting in a hybrid PCA-correlation filter which is dubbed CoreFaces. The performance of this is compared to a popular advanced correlation filter on the CMU PIE face database.

Part II on multisensor fusion emphasizes the use of visible and thermal infrared imaging for face recognition. Well known already are the advantages of complementary phenomenology of imaging in the visible and thermal infrared spectrums for face recognition. Human skin has high emissivity in the MWIR (3–5 μm) spectrum and even higher emissivity in the LWIR (8–14 μm) spectrum making face imagery by and large invariant to illumination variations in these spectrums. Illumination invariance is strongest under most indoor environments. In outdoor environments particularly in direct sunlight illumination invariance only holds true to good approximation for LWIR which fortunately is measured by the less expensive uncooled thermal infrared camera technology. Another advantage of thermal infrared imaging of the face is it is more direct relationship to underlying physical anatomy such as vasculature. On the other hand glass and most plastics are opaque in the thermal infrared; a significant hinderance to face imaging as approximately one-third of the US population wears glasses. For a large proportion of individuals this occludes the regions around the eyes a key area of the face for image registration and face discrimination. This is one aspect of why fusion with visible imaging is of great assistance. Visible imaging of the face is also less affected by activity variation than is thermal imaging. Visible and thermal sensors are well-matched candidates for image fusion as limitations of imaging in one spectrum seem to be precisely the strengths of imaging in the other. Research conducted over the past several years has shown that visible/thermal infrared image fusion for face recognition significantly boosts performance even when performance respective to each of the individual imaging spectrums is mediocre. An important issue for this very promising technology is the use of legacy image data such as passport and driver license photographs for training. The lack of any current standard for producing thermal infrared legacy imaging data has lead a number of researchers to speculate whether it is possible to synthesize a thermal image of the face of a subject from a visible image. Interestingly enough the ability to create a thermal image of a subject from its visible counterpart is at odds with the complementary nature of these two imaging spectrums which appear to be uncorrelated. Indeed if thermal signatures of the face could be derived from visible signatures this must use techniques that do not contradict the mounting number of ways that features from visible/thermal images can be fused to boost recognition performance. At present it seems that enrollment of *both* visible and thermal imagery is required to implement such systems.

The three chapters in Part II reflect an important trend in advancing the state-of-the-art of visible/thermal face recognition, namely explorations into what features to select and at what level(s) information fusion needs to occur to optimize performance. Section 12.4 of Chap. 12 is a good place to review the taxonomy of levels of information fusion. Chapter 6 innovates a match score level fusion scheme which adapts the weighting of visible and thermal image spectrum similarity scores according to how useful information is computed to be contained in the visible spectrum image. This is determined by how much variation in illumination exists between probe and gallery images and how much it is compensated for by preprocessing. Probability density functions estimated offline on the training corpus contribute to this weighting function. The component similarity scores for visible and thermal are in turn determined by a weighted combination of local mouth and eye image regions, and the entire face image region similarities. A glasses detector is used to determine whether the eye region should be weighted zero in the thermal image component similarity score due to occlusion. Optimal values for these weights are also computed on the training set. Prior to computing scores optimal band-pass filters are used to preprocess visible and thermal images. Chapter 7 proposes visible/thermal image fusion that uses two completely different techniques for deriving similarity scores separately for the visible and the thermal domains. On thermal images thermal minutia points (TMP) are extracted from segmentation of face vasculature and these points are matched between gallery and probe images to derive a similarity score. On visible images classical PCA is used. The similarity scores are combined although it is not specified how. Lastly, Chap. 8 compares performance of different levels of information fusion on a neighborhood-based feature selection method with a phase congruency feature method.

Part III of this book is on multimodal fusion, combining face with other biometric information. While biometric recognition technology using fingerprints or iris is more mature, multimodal fusion of face recognition with these other biometrics has enormous tradeoffs in practice. Face recognition from the start has claimed as an advantage that it is a stand-off, passive and relatively noninvasive technology at least in the days when only video cameras were used to acquire data. This still remains true with the use of multiple sensors passively imaging in different spectrums such as Near-IR and thermal. However fingerprint requires physical contact and iris a close-in view in turn requiring significant cooperation and restriction of the subject. This is compatible with access control systems where frontal face recognition by itself already works quite well under typically controlled environments. Multimodal fusion with face recognition seems to become of increasing value when additional biometric information complements face data under some of the most problematic conditions for face recognition such as oblique pose. Perhaps this is why the recent surge in interest of using multi modal fusion of face data with ear and gait biometrics as described in Chaps. 10 and 11 of Part III. Measurements of ear and/or gait are accessible at high angles away from frontal face position and can be obtained using passive imaging. Voice can also be passively measured and Chap. 9 in Part III describes fusion of face and voice from on a single small device.

Finally Part IV of the book consists of two chapters on generic techniques for data fusion with application to multibiometric systems. Chapter 12 expounds further on the general taxonomy of data fusion for biometric recognition. Chapter 13 describes a general framework for determining the performance of multibiometric fusion based upon likelihood models.

1.3 Evaluation of This Book

So how far does the work presented in this book advance the practical state-of-the-art? To try to answer this question in the concrete, put this into context of solving the following benchmark example problem which could be put to good use in many practical applications; Consider the problem of developing an end-to-end system for recognizing individuals outdoors, at different times of day, at a distance of several hundred feet, and who can be walking, standing, or sitting at arbitrary pose. To simplify matters the individuals are limited to a specific group of about 100 subjects. An easier version of this problem is having legacy data for each subject obtained under controlled conditions – the harder version is learning/training on-the-fly. Even under the easiest circumstances for this problem scenario it is fair to say that the ability of an end-to-end system to have a first rank match recognition performance of 90% or above is still well out of reach at present day. The chapters in this book can be effectively used to advance insights and address the right issues towards solving isolated components of this example problem. Analyzing different levels at which information fusion occurs, illumination normalization, synthesizing pose variations, and accounting for optical effects, are all critical components. Adding passive biometrics can augment an effective solution. These different components have been further advanced and should encourage more work in their respective aspects. Peak performance has not yet been nearly saturated and more study at the component level is required.

The major challenge ahead is in combining these component areas towards achieving more complete practical systems. Experimentation needs to evolve more away from environments typical for access control, where other existing biometrics already work well, to more remote stand-off regimes where face recognition involving fusion techniques can potentially apply the most unique leverage. For instance, attacking the problem of recognition under pose variations at large distances with multisensor fusion has yet to be aggressively studied. No doubt developing more complete fusion systems operating under a broader range of scenarios will risk lower performance, at least initially. But such risks will be necessary to bring face recognition more prominently into the realm of real world applications.

Part I

Space/Time Emerging Face Biometrics

2 Pose and Illumination Invariant Face Recognition Using Video Sequences

Amit K. Roy-Chowdhury and Yilei Xu

2.1 Introduction

Pose and illumination variations remain a persistent problem in face recognition, and has been documented in different studies [1, 2]. These two factors affect low-level tasks like face registration and tracking, which, in turn, reduce the final accuracy of the recognition algorithms. Also, it is often difficult to estimate illumination conditions accurately so as to factor them into the recognition strategies. Pose estimation problems are often made difficult by the fact that illumination is unknown. Therefore, it is extremely important to develop methods for face recognition that are robust to variations in pose and illumination.

It is believed by many that video-based systems hold promise in certain applications where motion can be used as a cue for face segmentation and tracking, and the presence of more data can increase recognition performance [1]. However, video-based face recognition systems have their own challenges such as low resolution of the face region, segmentation and tracking over time, 3D modeling, and developing measures for integrating information over the entire sequence. In this paper, we present a novel framework for video-based face tracking and recognition that is based on learning joint illumination and motion models from video, synthesizing novel views based on the learned parameters, and designing metrics that can compare two time sequences while being robust to outliers. We show experimentally that our method achieves high identification rates under extreme changes of pose and illumination.

2.1.1 Overview of the Approach

The underlying concept of this paper is a method for learning joint illumination and motion models of objects from video. The application focus is on video-based face recognition where the learned models are used to (1) automatically and accurately track the face in the video, and (2) synthesize novel views under different pose and illumination conditions. We can handle a variety of lighting conditions, including the presence of multiple and extended light sources, which is natural in outdoor environments (where face recognition performance is still poor [1–3]). We can also handle gradual and sudden changes of lighting patterns overtime. This is achieved using the spherical harmonics-based representation of illumination [4, 5] and our previous work that integrates motion and illumination models for video analysis [6]. In [4,5], the reflectance image was represented using a linear combination of spherical harmonics basis functions. For Lambertian objects, a ninth-order expansion was deemed sufficient to capture most of the energy in the signal, while non-Lambertian objects required higher order coefficients. In [6,7], we showed that the appearance of a moving object under arbitrary lighting could be represented as bilinear combination of 3D motion and the spherical harmonics coefficients for illumination.

This bilinear model of illumination and motion parameters allows us to develop an algorithm for tracking a moving object with arbitrary illumination variations. This is achieved by alternately projecting onto the appropriate motion and illumination bases of the bilinear space. In addition to the 3D motion estimates, we are also able to recover the illumination conditions as a function of time, which allows us to synthesize novel images under the same lighting conditions. The framework does not assume any model for the variation of the illumination conditions – lighting can change slowly or drastically and can originate from a combination of point and extended sources. The method *relies upon image differences and does not require computation of correspondences between images*. It leads to the development of an illumination invariant model-based tracking algorithm that is initialized by registering the model (e.g., a generic face model) to the first frame of the sequence.

The recognition algorithm proceeds as follows. We assume that a 3D model of each face in the gallery is available. (We later show experimentally that an approximate 3D model with the correct texture is often good enough). Given a probe sequence, we track the face automatically in the video sequence under arbitrary pose and illumination conditions (as explained above). During the process, we also learn the illumination model parameters. The learned parameters are used to synthesize video sequences for each gallery under the motion and illumination conditions in the probe. The distance between the probe and synthesized sequences is then computed for each frame. Next, the synthesized sequence that is at a minimum distance from the probe sequence is computed and is declared to be the identity of the person. Robust distance measures are studied for this purpose.

Experimental evaluation is carried out on a database of 32 people that we collected for this purpose. One of the challenges in video-based face recognition is the lack of a good dataset, unlike in image-based approaches [1]. The dataset in [8] is small and consists mostly of pose variations. The dataset described in [9] has large pose variations under constant illumination, and illumination changes in natural environments but mostly in fixed frontal/profile poses (these are essentially for gait analysis). An ideal dataset for us would be similar to the CMU PIE dataset [10], but with video sequences instead of discrete poses. This is the reason why we collected our own data, which has large, simultaneous pose and illumination variations. We are presently enlarging this dataset and adding expression variations.

2.1.2 Relation to Previous Work

We divide our survey of the relevant literature into two broad parts. First we look at face recognition, especially the problem of pose and illumination variations. Next, we compare our joint illumination and motion models with other some approaches that deal with illumination variations in motion analysis.

Face Recognition

Due to want of space, we refer the reader to a recent review paper for existing work on face recognition [1]. A recently edited book [11] also deals with many of well-known approaches for face processing, modeling, and recognition. For a comparison of the performance of various face recognition algorithms on standard databases, the reader can refer to [2, 3]. We will briefly review a few papers most directly related to this work.

Recently there have been a number of algorithms for pose and/or illumination-invariant face recognition, many of which are based on the fact that the image of an object under varying illumination lies in a lower-dimensional linear subspace. In [12], the authors propose to arrange physical lighting so that the acquired images of each object can be directly used as the basis vectors of the low-dimensional linear space. In [13], the authors proposed a 3D spherical harmonic basis morphable model (SHBMM) to implement a face recognition system given one single image under arbitrary unknown lighting. Another morphable model-based face recognition algorithm was proposed in [14], but they use the Phong illumination model, estimation of whose parameters can be more difficult than the spherical harmonics model in the presence of multiple and extended light sources. In [15], a method was proposed for using locality preserving projections (LPP) to eliminate the unwanted variations resulting from changes in lighting, facial expression, and pose. The authors in [16, 17] proposed to use Eigen light-fields and Fisher light-fields to do pose invariant face recognition. They used generic training data and gallery images to estimate the Eigen/Fisher light-field of the subject's head, and then compare the probe image and gallery light-fields to match the face. In [18], the authors used photometric stereo methods for face recognition under varying illumination and pose. Their method requires iteration overall the poses in order to find the best match. Correlation filters have been proposed for illumination-invariant face recognition from still images in [19]. A novel method for multilinear independent component analysis was proposed in [20] for pose and illumination-invariant face recognition. All of these methods deal with recognition in a single image or across discrete poses and do not consider continuous video sequences. The authors in [8] deal with the issue of video-based face recognition, but concentrate mostly on pose variations. A method for video-based face verification using correlation filters was proposed in [21]. The advantage of using 3D models in face recognition has been highlighted in [22], but their focus is on 3D models obtained directly from the sensors and not estimated from video. This paper provides a method for *learning* the pose and illumination conditions from video, using a 3D model that can be estimated from images.

Modeling Illumination Variations in Video

Learning the parameters of the *joint* illumination and motion space is a novel contribution of this paper and we briefly review some related work. One of the well-known approaches for 2D motion estimation is optical flow [23]. However, it involves the brightness constancy constraint, which is often violated in practice. Many researchers have tried overcoming this by introducing an illumination variation term within the standard optical flow formulation. In [24], the author coined the term "photometric motion" to define the intensity change of an image point due to object rotation, and applied it to solve for shape and reflectance. In [25], a parameterized function was proposed to describe the movement of the image points taking into account the illumination variation. In [26], the author combined the geometric and photometric effects for flow computation and highlighted the need for integrating the different variabilities in the process of image analysis. A method for shape reconstruction of a moving object under arbitrary, unknown illumination, assuming motion is known, was presented in [27]. Lighting changes were modeled by introducing illumination-specific parameters into the standard optical flow equations in [28]. Illumination-invariant optical flow estimation was also the theme of [29], where an energy function was proposed to account for illumination changes and optimized using graph cuts. Another well-known approach for 2D motion estimation in monocular sequences is the Kanade–Lucas–Tomasi (KLT) tracker [30], which selects features that are optimal for tracking, and its extensions to handle illumination variations [31]. All of these approaches deal with 2D motion estimation that can handle only small changes in the pose of the object.

Our approach is illumination-invariant 3D motion estimation, *while simultaneously learning the parameters of the model*. The 2D motion obtained by any of the above methods can be used along with the well-known structure from motion (SfM) methods [32] to compute 3D motion and structure. However, the accuracy of the 3D estimates will be limited by the accuracy of the 2D motion estimates in the case of lighting changes. As an alternative, model-based techniques have been used for direct 3D motion estimation from video [33]. Many 3D model-based motion estimation algorithms rely on optical flow for the 2D motion and most existing methods are sensitive to lighting changes. The authors in [34] use probabilistic models and particle filters within a Bayesian framework to robustly track the human body, thus accounting for moderate illumination variations indirectly. A related work is [35], which uses SfM with photometric stereo to estimate surface structure. However, all the frames are needed a priori and an orthographic camera is assumed. Illumination-invariant motion estimation is possible within the active appearance model framework [36, 37], but the method requires training images under different illumination conditions. While these methods can handle illumination variations within the video sequence, they are not able to explicitly recover the illumination conditions of each frame in the video.

In [4, 5], the authors independently derived a low order (9D) spherical harmonics-based linear representation to accurately approximate the reflectance images produced by a Lambertian object with attached shadows. This was an approximation of the infinite-dimensional convex cone representation derived in [38]. All of these methods work only for a single image of an object that is fixed relative to the camera, and do not account for changes in appearance due to motion.

We proposed a framework in [6, 7] for integrating the spherical harmonics-based illumination model with the motion of the objects leading to a bilinear model of lighting and motion parameters. This approach to illumination modeling takes into account the 3D shape of the object, which is in contrast to the 2D approaches for handling illumination variation, like gradient orientation histograms [39], scale-invariant feature transforms [40] and others [41, 42]. This is motivated by a number of reasons. Our final goal is to estimate the 3D motion and shape of the objects, in addition to the lighting conditions. Thus it makes sense to integrate the illumination models with the 3D shape models. Secondly, a number of authors have shown that 2D approaches to handle illumination variations have limited ability due to lack of knowledge of the underlying geometry of the object [43–45]. Thirdly, we not only want to achieve illumination invariance, but also learn the parameters of the illumination models from video sequences. The 3D approaches to illumination modeling allow this from video sequences of natural moving objects.

2.1.3 Organization of the Chapter

The rest of the paper is organized as follows. Section 2.2 presents a brief overview of the theoretical result describing the bilinear model of joint motion and illumination variables. Section 2.3 describes the algorithm for learning the parameters of the bilinear model. Section 2.4 describes our recognition algorithm. In Section 2.5 experimental results are presented. Section 2.6 concludes the paper and highlights future work.

2.2 Integrating Illumination and Motion Models in Video

The authors in [4, 5] proved that for a fixed Lambertian object, the set of reflectance images can be approximated by a linear combination of the first nine spherical harmonics, i.e.,

$$I(x,y) = \sum_{i=0,1,2} \sum_{j=-i,-i+1...i-1,i} l_{ij} b_{ij}(\mathbf{n}), \tag{2.1}$$

where I is the reflectance intensity of the image pixel (x, y), i and j are the indicators for the linear subspace dimension in the spherical harmonics representation, l_{ij} is the illumination coefficient determined by the illumination direction, b_{ij} are the basis images, and $\mathbf{n}$ is the unit norm vector at the reflection point. The basis images can be represented in terms of the spherical harmonics as

$$b_{ij}(\mathbf{n}) = \rho r_i Y_{ij}(\mathbf{n}), i = 0, 1, 2; j = -i, \ldots, i, \tag{2.2}$$

where ρ is the albedo at the reflection point, r_i is constant for each spherical harmonics order, and Y_{ij} is the spherical harmonics function. For brevity, we will refer to the work in [4] as the Lambertian reflectance linear subspace (LRLS) theory.

This result does not consider the relative motion between the object and the camera. In [6], it was shown that for moving objects it is possible to approximate the sequence of images by a bilinear subspace. We exploit this result for 3D motion estimation under arbitrarily varying illumination. We assume a perspective projection model for the camera, consider the focal length, f, of the camera as the only intrinsic parameter (can be relaxed), and assume the reference frame to be attached to the camera with the z-axis being along the optical axis. At time instance t_1, assume we know the 3D model of the object, its pose, and the illumination condition in terms of the coefficients $l_{ij}^{t_1}$. The ray from the optical center to the pixel (x, y) intersects with the surface at $\mathbf{P_1}$. Define the motion of the object in the above reference frame as the translation $\mathbf{T} = \begin{bmatrix} T_x & T_y & T_z \end{bmatrix}^T$ of the centroid of the object and the rotation $\mathbf{\Omega} = \begin{bmatrix} \omega_x & \omega_y & \omega_z \end{bmatrix}^T$ about the centroid. After the motion, $\mathbf{P_1}$ moves to $\mathbf{P_1}'$, and another point $\mathbf{P_2}$ moves to $\mathbf{P_2}'$. At the new time instance t_2, the direction of this ray does not change, and it intersects with the surface at $\mathbf{P_2}'$. The new illumination condition is represented in terms of the coefficients $l_{ij}^{t_2}$. This is represented pictorially in Fig. 2.1.

The authors in [6] proved that reflectance image at new time instance t_2 can be expressed as:

$$I(x, y, t_2) = \sum_{i=0,1,2} \sum_{j=-i,-i+1\ldots i-1,i} l_{ij}^{t_2} b_{ij}(\mathbf{n_{P_2'}}), \tag{2.3}$$

where

$$b_{ij}(\mathbf{n_{P_2'}}) = b_{ij}(\mathbf{n_{P_1}}) + \mathbf{AT} + \mathbf{B\Omega}. \tag{2.4}$$

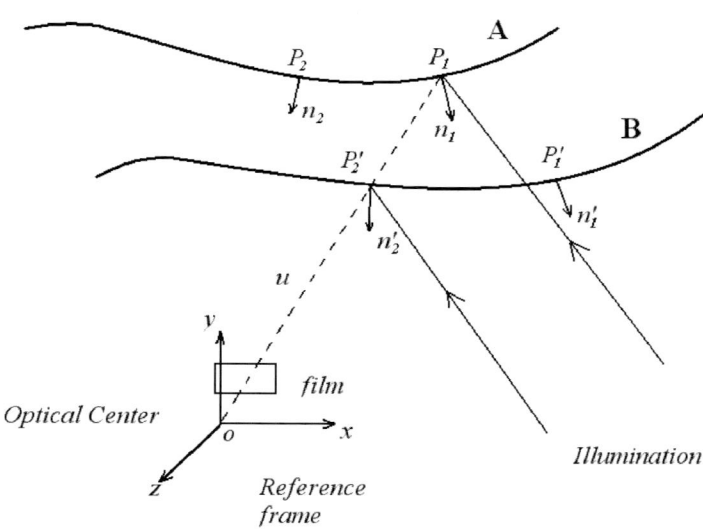

Fig. 2.1. Pictorial representation showing the motion of the object and its projection

In (2.3), $b_{ij}(\mathbf{n}_{\mathbf{P}_2'})$ and $l_{ij}^{t_2}$ are the basis images and illumination coefficients after motion. In (2.4), $b_{ij}(\mathbf{n}_{\mathbf{P}_1})$ are the original basis images before motion. $\mathbf{A}$ and $\mathbf{B}$ contain the structure and camera intrinsic parameters. Substituting (2.4) into (2.3), we see that the new image spans a bilinear space of six motion and approximately nine illumination variables (for Lambertian objects). The basic result is valid for general illumination conditions, but require consideration of higher order spherical harmonics.

When the illumination changes gradually, we can use the Talyor series to approximate the illumination coefficients as $l_{ij}^{t_2} = l_{ij}^{t_1} + \Delta l_{ij}$. Ignoring the higher order terms, the bilinear space now becomes a combination of two linear subspaces, as

$$I(x,y,t_2) = I(x,y,t_1) + \sum_{i=0,1,2} \sum_{j=-i,\ldots,i} l_{ij}^{t_1}(\mathbf{AT} + \mathbf{B}\Omega)$$
$$+ \sum_{i=0,1,2} \sum_{j=-i,\ldots,i} \Delta l_{ij} b_{ij}(\mathbf{n}_{\mathbf{P}_1}). \tag{2.5}$$

If the illumination does not change from t_1 to t_2 (often a valid assumption for a short interval of time), the new image at t_2 spans a linear space of the motion variables, since the third term in (2.5) is zero.

We can express the result in (2.3) succinctly using tensor notation as

$$\mathcal{I} = (\mathcal{B} + \mathcal{C} \times_2 \begin{pmatrix}\mathbf{T}\\ \Omega\end{pmatrix}) \times_1 \mathbf{l}, \tag{2.6}$$

where $\times_n$ is called the *mode-n product* [20], and $\mathbf{l} \in \mathbf{R}^9$ is the vector of l_{ij} components. The *mode-n product* of a tensor $\mathcal{A} \in \mathbf{R}^{I_1 \times I_2 \times \ldots \times I_n \times \ldots \times I_N}$ by a vector $\mathbf{V} \in \mathbf{R}^{1 \times I_n}$, denoted by $\mathcal{A} \times_n \mathbf{V}$, is the $I_1 \times I_2 \times \ldots \times 1 \times \ldots \times I_N$ tensor

$$(\mathcal{A} \times_n \mathbf{V})_{i_1 \ldots i_{n-1} 1 i_{n+1} \ldots i_N} = \sum_{i_n} a_{i_1 \ldots i_{n-1} i_n i_{n+1} \ldots i_N} v_{i_n}.$$

For each pixel (p,q) in the image, $\mathcal{C}_{klpq} = [\,A\ B\,]$ of size $N_l \times 6$, where N_l is the dimension of the illumination basis ($N_l \approx 9$ for Lambertian objects). Thus for an image of size $M \times N$, $\mathcal{C}$ is $N_l \times 6 \times M \times N$. $\mathcal{B}$ is a subtensor of dimension $N_l \times 1 \times M \times N$, comprising the basis images $b_{ij}(\mathbf{n}_{\mathbf{P}_1})$, and $\mathcal{I}$ is a subtensor of dimension $1 \times 1 \times M \times N$, representing the image. $\mathbf{l}$ is still the $N_l \times 1$ vector of the illumination coefficients.

These theoretical results can be used to synthesize video sequences of objects under different conditions of lighting and motion. This would rely on computing the basis images which are a function of the surface normal. In practice, the surface normals are computed by finding the intersection of the ray passing through a pixel with a 3D point, assuming that the 3D model is represented by a cloud of points. The normal is then calculated by considering neighboring points. If a mesh model of the object is used, the intersection of the ray with a triangular mesh is computed, and the normal to this mesh patch is calculated.

2.3 Learning Joint Illumination and Motion Models from Video

The joint illumination and motion space provides us with a novel method for 3D motion estimation under varying illumination. This is based on inverting the generative model for motion and illumination modeling. It can not only track the 3D motion under varying illumination, but also can estimate the illumination parameters.

Equation (2.3) provides us an expression relating the reflectance image I_{t2} with new illumination coefficients $l_{ij}^{t_2}$ and motion variables $\mathbf{m} = [\mathbf{T}, \mathbf{\Omega}]^T$, which lead to a method for estimating 3D motion and illumination as:

$$(\hat{\mathbf{l}}, \hat{\mathbf{T}}, \hat{\mathbf{\Omega}}) = \arg\min_{\mathbf{l},\mathbf{T},\mathbf{\Omega}} \|I_{t2} - \sum_{i=0,1,2}\sum_{j=-i}^{i} l_{ij}b_{ij}(\mathbf{n}_{\mathbf{P}_2'})\|^2,$$

$$= \arg\min_{\mathbf{l},\mathbf{T},\mathbf{\Omega}} \|\mathcal{I}_{t2} - (\mathcal{B}_{t1} + \mathcal{C}_{t1} \times_2 \begin{pmatrix} \mathbf{T} \\ \mathbf{\Omega} \end{pmatrix}) \times_1 \mathbf{l}\|^2, \quad (2.7)$$

where $\hat{x}$ denotes an estimate of x. The cost function is a square error norm, similar to the famous bundle-adjustment [32], but incorporates an illumination term. Motion and illumination estimates are obtained for each frame. Since the motion between consecutive frames is small, but illumination can change suddenly, we add a regularization term to the above cost function. It is of the form $\alpha\|\mathbf{m}\|^2$.

Since the image I_{t2} lies approximately in a bilinear space of illumination and motion variables (ignoring the regularization term for now), such a minimization problem can be achieved by alternately estimating the motion and illumination parameters by projecting the video sequence onto the appropriate basis functions derived from the bilinear space. Assuming that we have tracked the sequence upto some frame for which we can estimate the motion (hence, pose) and illumination, we calculate the basis images, b_{ij}, at the current pose, and write it in tensor form $\mathcal{B}$. Unfolding[1] $\mathcal{B}$ and the image $\mathcal{I}$ along the first dimension [46], which is the illumination dimension, the image can be represented as:

$$\mathcal{I}_{(1)}^T = \mathcal{B}_{(1)}^T \mathbf{l}. \quad (2.8)$$

This is a least square problem, and the illumination $\mathbf{l}$ can be estimated as:

$$\hat{\mathbf{l}} = (\mathcal{B}_{(1)}\mathcal{B}_{(1)}^T)^{-1}\mathcal{B}_{(1)}\mathcal{I}_{(1)}^T. \quad (2.9)$$

Keeping the illumination coefficients fixed, the bilinear space in (2.3) and (2.4) becomes a linear subspace, i.e.,

$$\mathcal{I} = \mathcal{B} \times_1 \mathbf{l} + (\mathcal{C} \times_1 \mathbf{l}) \times_2 \begin{pmatrix} \mathbf{T} \\ \mathbf{\Omega} \end{pmatrix}. \quad (2.10)$$

---

[1] Assume an Nth-order tensor $\mathcal{A} \in \mathbf{C}^{I_1 \times I_2 \times \ldots \times I_N}$. The matrix unfolding $\mathbf{A}_{(n)} \in \mathbf{C}^{I_n \times (I_{n+1}I_{n+2}\ldots I_N I_1 I_2 \ldots I_{n-1})}$ contains the element $a_{i_1 i_2 \ldots i_N}$ at the position with row number i_n and column number equal to $(i_{n+1} - 1)I_{n+2}I_{n+3}\ldots I_N I_1 I_2 \ldots I_{n-1} + (i_{n+2} - 1)I_{n+3}I_{n+4}\ldots I_N I_1 I_2 \ldots I_{n-1} + \cdots + (i_N - 1)I_1 I_2 \ldots I_{n-1} + (i_1 - 1)I_2 I_3 \ldots I_{n-1} + \cdots + i_{n-1}$.

Similarly, unfolding all the tensors along the second dimension, which is the motion dimension, and adding the effect of the regularization term, $\mathbf{T}$ and Ω can be estimated as:

$$\begin{pmatrix} \hat{\mathbf{T}} \\ \hat{\Omega} \end{pmatrix} = \left((\mathcal{C} \times_1 \mathbf{1})_{(2)} (\mathcal{C} \times_1 \mathbf{1})_{(2)}^T + \alpha \mathbf{I} \right)^{-1} (\mathcal{C} \times_1 \mathbf{1})_{(2)} (\mathcal{I} - \mathcal{B} \times_1 \mathbf{1})_{(2)}^T, \quad (2.11)$$

where $\mathbf{I}$ is an identity matrix of dimension 6×6. The above procedure for estimation of the motion should proceed in an iterative manner, since $\mathcal{B}$ and $\mathcal{C}$ are functions of the motion parameters. This should continue until the projection error $\|\mathcal{I} - \mathcal{B} \times_1 \hat{\mathbf{l}}\|^2$ does not decrease further. This process of alternate minimization leads to the local minimum of the cost function (which is quadratic in motion and illumination variables) at each time step. This can be repeated for each subsequent frame. We now describe the algorithm formally.

2.3.1 Algorithm

Consider a sequence of image frames $I_t, t = 0, \ldots, N - 1$.
Initialization. Take one image of the object from the video sequence, register the 3D model onto this frame and map the texture onto the 3D model. Calculate the tensor of the basis images $\mathcal{B}_0$ at this pose. Use (2.9) to estimate the illumination coefficients. Now, assume that we know the motion and illumination estimates for frame t, i.e., $\mathbf{T}_t, \Omega_t$ and $\mathbf{l}_t$.

- Step 1. Calculate the tensor form of the bilinear basis images $\mathcal{B}_t$ at the current pose using (2.4). Use (2.11) to estimate the new pose from the estimated motion.
- Step 2. Assume illumination does not change, i.e., $\hat{\mathbf{l}}_{t+1} = \hat{\mathbf{l}}_t$. Compute the motion $\mathbf{m}$ by minimizing the difference between an input frame and the rendered frame $\|\mathcal{I}_{t+1} - \left(\mathcal{B}_t + \mathcal{C}_t \times_2 \begin{pmatrix} \hat{\mathbf{T}}_{t+1} \\ \hat{\Omega}_{t+1} \end{pmatrix} \right) \times_1 \hat{\mathbf{l}}_{t+1}\|^2$, and estimate the new pose.
- Step 3. Using the new pose estimate, re-estimate the illumination using (2.9). Repeat Steps 1 and 2 with the new estimated $\hat{\mathbf{l}}_{t+1}$ for that input frame, till the error is below an acceptable threshold.
- Step 4. Set $t = t + 1$. Repeat Steps 1, 2, and 3.
- Step 5. Continue till $t = N - 1$.

In many practical situations, the illumination changes slowly within a sequence (e.g., cloud covering the sun). In this case, we use the expression in (2.5) instead of (2.3) and (2.4) in the cost function (2.7) and estimate Δl_{ij}.

2.3.2 Handling Occlusions

The optimization function (2.7) yields the maximum likelihood estimate under the assumption of additive Gaussian noise to the image observations. However, in the presence of occlusion, the optimization function can be used only if we can work with the unoccluded pixels, which will have to be estimated a priori. A simple way to do this is to set a threshold and discard those pixels that have an intensity change (with respect to the previous frame) greater than the threshold. However, a simple threshold strategy may eliminate the pixels that are not occluded, but whose intensity changes because of the change in illumination conditions. Therefore, we propose the following modification to our algorithm to handle occlusion.

Assume that we are able to obtain the tracking and illumination estimates upto some instance t. Then, we can calculate the bilinear basis images at the current pose, and project the frame at the next time instance, $t+1$, onto the linear subspace of the basis images. This gives an estimate of the illumination coefficients for the frame. Using the basis images, we can synthesize the image with the newly estimated illumination coefficients $\mathbf{l_{t+1}}$. In order to do this, the motion between I_{t+1} and I_t is assumed to be the same as between I_t and I_{t-1} (i.e., uniform motion). If the difference between the synthesized image and the observed one is larger than some threshold for some pixels, we will discard these pixels. By doing this, we store a mask for the pixels which are occluded. Note that the synthesized image has the new illumination condition, and thus is not affected by the problem noted above. Using the unoccluded pixels and the algorithm described in Sect. 2.3.1, we re-estimate the 3D motion as well as the new illumination coefficients $\mathbf{\hat{l}_{t+1}}$. For the image at time instance $t+2$, we will use the mask at time instance $t+1$ to estimate the illumination condition $\mathbf{\hat{l}_{t+2}}$, then repeat what we have done for $t+1$ frame and update the mask. This method works provided the occlusion happens slowly(most practical cases). For sudden occlusion, a RANSAC approach [32], that works with random subsets of feature points, will be adopted.

2.4 Face Recognition From Video

The generative framework for integrating illumination and motion models described in Sect. 13.1 and the method for learning the model parameters as described in Sect. 2.3 set the stage for developing a novel face recognition algorithm that is particularly suited to handling video sequences. The method is able to handle arbitrary pose and illumination variations and can integrate information over an entire video sequence.

In our method, the gallery is represented by a 3D model of the face. The model can be built from a single image [47], a video sequence [48] or obtained directly from 3D sensors [22]. In our experiments, the face model will be estimated from video. Given a probe sequence, we will estimate the motion and illumination conditions using the algorithms described in Sect. 2.3. Note that the tracking does not require a person-specific 3D model – a generic face model is usually sufficient. Given the motion and illumination estimates, we will then render images from the 3D models in the gallery. The rendered images can then be compared with the images in the probe sequence. Given the rendered images from the 3D models in the gallery and the probe images, we will design robust metrics for comparing these two sequences. A feature of these metrics will be their ability to integrate the identity over all the frames, ignoring some frames that may have the wrong identity. Since 3D shape modeling is done for the gallery sequences only, we avoid the issues of high computational complexity of 3D modeling algorithms in real time.

One of the challenges faced is to design suitable metrics capable of comparing two video sequences. This metric should be general enough to be applicable to most videos and robust to outliers. Let $P(f_i), i = 1, \ldots, N$ be N frames from the probe sequence. Let $SG_j(f_i), i = 1, \ldots, N$ be the frames of the synthesized sequence for galley j, where $j = 1, \ldots, M$ and M is the total number of individuals in the gallery. Note that the number of frames in the two sequences to be compared will always be the same in our method. By design, each corresponding frame in the two sequences will be under the same pose and illumination conditions, dictated by the accuracy of the estimates of these parameters from the probes and the synthesis algorithm. Let d_{ij} be the distance between the ith frames of P and G_j. We now compare two distance measures that can be used for obtaining the identity of the probe sequence.

1. $ID = \arg\min_j \min_i d_{ij}$
2. $ID = \arg\min_j \max_i d_{ij}$ (2.12)

The first alternative computes the distance between the frames in the probe and each synthesized sequence that are the most similar and chooses the identity as the individual with the smallest distance in the gallery. This can be looked upon as obtaining the identity of the probe from one image of it that is most similar to the gallery. The second distance measure can be interpreted as minimizing the maximum separation between the probe and synthesized gallery images. Both of these measures suffer from a lack of robustness, which can be critical for their performance since the correctness of the synthesized images depend upon the accuracy of the illumination and motion parameter estimates. For this purpose, we replace the $\max$ by the fth percentile and the $\min$ (in the inner distance computation of 1 in (2.12)) by the $(1 - f)$th percentile. In our experiments, we choose f to be 0.8 and use the first option.

A third possible option is to assign a weight to each image of each synthesized gallery that is inversely proportional to its distance from the corresponding probe image, sum all the weights and choose the gallery with largest weight as the identity. The problem with this method is that the recognition accuracy depends upon the choice of the weighting function, which in turn can vary with the probe and gallery sequences.

One point that still needs to be addressed is on how do we compute d_{ij}. Recall that a generic face model is used to track the face in the probe video and the estimated illumination and motion parameters are used to synthesize the videos for each person in the gallery using their 3D model. This sets up a mapping between the pixels in the synthesized images with the probe images through the 3D models. Also, the number of synthesized images is the same as the number of images in the probe, thus obviating any synchronization issues. Thus d_{ij} can be computed directly as the squared difference between the synthesized and probe image frames.

We now describe formally the video-based face recognition algorithm. Using the above notation, let $P(f_i), i = 1, \ldots, N$ be N frames from the probe sequence. Let $G_1, \ldots, G_M$ be the 3D models for each of M galleries.

- Step 1. Register a 3D generic face model to the first frame of the probe sequence. Estimate the illumination and motion model parameters for each frame of the probe sequence using the method described in Sect. 2.3.
- Step 2. Using the estimated illumination and motion parameters, synthesize, for each gallery, a video sequence using the generative model of (2.4). Denote these as $SG_j(f_i), i = 1, \ldots, N$ and $j = 1, \ldots, M$.
- Step 3. Compute d_{ij} in (2.12).
- Step 4. Obtain the identity using a suitable distance measure from (2.12), modifying it for robustness as necessary (see discussion above).

2.5 Experimental Results

Since the tracking and synthesis algorithms are the foundation for the recognition strategy, we first present results on these two aspects highlighting the accuracy of the methods in a controlled environment. We then describe our face video database and the results of the recognition algorithms.

2.5.1 Tracking and Synthesis Results

We synthesized a video sequence of a face with known motion and lighting. A generic 3D model was registered to the first frame of the sequence manually and tracked using the algorithm described in Sect. 2.3. Figures 2.2–2.4 show the results of our tracking algorithm on this sequence. The images in Fig. 2.2 are synthesized from a 3D model, and thus the motion and illumination are known. The face is rotating along y-axis from $-30°$ to $+30°$, and the illumination is changing such that the light always comes from the front of the face. The resolution of the image is

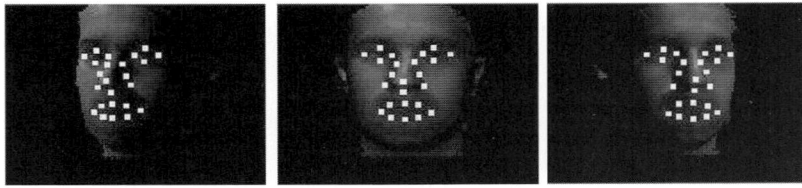

Fig. 2.2. The back projection of the mesh vertices of the 3D face model using the estimated 3D motion onto some input frames. Face is rotating about the y-axis, and illumination is changing in the same way as pose

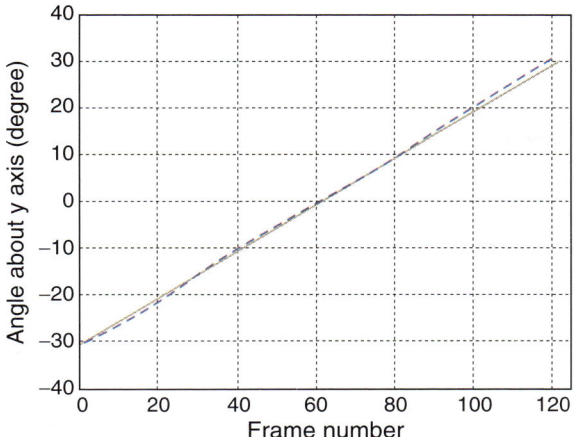

Fig. 2.3. The *solid line* shows the true pose (represented by the angle of face about y-axis) and the broken line is the estimated pose

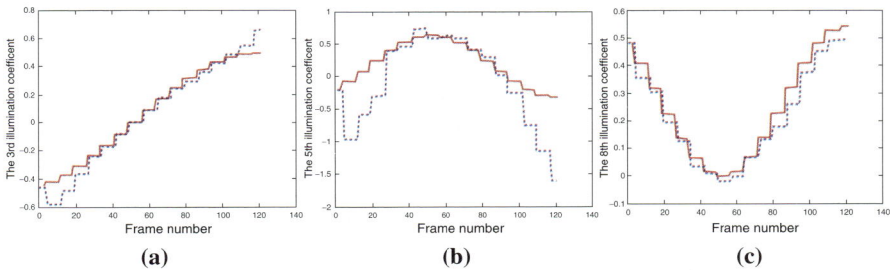

Fig. 2.4. (**a**), (**b**), and (**c**) are the estimates of the third, fifth, and eight illumination coefficients, respectively. The *solid line* shows the true illumination coefficients using the LRLS method, and the dotted line shows the estimated illumination coefficients

240 by 320. Figures 2.3 and 2.4 show plots of the estimated motion and illumination against the true values.

We also show results of the synthesis algorithm on a real-life video sequence. Frames from a synthesized video sequence using learned motion and illumination parameters are shown in Fig. 2.5. Motion and illumination are learned from the frames in the first and second row, respectively, and images in the third row are synthesized with the motion and illumination parameters learned from the corresponding frames in the same column. The reader can visually compare the synthesized images for accuracy of pose and illumination estimates.

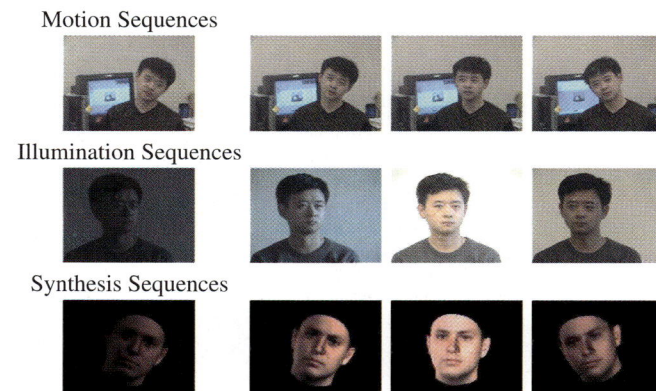

Fig. 2.5. An example of video synthesis with learned motion and illumination models. Motion and illumination are learned from the frames in the first and second row, respectively, and images in the third row are synthesized with the motion and illumination parameters learned from the corresponding frames in the same column

2.5.2 Face Recognition Results

Face Database

Our database consists of videos of 32 people. Each person was asked to move his/her head as they wished and the illumination was changed randomly. The illumination consisted of ceiling lights, lights from the back of the head and sunlight from a window on the left side of the face. Random combinations of these were turned on and off and the window was controlled using dark blinds. An example of some of the images in the video database is shown in Fig. 2.6. The resolution of the face varied depending on the person and the movement. A statistical analysis showed that the average size was about 70×70, with the minimum size being 50×50. Each sequence was divided into two parts – gallery and probe. The frames in Fig. 2.6 are arranged in the same order as in the original video, with the first column representing a frame from the gallery, the third column representing the image in Experiment 1 (see below), and the fifth column representing the image in Experiment 3 (see below).

A 3D model of each face was constructed from the gallery sequence. In the set of experiments shown, a generic model was registered to one approximately frontal image in the gallery manually by choosing seven points on the face. Thereafter the texture of the face was mapped onto the model. The shape was not changed from the generic model. We would like to emphasize that any other 3D modeling algorithm would also have worked and we plan to integrate our previous work in [49] with this system.

From the portion of each sequence designated as probe, we designed five experiments by choosing different parts of it, as described below:

Fig. 2.6. Sample frames from the video sequences collected for our database

- Experiment 1: A single image, some examples of which are shown in the third column of Fig. 2.6, was used as the probe.
- Experiment 2: A video sequence starting with the frame in Experiment 1 was used as the probe. Examples of these frames can be seen from the third column and beyond in Fig. 2.6.
- Experiment 3: A single image, some examples of which are shown in the fifth column of Fig. 2.6, was used as the probe.
- Experiment 4: A video sequence starting with the frame in Experiment 3 was used as the probe. Examples of which can be seen from the third column and beyond in Fig. 2.6.
- Experiment 5: A video sequence that has a portion with frontal face and illumination similar to the gallery was used as the probe. This is achieved by considering the probe sequence to start immediately after the gallery sequence ends in our collected data.

As can be seen from Fig. 2.6, the pose and illumination varies randomly in the video. The reason for choosing the experiments in this way are the following (1) to study the advantage video provides over image-based recognition, (2) how sensitive recognition rates are with respect to the actual frames in the video (hence the change in the starting frame in Experiment 4 compared to Experiment 2), and (3) how recognition rates are affected if there is a small portion of the video in the probe very similar to the gallery, even though the other frames may not be.

The results on tracking and synthesis on three of the probes are shown in Fig. 2.7. We plot the cumulative match characteristic (CMC) [1, 2] for all the experiments

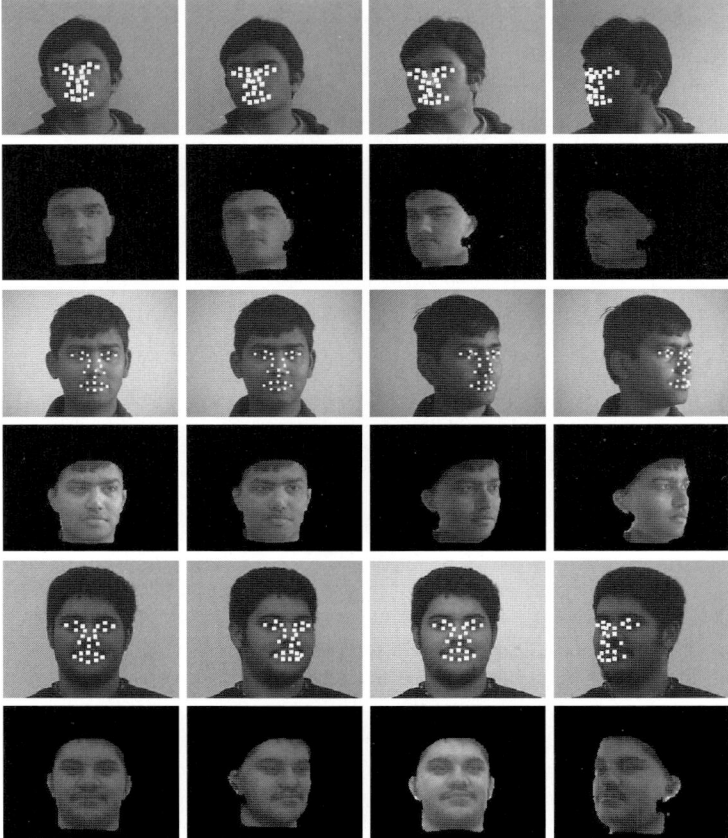

Fig. 2.7. Tracking and synthesis results are shown in alternating rows for three of the probes

in Fig. 2.8. The following are the main conclusions that we can draw from our experiments:

- Our proposed algorithm gives relatively high performance (about 90% on the average for Experiments 1, 3, and 5 that deal with video sequences) on videos with large and arbitrary variations of pose and illumination.
- There is a significant change increase in performance in considering a video sequence compared to a single image, as evidenced by the improvements between Experiments 1 and 2, and between Experiments 3 and 4. Between Experiments 1 and 2 there is a 10% increase in the Rank 1 identification rate, as well as a significant increase in the slope of the CMC curve. Between Experiments 3 and 4, there is again a 10% increase in the identification rate. However, the recognition rates between Experiments 2 and 4 are different, demonstrating

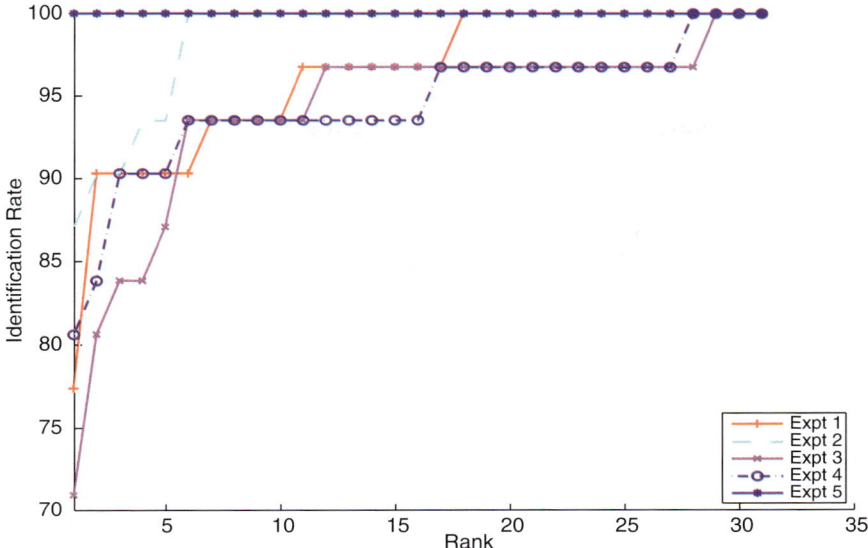

Fig. 2.8. CMC curve for video-based face recognition experiments

the sensitivity of the algorithm to the actual frames in the sequence (which is to be expected).
– When a part of the video sequence has overlap with the gallery (even one frame), our system gives a 100% recognition rate (Experiment 5).

All these experiments demonstrate the effectiveness of video-based face recognition methods over still image-based approaches. However, the recognition rate is affected significantly by the actual conditions under which the video was captured.

2.6 Conclusions

In this chapter, we have proposed a method for video-based face recognition that relies upon a novel theoretical framework for integrating illumination and motion models for describing the appearance of a video sequence. We started with a brief exposition of this theoretical result, followed by methods for learning the model parameters. Then, we described our recognition algorithm that relies on synthesis of video sequences under the conditions of the probe. Finally, we demonstrated the effectiveness of the method on video databases with large and arbitrary variations in pose and illumination. In future, we will work on improving the tracking and synthesis algorithms (which we believe will improve recognition performance), performing thorough experimentation to understand the effect of the different variabilities, and analyzing performance on larger datasets.

3 Recognizing Faces Across Age Progression

Narayanan Ramanathan and Rama Chellappa

3.1 Introduction

Perceiving human faces and modeling the distinctive features of human faces that contribute most toward face recognition, are some of the challenges faced by computer vision and psychophysics researchers. Human faces comprise a special class of 3D objects, modeling of which involves developing accurate characterizations that account for illumination variations, head pose variations, facial expressions, etc. Moreover, human faces also undergo growth related changes that are manifested in the form of shape variations and textural variations. Hence, the robustness of a face recognition system to variations due to illumination, pose, facial expressions, aging, etc., forms a significant evaluation criterion for the system. In this chapter, we discuss methods to characterize facial variations due to aging effects and propose methods to verify face images across age progression.

Facial aging effects are manifested in different forms in different age groups. While aging effects are often manifested in the form of shape variations from infancy to teenage years (due to craniofacial growth) [50], they are observed in the form of textural variations such as facial wrinkles and other skin artifacts, during adulthood. Apart from biological factors, factors such as climatic conditions, ethnicity, mental stress, etc., are often attributed to play a role in the process of aging. Some of the interesting applications of studying age progression in human faces are:

- Face recognition/face verification across age
- Automatic age progression systems
- Age-based human–computer interaction systems
- Automatic age identification systems

Next, we provide a brief overview on some of the previous works on age progression in human faces.

3.1.1 Previous work on Age Progression

D'arcy Thompson's study of morphogenesis [51] largely laid the foundations for studies related to the perception of growing faces. Thompson pioneered the use of geometric transformations in the study of morphogenesis. Biological forms were embedded within coordinate systems and different morphogenetic events were described by means of global geometric transformations within the coordinate system. All through his study, he maintained that morphological changes were a result of the physical forces such as biomechanical stress and gravity that act on biological forms. From the perspective of modeling craniofacial growth, Thompson's approach translates into identifying those physical forces that are responsible for the remodeling of the human head with growth.

Pittenger and Shaw [52] studied facial growth as a visco-elastic event defined on the craniofacial complex. They applied strain and shear transformations on face profiles and studied the relative significance of each of the applied transformations, in accounting for facial growth. They observed that, shape changes in facial profiles induced by cardioidal strain transformations formed the primary source of perceptual information for relative age judgements. Interestingly, when cardioidal strain transformations were applied on inanimate objects, inanimate objects were perceived to undergo growth related transformations. Pittenger and Shaw [53] observed this behavior on cartoon drawings of dogs, monkeys, Volkswagen "beetles," etc. Mark et al. [54] identified geometric invariants associated with growth related transformations and proposed that only transformations that preserve such geometric invariants in human faces would be perceived as growth-related transformations. The geometric invariants were described as follows (1) angular coordinates of every point on an object in a polar coordinate system, being preserved (2) bilateral symmetry about the vertical axis being maintained, and (3) continuity of object contours being preserved. Table 3.1 illustrates some of the transformation functions that were proposed to model craniofacial growth [54,55] and further illustrates the geometric invariants that are preserved under each of the transformations. Figure 3.1 illustrates the effect of different geometric transformations on a human face profile. The "revised" cardioidal strain transformation model proposed by Todd et al. [56], the model that draws analogy between the remodeling of the human head with growth and the remodeling of a fluid-filled spherical object with pressure, was considered very effective in modeling craniofacial growth.

Table 3.1. Some geometric tranformations that were proposed to model craniofacial growth. (The geometric invariants designated as I, II, and III are explained Sect. 1.1)

applied transformation	model	geometric invariants		
		(I)	(II)	(III)
cardioidal strain (polar coordinates)	$\theta_{t+1} = \theta_t$ $R_{t+1} = R_t(1 - k\cos(\theta_t))$	✓	✓	✓
spiral strain (polar coordinates)	$\theta_{t+1} = \theta_t$ $R_{t+1} = R_t(1 + k|\theta_t|)$	✓	✓	✓
affine shear (cartesiancoordinates)	$Y_{t+1} = Y_t$ $X_{t+1} = X_t + kY_t$	✗	✓	✓
revised cardioidal strain (polar coordinates)	$\theta_{t+1} = \theta_t$ $R_{t+1} = R_t(1 + k(1 - \cos(\theta_t)))$	✓	✓	✓

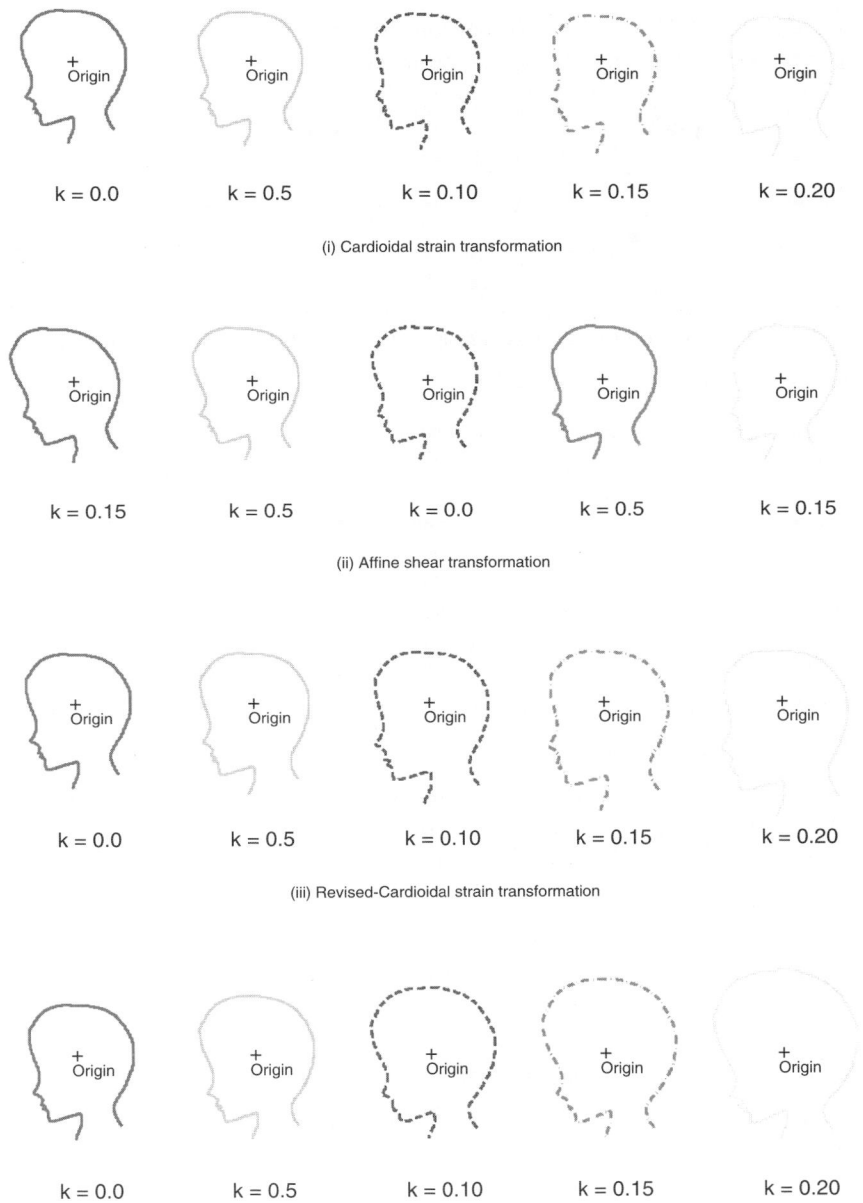

Fig. 3.1. Different transformation models proposed to model facial growth

Table 3.2. Age progression: computer vision approaches

reference	operation	approach: overview
Kwon and da Lobo [57, 58]	age-based classification of face images (age group: all ages)	study face anthropometric ratios and propose facial wrinkle analysis methods
Tidderman et al. [59] Burt and Perrett [60]	automatic age progression; age perception (age group: 20–60 years)	create composite faces for different age groups and employ wavelet-based approaches
Lanitis et al. [61]	automatic age progression; face recognition across age (age group: 0–30 years)	build a shape-intensity model for faces and propose PCA based age transformation function
Lanitis et al. [62]	age-based classification of face images (age group: 0–30 years)	compare performance of neural network-based classifiers and hierarchical classifiers
Gandhi et al. [63]	automatic age progression; age estimation (age group: 15 years +)	propose SVM-based approach for age estimation and modify IBST approach for age progression
Ramanathan and Chellappa [64] and [65]	face verification and facial similarity across age (age group: 20–70 years)	propose a Bayesian age difference classifier built on probabilistic eigenspaces framework
Ramanathan and Chellappa [66]	automatic age progression; face recognition across age (age group: 0–18 years)	propose a craniofacial growth model incorporating age-based face anthropometric data

In the computer vision literature, age progression in human faces has been addressed from two perspectives: one towards automatic age estimation and age-based classification from face images and the other towards automatic age progression systems. Table 3.2 summarizes some of the recent computer vision approaches towards studying age progression in human faces.

3.1.2 Problem Statement

While face images have traditionally been used in identification documents such as passports, driver's licenses, voter ID, etc., in recent years, face images are being increasingly used as additional means of authentication in applications such as credit/debit cards and in places of high security. Since faces undergo gradual variations due to aging, periodically updating face databases with more recent images of subjects might be mandatory for the success of face recognition systems. Since periodically updating such large databases would be a tedious task, a better alternative would be to develop face recognition systems that verify the identity of individuals from a pair of age separated face images. Further, understanding the role of age progression in affecting the similarity between two face images of an individual would be important for such a task.

Table 3.3. Database of passport images

age difference (years)	no. of image pairs
1–2	165
3–4	104
5–7	81
8–9	115

Fig. 3.2. A few sample age separated images of individuals retrieved from their passports

We wish to address the following problems: how similar are a pair of age separated face images of an individual? How do inherent changes in a human face due to aging effects affect facial similarity? Given a pair of age separated face images of an individual, what is the confidence measure associated with verifying the identity? Our database is comprised of pairs (younger and most recent) of face images retrieved from the passports of 465 individuals. Table 3.3 summarizes the database. The age span of individuals in this database was 20–70 years. Figure 3.2 shows a few sample images from our database.

3.2 Age Difference Classifier

We propose a Bayesian age-difference classifier built on a probabilistic eigenspaces framework, to verify the identity of individuals from pairs of age separated face images. Further, the classifier is also designed to estimate the age-difference between intrapersonal face images (face images of the same individual) that are age separated. Since our database is comprised of adult face images, the textural variations (wrinkles and other skin artifacts) commonly observed due to age progression, form the primary basis for classification. Across each pair of face images, we compute the difference image by subtracting the more recent image from the older image. The difference image, when computed between age separated images of the same individual (intrapersonal images), captures facial variations due to aging effects. Intuitively, the difference images obtained from the intrapersonal image pairs (image

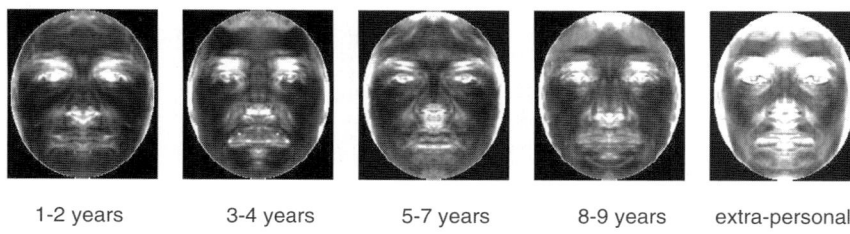

1-2 years   3-4 years   5-7 years   8-9 years   extra-personal

Fig. 3.3. Average difference images from the intrapersonal (under each of the four age-difference categories) and extrapersonal classes

pairs of the same individual) with lesser age separation would be less exaggerated than that obtained from the intrapersonal image pairs with larger age separation. Further, the extrapersonal difference images would differ significantly from the intrapersonal difference images, due to the large mismatch in facial features that is observed between extrapersonal image pairs. Figure 3.3 illustrates the average difference images computed from the intrapersonal image pairs (with an age separation of 1–2 years, 3–4 years, 5–7 years, and 8–9 years) and that computed from the extrapersonal image pairs. The sagging of facial features getting more prominent with an increase in the age separation between the intrapersonal image pairs is evident from Fig. 3.3.

3.2.1 Bayesian Framework

The framework proposed in [67] was adopted primarily to estimate complex density functions in high-dimensional image spaces and subsequently to compute class conditional density functions. The classification of pairs of face images based on their age-differences, consists of two stages. In the first stage of classification, the identity between the pair of face images is established. In the second stage, the pairs of age separated face images that were identified as ones belonging to the same individual, are further classified based on their age-differences.

Let Ω_I denote the intrapersonal space and let Ω_E denote the extrapersonal space. Let $I_{11}, I_{12}, I_{21}, I_{22}, \ldots, I_{M1}, I_{M2}$ be the set of N x 1 vectors formed by the lexicographic ordering of pixels in each of the M pairs of faces images of M individuals, respectively. The intrapersonal image differences $\{\mathbf{x}_i\}_{i=1}^{M}$ are obtained by computing the differences between pairs of age separated face images of individuals.

$$\mathbf{x}_i = I_{i1} - I_{i2}, \; 1 \leq i \leq M \tag{3.1}$$

The extrapersonal image differences $\{\mathbf{z}_i\}_{i=1}^{M}$ are obtained by computing the difference between face images of different individuals.

$$\mathbf{z}_i = I_{i1} - I_{j2} \, , \; j \neq i, \; 1 \leq i,j \leq M \tag{3.2}$$

Firstly, from a set of intrapersonal image differences $\{\mathbf{x}_i\}_{i=1}^{M} \in \Omega_I$ we estimate the likelihood function for the data $P(\mathbf{x}_i|\Omega_I)$. We assume the intrapersonal difference images to be Gaussian distributed. Upon performing a Karhunen–Loeve transform [68] on the training data we obtain the basis vectors $\{\Phi_i\}_{i=1}^{N}$ that span the intrapersonal space. But due to the high dimensionality of data such a computation is infeasible. We perform PCA [69], and extract the k basis vectors $\{\Phi_i\}_{i=1}^{k}$ that capture 99% variance in the data. The space spanned by $\{\Phi_i\}_{i=1}^{k}$ corresponds to the principal subspace or the feature space F. The remaining basis vectors $\{\Phi_i\}_{i=k+1}^{N}$ span the orthogonal complement space or the error space $\bar{F}$. The likelihood function $P(x_i|\Omega_I)$ is estimated as

$$P(\mathbf{x}\,|\,\Omega_I) = \frac{\exp(-\frac{1}{2}(\mathbf{x}-\bar{\mathbf{x}})^T \Sigma^{-1}(\mathbf{x}-\bar{\mathbf{x}}))}{(2\pi)^{N/2}|\Sigma|^{1/2}}$$

$$= \frac{\exp\left(-\frac{1}{2}\sum_{i=1}^{N}\frac{y_i^2}{\lambda_i}\right)}{(2\pi)^{N/2}\prod_{i=1}^{N}\lambda_i^{1/2}}$$

$$\simeq \left[\frac{\exp\left(-\frac{1}{2}\sum_{i=1}^{k}\frac{y_i^2}{\lambda_i}\right)}{(2\pi)^{k/2}\prod_{i=1}^{k}\lambda_i^{1/2}}\right] \cdot \left[\frac{\exp\left(-\frac{\epsilon^2(\mathbf{x})}{2\rho}\right)}{(2\pi\rho)^{(N-M)/2}}\right]$$

$$= P_F(\mathbf{x}\,|\,\Omega_I) \cdot \hat{P}_{\bar{F}}(\mathbf{x}\,|\,\Omega_I) \qquad (3.3)$$

where $y_i = \Phi_i^T(x-\bar{x})$, are the principal component feature vectors, λ_i are the eigenvalues. The marginal density in the orthogonal complement space $\hat{P}_{\bar{F}}(\mathbf{x}\,|\,\Omega_I)$ is estimated using the error in PCA reconstruction $\epsilon^2(\mathbf{x}) = \|\tilde{\mathbf{x}}^2\| - \sum_{i=1}^{k} y_i^2$ and the estimated variance along each dimension in the orthogonal subspace, $\rho = \frac{1}{N-k}\sum_{i=k+1}^{N}\lambda_i$. The sum $\sum_{i=k+1}^{N}\lambda_i$ is estimated by fitting a cubic spline function on the computed eigenvalues $\{\lambda_i\}_{i=1}^{k}$ and subsequently extrapolating the function.

Next, from a set of extrapersonal image differences $\{\mathbf{z}_i\}_{i=1}^{M} \in \Omega_E$, we estimate the likelihood function for the data $P(\mathbf{z}_i\,|\,\Omega_E)$. Adopting a similar approach as earlier, the extrapersonal space is decomposed into two complementary spaces: the feature space and the error space. Since the assumption of Gaussian distribution of extrapersonal image differences may not hold, we adopt a parametric mixture model (mixture of Gaussian) to estimate the marginal density in the feature space and follow a similar approach as earlier to estimate the marginal density in the orthogonal complement space. We estimate the likelihood for the data as

$$\hat{P}(\mathbf{z}\,|\,\Omega_E) = P(\mathbf{y}\,|\,\Theta^*) \cdot \hat{P}_{\bar{F}}(\mathbf{z}\,|\,\Omega_E)$$

where

$$P(\mathbf{y}\,|\,\Theta) = \sum_{i=1}^{N_c} w_i N(\mathbf{y}; \mu_i, \Sigma_i) \qquad (3.4)$$

$$\Theta^* = argmax\left[\prod_{i=1}^{M} P(\mathbf{y}_i\,|\,\Theta)\right] \qquad (3.5)$$

$N(\mathbf{y};\mu_i,\Sigma_i)$ is Gaussian with parameters (μ_i,Σ_i) and w_i correspond to the mixing parameters such that $\sum_{i=1}^{N_c} w_i = 1$. We solve the estimation problem using the expectation-maximization algorithm [70]. N_c, the number of components that comprise the gaussian mixture model, was selected based on the distribution of the top three principal components in the feature space.

During the first stage of classification given a pair of age separated face images, we compute the difference image $\mathbf{x} = I_1 - I_2$. The a posteriori probability $P(\Omega_I \,|\, \mathbf{x})$ is computed using the Bayes rule.

$$P(\Omega_I \,|\, \mathbf{x}) = \frac{P(\mathbf{x} \,|\, \Omega_I)P(\Omega_I)}{P(\mathbf{x} \,|\, \Omega_I)P(\Omega_I) + P(\mathbf{x} \,|\, \Omega_E)P(\Omega_E)} \qquad (3.6)$$

The classification of the image difference as intrapersonal or extrapersonal is based on a maximum a posteriori (MAP) rule. For operational conditions, $P(\Omega_I)$ and $P(\Omega_E)$ are set equal and the difference image $\mathbf{x}$ is classified as intrapersonal if $P(\Omega_I \,|\, \mathbf{x}) > \frac{1}{2}$.

The second stage of classification deals with classifying the image pairs that were identified as intrapersonal, further based on their age-differences. We build the following age-difference based intrapersonal spaces $\Omega_1, \Omega_2, \Omega_3, \Omega_4$ for the age-difference categories 1–2 years, 3–4 years, 5–7 years, and 8–9 years, respectively. Next, from a set of age-difference based intrapersonal difference images we estimate the likelihood function $P(\mathbf{x} \,|\, \Omega_j)$, $j \in 1,2,3,4$ for each of the four age-difference categories. Given a difference image $\mathbf{x}$ that has been classified as intrapersonal, we compute the *a posteriori* probability $P(\Omega_i \,|\, \mathbf{x})$ with $i = 1,2,3,4$ as:

$$P(\Omega_i \,|\, \mathbf{x}) = \frac{P(\mathbf{x} \,|\, \Omega_i)P(\Omega_i)}{\sum_{j=1}^{4} P(\mathbf{x} \,|\, \Omega_j)P(\Omega_j)} \qquad (3.7)$$

Thus if $P(\Omega_i \,|\, \mathbf{x}) > P(\Omega_j \,|\, \mathbf{x})$ for all $j \neq i$, $i,j = 1,2,3,4$, then Ω_i is identified to be the class to which the difference image $\mathbf{x}$ belongs. Figure 3.4 gives a complete overview of the age-difference classifier.

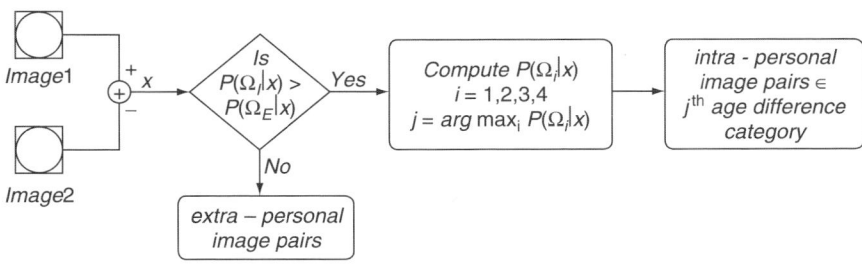

Fig. 3.4. An Overview of the Bayesian age-difference classifier

3.2.2 Experiments and Results

Using the above formulation, we performed classification experiments on the passport database. We selected pairs of face images of 200 individuals from our database. Using their intrapersonal image differences, we created the intrapersonal space Ω. Computing the extra personal difference images (by randomly selecting two images of different individuals from the 200 pairs of images) we created the extrapersonal space Ψ. We created two sets of image differences: set I is comprised of intrapersonal difference images computed from the face images of 465 image pairs and Set II is comprised of 465 extrapersonal difference images computed by a random selection of face images of different individuals. The image pairs from Set I and Set II were classified as either intrapersonal or extrapersonal.

During the second stage of classification, 50 pairs of face images from each of the following age-difference categories 1–2 years, 3–4 years, 5–7 years, and 8–9 years were randomly selected and their corresponding difference image subspaces namely Ω_1, Ω_2, Ω_3, Ω_4 were created. The image pairs from Set I that were classified as intrapersonal, were further classified into one of the four age-difference categories using the formulation discussed previously. The classification experiment was repeated many times using different sets of images from each age-difference category to create the intrapersonal spaces. The classification results are reported in Table 3.4 in the form of percentage of images under each category that were classified into one of the four classes. The mean and the standard deviations of the classification results generated from the many iterations are reported in Table 3.4. The bold entries in the table correspond to the percentage of image pairs that were correctly classified to their age-difference category. The entries within parenthesis denote the standard deviations.

The classification results are as follows :

- At the operating point, 99 % of the difference images from Set I were correctly classified as intrapersonal. 83 % of the difference images from Set II were correctly classified as extrapersonal. It was observed that the image pairs from Set I that were misclassified as extrapersonal differed from each other significantly either in facial hair or glasses. Moreover, the average age-difference of intrapersonal images that were misclassified was 7.4 years. The ROC plot in Fig. 3.5 was generated by varying the thresholds adopted for classification. The equal error rate was 8.5%.

Table 3.4. The overall results of the Bayesian age-difference classifier

type	class	1–2 years	3–4 years	5–7 years	8–9 years
	Ω_1	**41.0 (1.1)**	12.0 (6.9)	9.0 (5.0)	38.0 (7.2)
original set	Ω_2	8.0 (5.0)	**46.0 (5.6)**	8.0 (4.9)	37.0 (9.2)
of images	Ω_3	10.0 (3.3)	8.0 (6.3)	**53.0 (4.4)**	28.0 (6.9)
	Ω_4	10.0 (2.3)	12.0 (7.3)	5.0 (5.4)	**73.0 (8.2)**

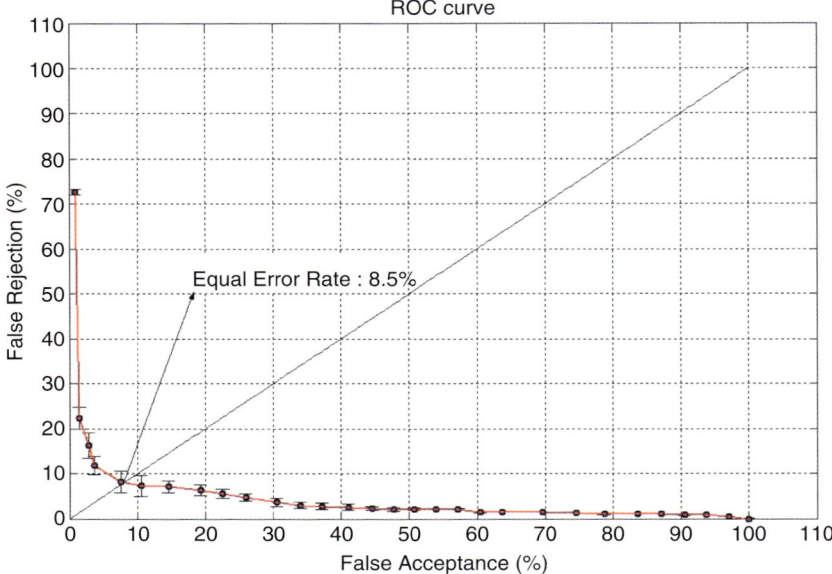

Fig. 3.5. Face verification results: ROC curve

- When the image pairs from Set I that were correctly classified as intrapersonal were classified further based on age-differences, it was observed that image pairs with little variations due to factors such as facial expressions, glasses and facial hair were more often classified correctly to their respective age-difference category.
- Image pairs, that belong to the age-difference categories 1–2 years or 3–4 years or 5–7 years, with significant differences in facial hair or expressions or glasses, were misclassified under the category 8–9 years. Since Ω_4 was built using images from the age-difference category 8–9 years, it spans more intra pair variations than that compared with the other three age-difference categories and hence the above trend is observed.

The eigenspace decomposition which forms an inherent part of the density estimation process reduces computational complexities significantly. Further, since the estimation of the class conditional density functions is an off-line process, the real-time computations involved in classifying image pairs based on age-differences are simple.

3.3 Facial Similarity

We designed the following experiment to study how age progression affects the measure of facial similarity. We created an eigenspace using 200 faces retrieved from the database of passport images. The 465 pairs of faces were projected onto the space of eigenfaces and were represented by the projections along the eigenfaces that correspond to 95% of the variance. We adopt the similarity measure as proposed in Sect. 3.2. Since illumination variations and pose variations across each pair of faces is minimal, the similarity score between each pair would be affected by factors such as age progression, facial expression variations, and occlusions due to facial hair and glasses. We divided our database into two sets: the first set is comprised of those images where each pair of passport images had similar facial expressions and similar occlusions if any, due to glasses and facial hair. The second set is comprised of those pairs of passport images where differences due to facial expressions or occlusions due to glasses and facial hair were significant.

The distribution of similarity scores across the age-difference categories namely 1–2 years, 3–4 years, 5–7 years, and 8–9 years is plotted in Fig. 3.6. The statistical variations in the similarity scores across each age-difference category and across each set of passport images are tabulated in Table 3.5.

- From Fig. 3.6 we note that as the age-difference between the pairs of images increases, the proportion of images with high similarity scores decreases.
- From Table 3.5 we note that as the age-difference increases, across both the sets of images and across all the variations such as expression, glasses, and

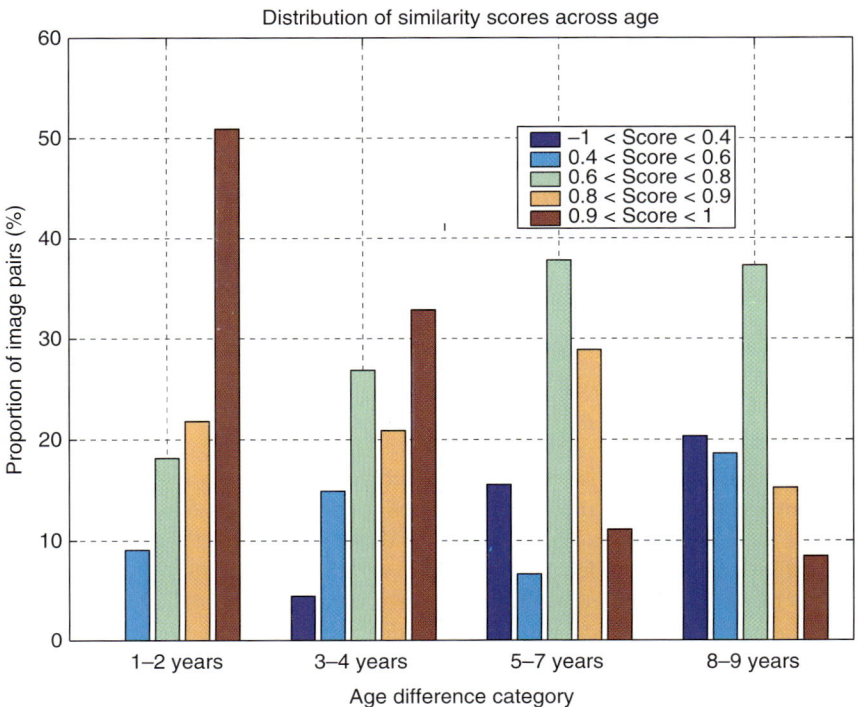

Fig. 3.6. Facial similarity across time: distribution of similarity scores across age

Table 3.5. Similarity measure

age-difference (years)	age-based similarity measure							
	first set		second set					
			expression		glasses		facial hair	
	μ	σ^2	μ	σ^2	μ	σ^2	μ	σ^2
1–2	0.85	0.02	0.70	0.021	0.83	0.01	0.67	0.04
3–4	0.77	0.03	0.65	0.07	0.75	0.02	0.63	0.01
5–7	0.70	0.06	0.59	0.01	0.72	0.02	0.59	0.10
8–9	0.60	0.08	0.55	0.10	0.68	0.18	0.55	0.10

facial hair, the mean similarity score drops gradually and the variance of the similarity scores increases.
– Within each age-difference category, we see a notable drop in similarity scores when variations due to expressions and facial hair are more pronounced.

3.4 Craniofacial Growth Model

In this section, we propose a craniofacial growth model that characterizes growth related shape variations commonly observed in children's faces across years. The model draws inspiration from the "revised" cardioidal strain transformation model and further accounts for anthropometric evidences collected on facial growth. Mathematically, the "revised" cardioidal strain transformation model is expressed as follows [56]. Let P denote the pressure at the particular point on the object surface acting radially outward. Let (R_0, θ_0) and (R_1, θ_1) denote the angular co-ordinates of a point on the surface of the object, before and after the transformation. Let k denote a growth related constant. Figure 3.7a illustrates the pressure distribution inside a fluid-filled spherical object. (A similar illustration appears in [71].)

$$\begin{aligned} P &\propto R_0(1 - cos(\theta_0)) \\ R_1 &= R_0 + k(R_0 - R_0 \cos(\theta_0)) \\ \theta_1 &= \theta_0 \end{aligned} \quad (3.8)$$

Face anthropometric studies report that different facial regions reach maturation at different years and hence a few facial features change relatively less when compared to other facial features, as age increases. In the context of the "revised" cardioidal strain transformation model, this observation translates into the fact that different regions of human faces have different growth parameters across age. Hence, it is important to incorporate anthropometric evidences collected on facial growth while developing the model, whereby we can reliably estimate the growth parameters for different regions of the human face across age. Farkas [72] provides a comprehensive overview of face anthropometry and its many significant applications. He defines face anthropometry in terms of measurements taken from 57 landmarks

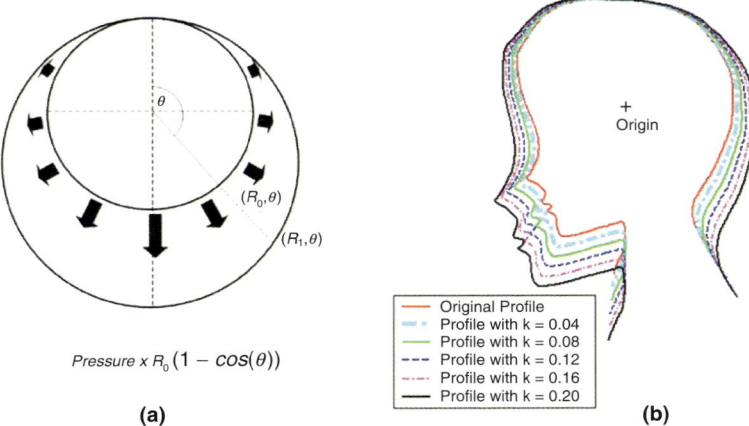

Fig. 3.7. (**a**) Remodeling of a fluid-filled spherical object. (**b**) Facial growth simulated on the profile of a child's face using the "revised" cardioidal strain transformations

on human faces. We use the age-based facial measurements and proportion indices (ratios of distances between facial landmarks) provided in [72, 73] to build the craniofacial growth model. Figure 3.8 illustrates the 24 facial landmarks and some of the important facial measurements that were used in our study.

3.4.1 Model Computation: An Optimization Problem

The origin of reference for the craniofacial growth model is located between landmarks "tr" and "n" along the facial midline axis [66]. Let the facial growth parameters of the "revised" cardioidal strain transformation model, that correspond to facial landmarks designated by [n, sn, ls, sto, li, sl, gn, en, ex, ps, pi, zy, al, ch, go] be [k_1, k_2, $\cdots$ k_{15}], respectively. The facial growth parameters for different age transformations can be computed using anthropometric constraints on facial proportions. The computation of facial growth parameters is formulated as a nonlinear optimization problem. We identified 52 facial proportions that can be reliably estimated using the photogrammetry of frontal face images. Anthropometric constraints based on proportion indices translate into linear and nonlinear constraints on selected facial growth parameters. While constraints based on proportion indices such as the *intercanthal index*, *nasal index*, etc., result in linear constraints on the growth parameters, constraints based on proportion indices such as *eye fissure index*, *orbital width index*, etc., result in nonlinear constraints on the growth parameters.

Let the constraints derived using proportion indices be denoted as $r_1(\mathbf{k}) = \beta_1$, $r_2(\mathbf{k}) = \beta_2, \cdots, r_N(\mathbf{k}) = \beta_N$. The objective function $f(\mathbf{k})$ that needs to be minimized w.r.t $\mathbf{k}$ is defined as

$$f(\mathbf{k}) = \frac{1}{2}\sum_{i=1}^{N}(r_i(\mathbf{k}) - \beta_i)^2 \qquad (3.9)$$

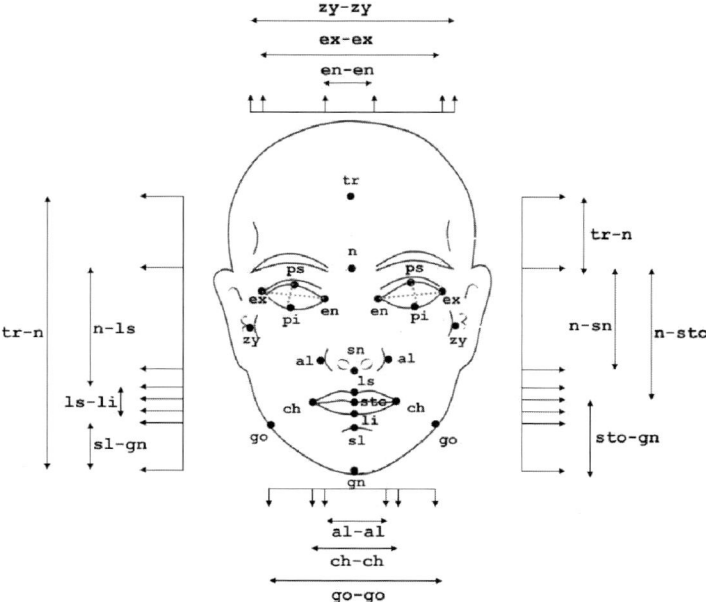

Fig. 3.8. Face anthropometry: of the 57 facial landmarks defined in [72], we choose 24 landmarks illustrated above for our study. We further illustrate some of the key facial measurements that were used to develop the growth model

The following equations illustrate the constraints that were derived using different facial proportion indices.

$$r_1 : \left[\frac{n-gn}{zy-zy} = c_1\right] \equiv \alpha_1^{(1)} k_1 + \alpha_2^{(1)} k_7 + \alpha_3^{(1)} k_{12} = \beta_1$$

$$r_2 : \left[\frac{al-al}{ch-ch} = c_2\right] \equiv \alpha_1^{(2)} k_{13} + \alpha_2^{(2)} k_{14} = \beta_2$$

$$r_3 : \left[\frac{li-sl}{sto-sl} = c_3\right] \equiv \alpha_1^{(3)} k_4 + \alpha_2^{(3)} k_5 + \alpha_3^{(3)} k_6 = \beta_3$$

$$r_4 : \left[\frac{sto-gn}{gn-zy} = c_4\right] \equiv \alpha_1^{(4)} k_5 + \alpha_2^{(4)} k_7 + \alpha_3^{(4)} k_{12} + \alpha_4^{(4)} k_4^2 + \alpha_5^{(4)} k_7^2$$
$$+ \alpha_6^{(4)} k_{12}^2 + \alpha_7^{(4)} k_4 \, k_7 + \alpha_8^{(4)} k_7 \, k_{12} = \beta_4$$

(α_j^i and β_i are constants. c_i is age-based proportion index obtained from [72].)

We use the Levenberg–Marquardt nonlinear optimization algorithm [74] to compute the growth parameters that minimize the objective function in an iterative fashion. We use the craniofacial growth model defined in (3.8) to compute the initial estimate of the facial growth parameters. The initial estimates are obtained using

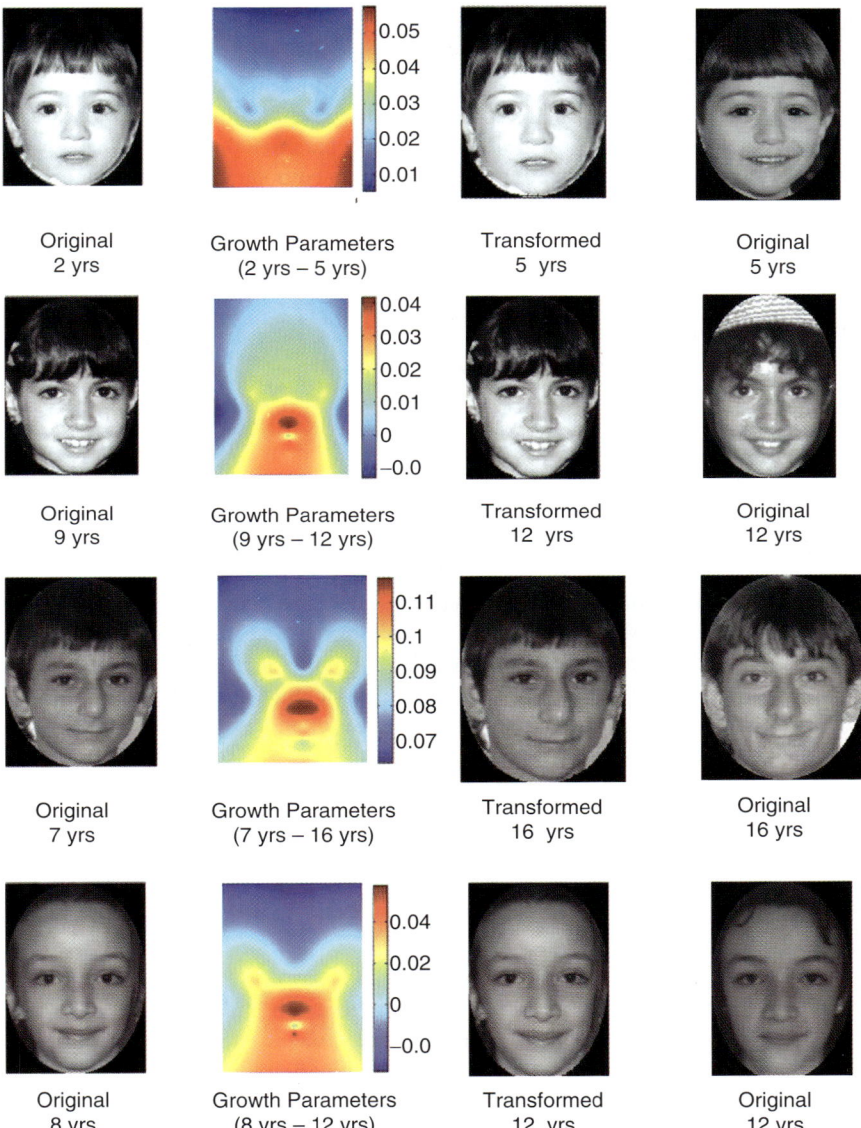

Fig. 3.9. Age transformation results on different individual (the original images shown above were taken from the FG-Net database [76])

the age-based facial measurements provided for each facial landmark, individually. The iterative step involved in the optimization process is defined as

$$\mathbf{k}_{i+1} = \mathbf{k}_i - (\mathbf{H} + \lambda diag[\mathbf{H}])^{-1} \nabla f(\mathbf{k}_i) \qquad (3.10)$$

where $\nabla f(\mathbf{k}_i) = \sum_{i=1}^{N} r_i(\mathbf{k}) \nabla r_i(\mathbf{k})$ and **H** corresponds to the Hessian matrix evaluated at $\mathbf{k}_i$. At the end of each iteration, λ is updated as illustrated in [74]. Next, using the growth parameters computed over selected facial landmarks, we compute the growth parameters over the entire face region. This is formulated as a scattered data interpolation problem [75]. Thus, the growth parameters computed at selected facial features using face anthropometry are used to compute the growth parameters over the entire facial region. Upon computing the growth parameters, the proposed craniofacial growth model can be applied to automatically age the face image. For an age transformation from "p" years to "q" years $(q > p)$, the model takes the form similar to the one defined in (3.8). On a polar coordinate framework, the transformation is defined as

$$R_q^i = R_p^i(1 + k_{pq}^i(1 - \cos(\theta_p^i)))$$
$$\theta_q^i = \theta_p^i \tag{3.11}$$

where i corresponds to the i'th facial feature and k_{pq}, the growth parameters for a transformation from "p" years to "q" years.

Figure 3.9 shows some of the age transformation results obtained using the proposed model. The growth parameters illustrate the different growth rates that are observed over different facial regions, across ages. The proposed appearance prediction model serves well the purpose of performing face recognition across age progression on children's face images [66] as the model facilitates better facial feature alignment between age separated face images of children. The model implicitly accounts for the different growth spurts observed in adolescent boys and girls, as the model is built using anthropometric data that pertains to the growth observed in boys and girls, separately. The anthropometric measurements provided in [72] were retrieved from Caucasian faces and hence the model should perform commendably well in predicting Caucasian faces across age progression.

3.5 Conclusions

While there have been considerable amounts of research in developing face recognition systems that are robust to variations in face images due to illumination, head pose orientations, facial expressions, etc., research pertaining to the natural process of facial aging is still in its nascent stages. Developing automatic age progression systems for children and adults, would further enhance the performance of face recognition systems. Developing systems that can automatically detect the age of individuals from their face images would have a significant impact on face processing applications. In future, we wish to develop automatic age estimation systems and automatic age progression systems for adult faces.

4 Quality Assessment and Restoration of Face Images in Long Range/High Zoom Video

Yi Yao, Besma Abidi, and Mongi Abidi

4.1 Introduction

Over the last two decades, substantial developments have been made in face recognition research. However, most of the efforts are limited to close range scenarios, which are well suited for applications with controlled distances, such as identity verification at access points. Little attention is paid to long range face related research. The rapidly increasing need for long range surveillance and wide area monitoring calls for a passage in face recognition from close-up distances to long distances and accordingly from low and constant camera zoom to high and adjustable zooms. The research work described herewith serves this purpose and establishes the foundation for long range face related research.

Before continuing our discussion, we first give in Table 4.1 the designation of different ranges of system magnifications and observation distances for near-ground surveillance (both indoor and outdoor).

4.1.1 Scope

In this effort, we first describe our long range/high magnification face video database. Both indoor and outdoor sequences are collected under uncontrolled surveillance conditions. The significance of this database lies in the fact that it is the first database to provide face images from long distances (indoor: 10–20 m and outdoor: 50–300 m) and high magnifications (indoor: $10\times$ to $20\times$ and outdoor: $60\times$ to $375\times$). The database has applications in experimentations with human identification and authentication in long range surveillance and wide area monitoring. The database will be made public to the research community for perusal towards long range face related research.

Our database has the following distinguishing characteristics. (1) Although many face databases have been collected in the past, such as XM2VTS [77], BANCA [78], FERET [79], CAS-PEAL [80], and CMU PIE [81], according to the definitions in Table 4.1, most of the existing face databases fall into the category of low magnification with very few achieving medium magnification. The database collected by the University of Texas at Dallas involves medium distance sequences [82]. Their parallel walking videos start from a distance of 13.6 m and their perpendicular walking videos are collected from a distance of 10.4 m. However, their camera zoom remains low and constant. In comparison, our database

Table 4.1. Designation of magnification/distance ranges

range	low/short	medium	high/long	extreme
magnification (×)	1–3	3–10	10–30	>30
distance (m)	< 3	3–10	10–100	>100

aims at high to extreme magnifications and long to extreme distances. For indoor sequences, high magnifications (10× to 20×) are used while for outdoor sequences extreme magnifications are obtained with a maximum of 375×. As a result, degradations induced by high magnification and long distance, such as magnification blur, are systematically present in the data. (2) Our database closely resembles the real near-ground surveillance conditions (illumination changes caused by nonuniform roof light, air turbulence, and subject motion) and, more importantly, includes the effect of camera zooming, which is commonly ignored by other existing databases. Furthermore, sequences with various combinations of still/moving subjects and constant/variable camera zoom are collected for the study of individual and combined effects of target and camera motions.

Apart from illumination, pose, and expression, magnification blur is identified as an additional major deteriorating source for long range face recognition. To describe the corresponding degradations, a face image quality measure is developed. Since imaging conditions, such as image noise level and available incoming lights, fluctuate considerably with respect to system magnifications and observation distances, conventional sharpness measures, sensitive to image noise and brightness, are not useful. Developed from gradient-based sharpness measures, a class of adaptive sharpness measures is proposed, where special weight functions are employed to suppress artificially elevated sharpness values from increased image noise. The proposed adaptive sharpness measures have been successfully applied to high magnification systems by the authors [83] to quantify magnification blur. In this chapter, adaptive sharpness measures are used to evaluate face image quality, predict overall FRR, and determine whether subsequent enhancement is necessary.

To compensate for the decrease in FRR caused by high magnification and long observation distance, several image enhancement algorithms are implemented and their performances compared. Wavelet-based approaches, capable of multiscale processing, are selected. Unsharp masking (UM) and regularized deconvolution are exploited to enhance facial features from the approximation coefficients transformed by the Harr wavelet. Both methods prove efficient and result in a substantially improved FRR.

4.1.2 Related Work

According to the scope of this effort, a brief review of related work in literature is given in the following two aspects: face image quality measures and face image deblurring algorithms. In addition, since the proposed face image quality measure is developed from gradient-based sharpness measures [83], an overview of existing gradient-based sharpness measures is also conducted.

Face Quality Measures

Cost functions have been used in literature to describe the probability of an area being a face image. The term face quality assessment was first explicitly introduced by Identix [84], where a face image is evaluated according to the confidence of detectable eyes, frontal face geometry, resolution, illumination, occlusion, contrast, focus, etc. Kalka et al. applied the quality assessment metrics originally proposed for iris [85] to face images. Criteria such as lighting (illumination), occlusion, pixel count between eyes (resolution), and image blurriness caused by both out-of-focus and motion are considered. Xiong et al. developed a metric based on bilateral symmetry, color, resolution, and expected aspect ratio (frontal face geometry) to determine whether a detected face image in a surveillance video is suitable to be added to an on-the-fly database [86].

Face Image Deblurring

Apart from numerous image deblurring algorithms, such as adaptive unsharp masking [87, 88] and regularized image deconvolution [89], algorithms are proposed especially for face deblurring by making use of known facial structures. Fan et al. incorporated prior statistical models of the shape and appearance of a face into the regularized image restoration formulation [90]. A hybrid recognition and restoration architecture was described by Stainvas and Intrator [91], where a neural network is trained by both clear and blurred face images. Liao et al. applied Tikhonov regularization to eigenface subspaces to overcome the algorithm's sensitivity to image noise [92].

Sharpness Measures

Sharpness measures are traditionally used to evaluate out-of-focus blur and can be grouped into the following categories [93]: gradient-based, variance-based, correlation-based [94], histogram-based [95], and frequency domain-based methods [96–98]. With the development of practical edge detectors, edge-based sharpness measures have attracted increasing attention [99–101]. Meanwhile, sharpness measures using wavelet transform also came into view [102, 103]. A detailed survey regarding existing sharpness measures along with performance comparisons can be found in [104]. We will focus on gradient-based methods herewith, from which the proposed adaptive measures are developed.

Gray level differences among neighboring pixels provide a reasonable representation of image sharpness. Image gradient obtained by differencing or using high pass filters are abundant in literature. Different forms of gradients can be used [93] (1) the absolute gradient defined as

$$S = \sum_M \sum_N |I(x, y+n) - I(x,y)| + |I(x+n, y) - I(x,y)|,$$

(2) the squared gradient given by

$$S = \sum_M \sum_N \sqrt{|I(x, y+n) - I(x,y)|^2 + |I(x+n, y) - I(x,y)|^2},$$

and (3) the maximum gradient formulated as

$$S = \sum_M \sum_N max\{I(x, y+n) - I(x,y)|, |I(x+n, y) - I(x,y)|\},$$

where $I(x, y)$ represents the image intensity, M/N denotes the total number of image rows/columns, and n is the differencing step. The absolute gradient with $n = 1$ is also called Sum-Modulus-Difference (SMD) and the case with $n = 2$ is commonly referred to as the Brenner measure [93]. The most well-known measure based on high pass filters is the Tenengrad measure [95] which is given by:

$$S = \sum_M \sum_N \left[I_x(x,y)^2 + I_y(x,y)^2\right], while \sqrt{I_x(x,y)^2 + I_y(x,y)^2} \geq T,$$

with the horizontal and vertical gradients, $I_x(x,y)$ and $I_y(x,y)$, obtained using the Sobel filters. The Laplacian filter is another popular choice [95], where the sharpness is defined as

$$S = \sum_M \sum_N |I_L(x,y)|, while |I_L(x,y)| \geq T,$$

with $I_L(x,y) = I(x,y) * h(x,y)$ and $h(x,y)$ a Laplacial filter. Choi et al. utilized a linear combination of multiple median filters, referred to as frequency selective weighted median (FSWM) filter [105].

4.1.3 Chapter Organization

The remainder of this chapter is organized as follows. Section 4.2 describes our high magnification face video database. The face quality measure and enhancement algorithms are discussed in Sects. 4.3 and 4.4. The efficiency of the proposed algorithms is validated via experimental results in Sect. 4.5. Section 4.6 concludes this chapter.

4.2 Database Acquisition

Our database collection, including indoor and outdoor sessions, began in February 2006 and is scheduled to be finished in October 2006. The final deliverable contains frontal view face images collected with various system magnifications ($10\times$ to $375\times$), different observation distances (10–300 m), indoor (office roof light and side light) and outdoor (sunny and partly cloudy) illuminations, still/moving subjects, and constant/varying camera zooms. Small expression and pose variations are also included in the video sequences of our database, closely resembling the variations encountered in uncontrolled surveillance applications.

The indoor and outdoor sessions share the same gallery, which is collected by a Canon A80 camera under controlled indoor environment, as shown in Fig. 4.2a. The observation distance for the gallery images is 0.5 m. The image resolution is $2,272 \times 1,704$ pixels and the camera's focal length 114 mm (magnification: $2.28\times$).

4.2.1 Indoor Sequence Acquisition

For the indoor sequence collection, the observation distance is varied from 10 to 16 m. Given this distance range and an image resolution of 640×480, a $22\times$ system magnification is sufficient to yield a face image with 60 pixels between the subject's eyes.[1] Therefore, a commercially available PTZ camera (Panasonic WV-CS854) is used. An ImperX VCE-PRO grabber is employed to capture the analog video signals from the Panasonic camera. A graphical user interface (GUI), shown in Fig. 4.1b, is developed for real-time video collection and camera control including camera zoom and focus.

Our indoor database includes both still images (7 images per subject) and video sequences (6 sequences per subject). Still images are collected at uniformly distributed distances in the range of 10–16 m with an interval of 1 m approximately. The corresponding system magnifications vary from $10\times$ to $20\times$ with an increment of $2\times$, achieving an approximately constant face image size to eliminate the influence of resolution. A still image is also taken with low magnification ($1\times$) and from close distance (1 m), as shown in Fig. 4.2b. The performance of face recognition algorithms on the $1\times$ images is used as a reference for higher magnification images. Still images will be used for the study of degradations from high magnifications.

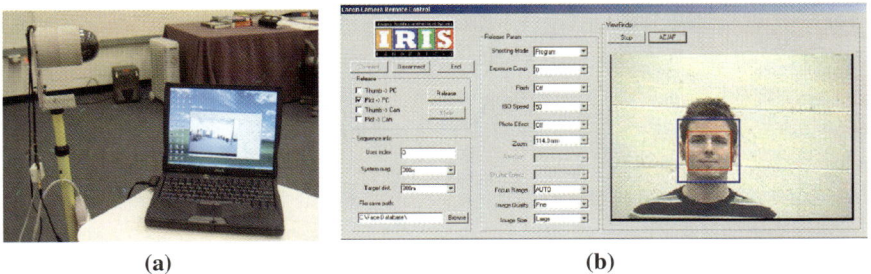

(a) (b)

Fig. 4.1. The indoor sequence collection system. (**a**) Panasonic PTZ camera and (**b**) GUI. The *blue and red rectangles* depict the position and image area of the subject's face and T-zone (including eyes, nose, and mouth), respectively

---

[1] A minimum distance of 60 pixels between eyes is recommended by FaceIt® for successful face recognition.

The observation distance and system magnification are two major factors, to which this effort is devoted. Meanwhile, the effect of composite target and camera motions are included to achieve a close resemblance to practical surveillance scenarios. Therefore, indoor video sequences are recorded under the following three conditions (1) constant distance and varying system magnification, (2) varying distance (the subject walks at a normal speed towards the observation camera) & constant system magnification, and (3) varying distance and varying system magnification. Conditions 1 and 2 concentrate on the individual effect of camera zooming and subject motion while the combined effect can be observed in condition 3. In addition, system magnification is varied so that a constant face image size is obtained in condition 3. These video sequences can be used for studies of resolution, target motion, and camera zooming.

The above still images and video sequences are collected under fluorescent roof lights with full intensity (approximately 500 lx) and include a certain degree of illumination changes caused by the nonuniformly distributed roof lights. Our indoor database also considers a large amount of illumination changes under high magnification. A halogen side light (approximately 2,500 lx) is added and a sequence is recorded as the intensity of the roof lights is decreased from 100 to 0%, which creates a visual effect of a rotating light source. Figure 4.2 illustrates the still images of one data record in the database and Table 4.2 summarizes the specifications of the collected video sequences.

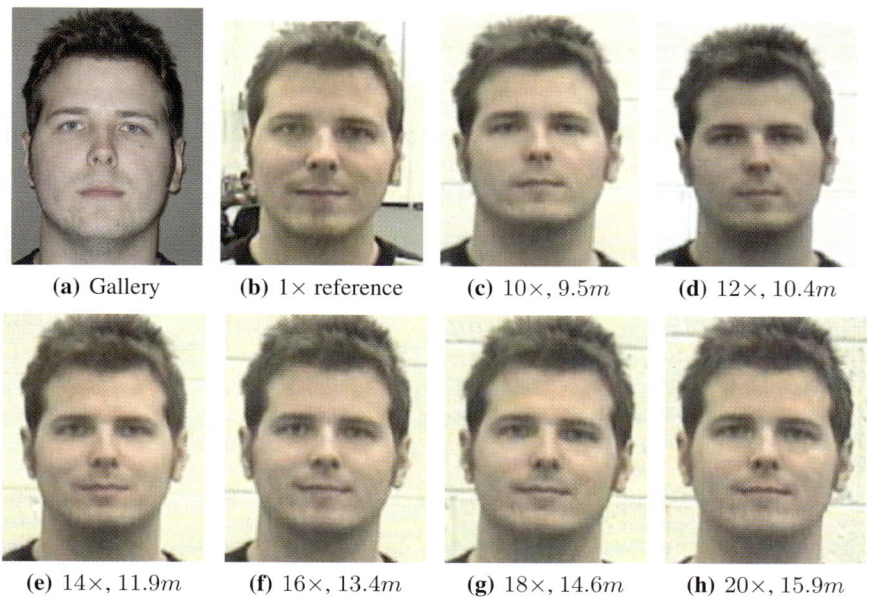

(a) Gallery (b) 1× reference (c) 10×, 9.5m (d) 12×, 10.4m

(e) 14×, 11.9m (f) 16×, 13.4m (g) 18×, 14.6m (h) 20×, 15.9m

Fig. 4.2. A set of still images in one data record

Table 4.2. Indoor video sequence specifications

conditions	system magnification ($\times$)	distance (m)
constant distance and varying system mag.	$10 \rightarrow 20$	13.4 and 15.9
varying distance and constant system mag.	10 and 15	$9.5 \rightarrow 15.9$
varying distance and constant system mag.	$10 \rightarrow 20$	$9.5 \rightarrow 15.9$
illumination changes	20	15.9

The indoor session has 55 participants (78% male and 22% female). The ethnic diversity is defined as a collection of 73% Caucasian, 13% Asian, 9% Asian Indian, and 5% of African Descent.

4.2.2 Outdoor Sequence Acquisition

For the outdoor sequence acquisition, a composite imaging system is built where a Meade ETX-90 telescope (focal length: 1,250 mm) is coupled with a Canon A80 camera (35 mm equivalent focal length: 38–114 mm) for still image collection and a Panasonic VDR-M53 camcorder (focal length: 2.1–50.4 mm) for video sequence collection via various eyepieces following an afocal connection (Fig. 4.3a). In order to achieve the required resolution (60 pixels between the subject's eyes) at long to extreme distances (50–300 m), three eyepieces are used: Meade 4.7 mm, Meade 26 mm, and Celestron 40 mm. Accordingly, the achievable system magnification range is $24\times$ to $800\times$. Still images and video sequences are collected at uniformly distributed distances in the range of 50–300 m with an interval of 50 m. The corresponding system magnification varies from $60\times$ to $375\times$ with an increment of $60\times$ approximately . Our outdoor database is still in the collection phase and is scheduled to be finished in October 2006.

4.3 Face Image Quality Assessment

Long distance and high magnification introduce severe and nonuniform blur, which is unique to our database in comparison to most existing databases usually collected from close distances and with low magnifications. Our first priority is to examine the effect of high magnification blur on face image quality and the overall FRR.

4.3.1 Face Recognition Rate vs. System Magnification

We employ FaceIt® [106] as an evaluation tool and focus on the first rank performance and the cumulative match characteristic (CMC) measure. The CMC measure is a quantified measurement of a CMC curve defined as $Q_{CMC} = \sum_{i=1}^{K} C_i/i$, where K is the number of ranks considered and C_i denotes the percentage of probes correctly recognized at rank i [107]. In the following experiments, $K = 10$ is used.

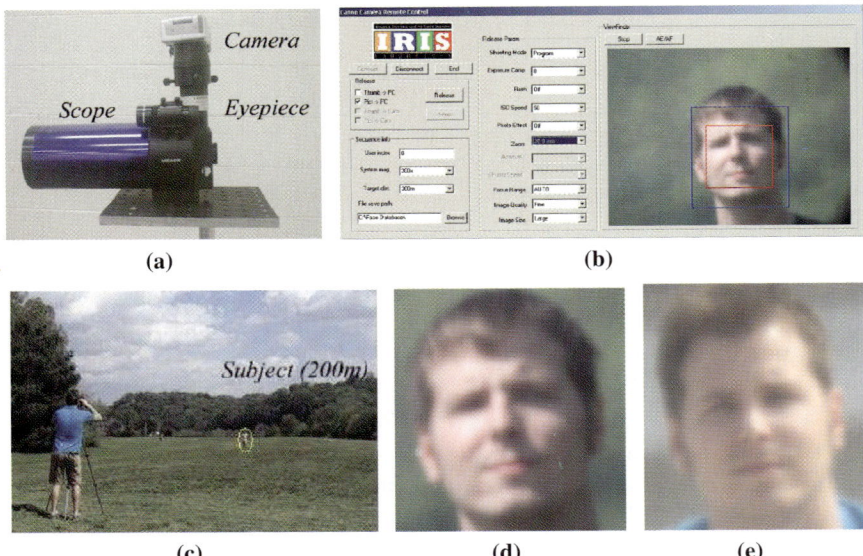

Fig. 4.3. The outdoor sequence collection system: (**a**) composite imaging system and (**b**) GUI. (**c**) Data collection scene. Still images: (**d**) 250×, 200 m and (**e**) 375×, 300 m, severely blurred by high magnification and air turbulences

The relationship between face recognition performance, characterized by CMC, and system magnification is illustrated in Fig. 4.4 and Table 4.3. It is obvious that deterioration from limited available fine details causes the FRR to drop gradually as the system magnification increases. From magnification 10× to 20×, the CMC measure declines from 74.3 to 58.8%. Although, the face resolution is kept constant, there exists a significant performance gap between the low (1×) and high (20×) magnification images, which reveals that magnification blur is an additional major degrading factor in long distance face recognition. This performance degradation is to be quantified by a face image quality measure and compensated for by image postprocessing.

4.3.2 Adaptive Sharpness Measures

Sharpness measures have been traditionally proposed to evaluate out-of-focus blur. Nevertheless, their extension to quantify magnification blur is nontrivial. Since image noise level increases with system magnification, conventional sharpness measures, sensitive to image noise, are not applicable. To avoid artificially elevated sharpness values due to image noise, adaptive measures are proposed [83]. In order to differentiate variations caused by actual image edges from those introduced by image noise and artifacts, adaptive sharpness measures assign different weights

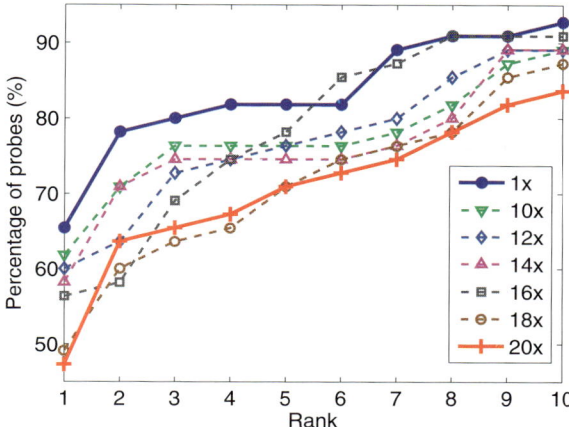

Fig. 4.4. CMC comparison across system magnifications. The gallery consists of face images with high resolution and collected from a close distance. The probe images differ only in system magnification and observation distance

Table 4.3. CMC measure and rank 1 performance comparison across system magnifications

system magnification ($\times$)	1	10	12	14	16	18	20
CMC measure (%)	74.3	69.7	67.3	67.5	64.9	59.4	58.8
CMC at rank 1 (%)	65.5	61.8	60.0	58.2	56.4	49.1	47.3

to pixel gradients according to their local activities. For pixels in smooth areas, small weights are used. For pixels adjacent to strong edges, large weights are allocated.

The definition of local activity and the selection of weight functions are two major factors in constructing adaptive sharpness measures. According to the description of local activity, sharpness measures can be divided into two groups: separable and nonseparable. As the name suggests, separable methods consider horizontal and vertical edges independently, while nonseparable methods include the contributions from diagonal edges. For separable measures, two signals are constructed, a horizontal

$$g_x(x, y) = I(x + 1, y) - I(x - 1, y)$$

and a vertical

$$g_y(x, y) = I(x, y + 1) - I(x, y - 1).$$

For nonseparable methods, the local activities are given by

$$g(x, y) = I(x - 1, y) + I(x + 1, y) - I(x, y - 1) - I(x, y + 1).$$

Different forms of weights can be used, among which polynomial and rational functions are two popular choices. The polynomial, to be more specific cubic, and rational functions are also exploited in adaptive unsharp masking [87,88]. The polynomial weights suppress small variations, mostly introduced by image noise, and have been proved efficient in evaluating the sharpness of high magnification images [83]. The rational weights emphasize a particular range of image gradients. Applying the nonseparable $g(x,y)$ for example, the polynomial weights can be written as

$$\omega(x,y) = g(x,y)^{p_\omega},$$

where p_ω is a power index determining the degree of noise suppression. The rational weights are given by:

$$\omega(x,y) = \frac{(2k_0 + k_1)g(x,y)}{g^2(x,y) + k_1 g(x,y) + k_0},$$

where k_0 and k_1 are coefficients associated with the peak position L_0 and width ΔL of the corresponding function, respectively, and comply with the following equations:

$$k_0 = L_0$$
$$k_1^2 + 8k_0 k_1 + 12k_0^2 - \Delta L^2 = 0.$$

Figure 4.5 illustrates the comparison of different forms of weight functions.

The newly developed weights are then applied to gradient-based sharpness measures to construct the adaptive sharpness measures. Considering the Tenengrad measure for instance, the resulting separable version is given by:

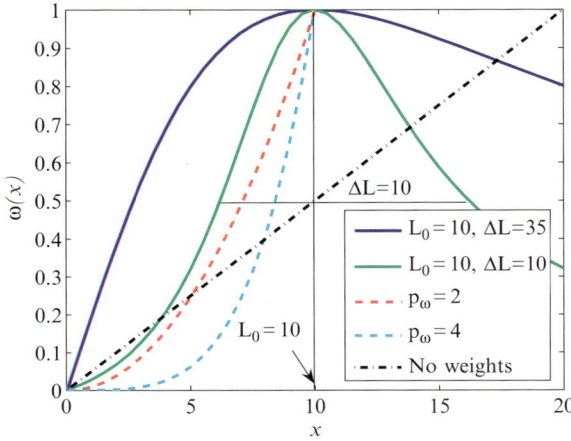

Fig. 4.5. Illustration of weight functions in the 1D case. *Solid curves*: rational functions and *dashed curves*: polynomial functions

$$S = \sum_M \sum_N [\omega_x(x,y) I_x(x,y)^2 + \omega_y(x,y) I_y(x,y)^2],$$

where $\omega_x(x,y)/\omega_y(x,y)$ denotes the weights obtained from $g_x(x,y)/g_y(x,y)$. For nonseparable methods, the corresponding adaptive Tenengrad is formulated as

$$S = \sum_M \sum_N \omega(x,y) [I_x(x,y)^2 + I_y(x,y)^2].$$

4.3.3 Image Sharpness and System Magnification

To validate the need for adaptive sharpness measures, noise levels of face images at various magnifications are first studied. The standard deviation of a uniform background patch closely describes the behavior of image noise and is computed with respect to system magnification, as shown in Fig. 4.6. Image noise level increases as system magnification increases. Therefore, to exclude the artificially elevated sharpness values from increased noise levels, adaptive sharpness measures are preferred.

Figure 4.7 shows the computed sharpness values (nonseparable Tenengrad with polynomial weight of degree 2) for all still face images in our database. The mean sharpness values give a clearer view of the overall performance with respect to system magnification. As expected, image sharpness decreases as system magnification increases. The decrease in FRR caused by magnification blur is consistent with the behavior of image sharpness measures. Therefore, we could use sharpness measures as an indicator not only for the degree of magnification blur but also for achievable recognition rates.

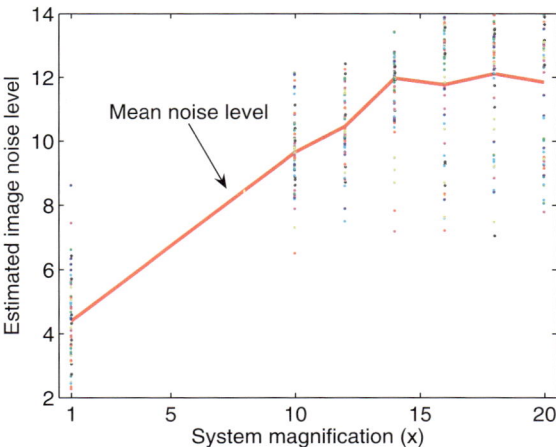

Fig. 4.6. Face image noise level. Image gray level: 0–255. *Dots* represent the noise variances computed from face images of different subjects and system magnifications

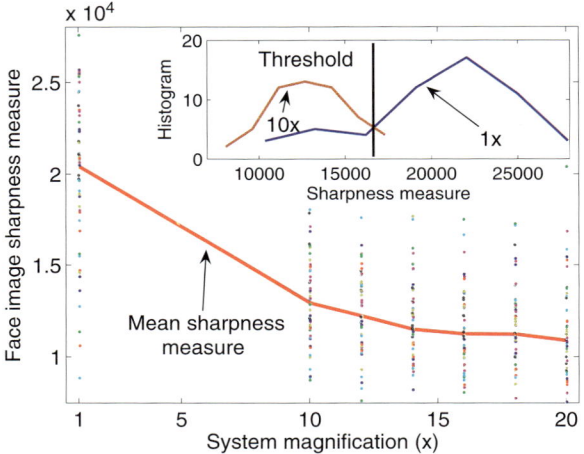

Fig. 4.7. Face image sharpness measures. *Dots* represent the sharpness measures computed from face images of different subjects and system magnifications

From the distribution of these sharpness values, especially those of 1× and 10× images (Fig. 4.7), a threshold (the intersection point) can be derived, $S_{th} = 16,600$, which separates the tested face images into two groups: one with acceptable sharpness ($S \geq S_{th}$) and the other ($S < S_{th}$) severely degraded by magnification blur. Images in the first group contain sufficient facial features and thus will not deteriorate the overall FRR. On the contrary, images in the second group, deficient in necessary facial features, require image enhancement and/or restoration so that the overall FRR can be maintained.

The threshold $S_{th} = 16,600$ is obtained empirically and is application dependent. In practice, the sharpness measures of low magnification images, usually the gallery images, can be computed and their statistics, such as the mean S_0 and the standard deviation σ, can be estimated. The threshold can then be defined as $S_{th} = S_0 - \sigma$. The threshold can also be estimated and updated on-the-fly by studying the distributions of face image sharpness at various magnifications.

4.4 Face Image Enhancement

As illustrated in the Sect. 4.3.3, high magnification images suffer from both increased image blur and noise levels. In general, deblurring algorithms are prone to aggravated image noise, while denoising algorithms usually smooth out image details. The resulting images are either short of details or overwhelmed by elevated image noise. Since FaceIt® is sensitive to both degradations, a good balance needs to be found for an optimal FRR. Multiscale processing based on wavelet transform proves to be the most promising candidate.

For each probe, the face image sharpness is computed and its value is compared with a predefined threshold S_{th}. If the current sharpness value is smaller than S_{th}, image enhancement is performed. Otherwise, no postprocessing is conducted. In so doing, only those images which may deteriorate the overall FRR are processed. Images with acceptable sharpness are fed to the face recognition engine directly to prevent a possible increase in image noise from unnecessary enhancement. The importance of choosing an efficient measure of face image quality becomes evident. Another advantage of using a face quality measure is attributed to the reduced computational complexity, which is also crucial to real-time applications. The block diagram of the proposed algorithm is depicted in Fig. 4.8.

The proposed algorithm proceeds as follows. (1) Compute the sharpness measure of the input face image: S. (2) If $S < S_{th}$, go to step (3). Otherwise, go to step (1) and wait for the next probe. (3) Decompose the face image via the Harr wavelet transform of level 1. (4) Apply deblurring algorithms to the approximation image and denoising algorithms to the vertical/horizontal/diagonal detail images. (5) Apply adaptive grey level contrast stretching. (6) Reconstruct the output image via the Harr wavelet transform.

A straightforward thresholding is applied for denoising all detail images. Two types of deblurring algorithms, UM, and regularized deconvolution, are implemented to enhance the approximation image. The UM method follows its traditional implementation and a Laplacian filter is used. Our regularized deconvolution utilizes the Lasso regularization [108].

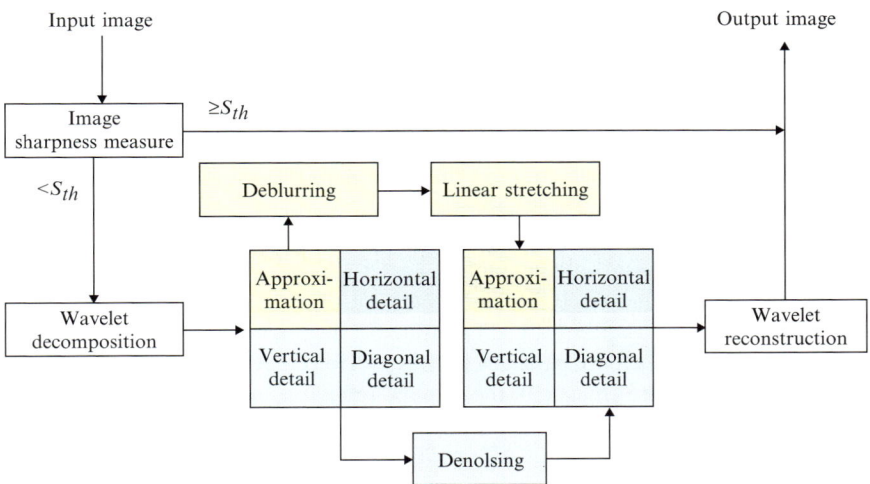

Fig. 4.8. Block diagram of high magnification image enhancement

A typical regularized deconvolution solves the following minimization problem:

$$f_\lambda = argmin\{||Af - f_b||^2_{\mathcal{L}_2} + \lambda||Lf||^2_{\mathcal{L}_2}\},$$

where f and f_b are the original and blurred images in vector format, A and L represent the blurring filter and a predefined mask in vector format, and λ denotes the regularization parameter. Various forms of L can be found in literature, among which the identity matrix and Laplacian filter are two popular choices [109, 110]. The Tikhonov regularization uses norm-2 definition and does not allow discontinuities in the solution, leading to overall smoothed edges in the restored images. The total variation (TV) regularization is proposed to preserve edges in the reconstructed images [111], where a $\mathcal{L}_1$ definition is adopted. The corresponding regularization term is $||\sqrt{f_x^2 + f_y^2}||_{\mathcal{L}_1}$, where f_x and f_y denote the vertical and horizontal image gradients in vector format. The TV regularization is capable of preserving edges but suffers from significantly increased computational complexity. In our implementation, we utilize the Lasso regularization and design the regularization term as $||f||_{\mathcal{L}_1}$ [108]. The Lasso regularization achieves similar edge preservation as the TV regularization with substantially reduced computations.

4.5 Result Validation

For the purpose of performance comparison, we chose adaptive UM and Lucy–Richardson deconvolution (LRD) from general deblurring methods. Also implemented is the algorithm proposed by Liao and Lin [92] for face deblurring. Fan's algorithm [90] is not selected due to the difficulties in establishing an accurate face statistical model based on the limited number of available samples in our database. The hybrid recognition and restoration structure [91] is not feasible since our approach regards recognition and restoration as two separate stages. The face recognition is conducted using an existing commercial tool, FaceIt®, keeping our focus on restoration only. Experiments are conducted for images at all magnifications and similar observations are obtained. In the interest of space, only the comparisons at $20\times$ (Fig. 4.9 and Table 4.4) and $10\times$ (Fig. 4.10 and Table 4.5) are demonstrated. Note that since the major concern of our work is the ability of the tested enhancement algorithms to compensate for degradations caused by high system magnifications, the performances of these various enhancement algorithms are compared with the performance of the original reference images at $1\times$.

Wavelet-based methods are able to achieve the most improvement with a relative increase of 25 and 26% in CMC measure for the UM and Lasso regularized deconvolution approaches, respectively, yielding a performance comparable to the $1\times$ reference. With proper postprocessing, the degradation in FRR caused by magnification blur can be successfully compensated for. Wavelet-based methods, capable of multiscale processing, can reduce image noise and enhance image details simultaneously, thus resulting in better FRR performances. Compared with the UM-based

4 Quality Assessment and Restoration of Face Images 57

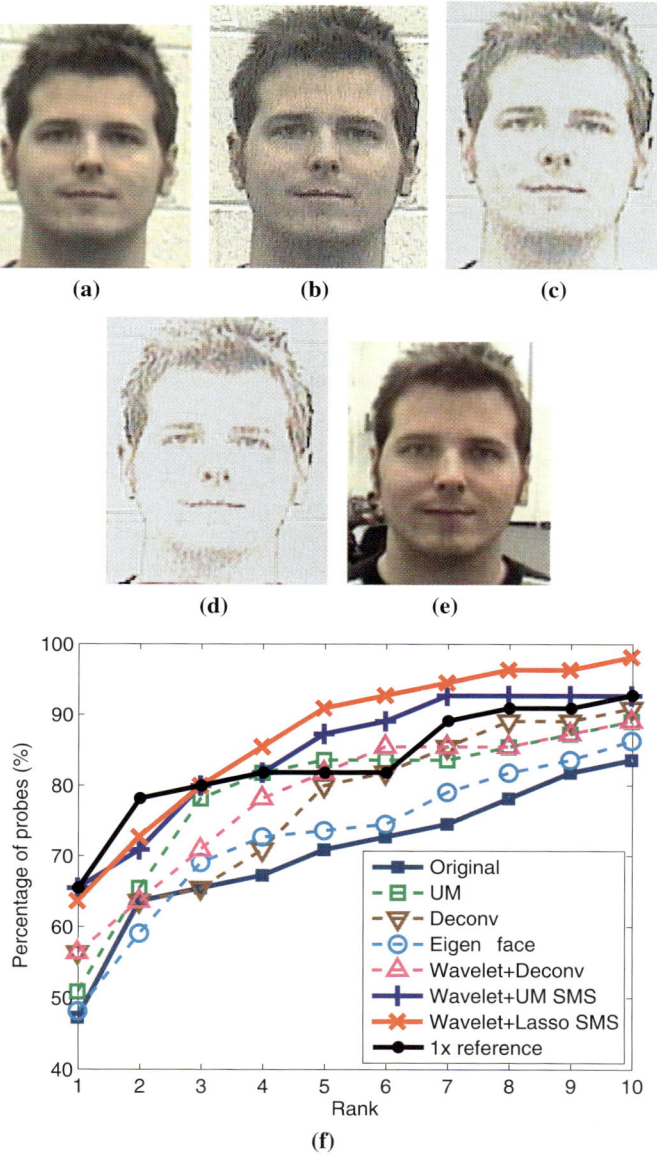

Fig. 4.9. (**a**) 20× original image. (**b**) Enhanced by UM. (**c**) Enhanced by wavelet transform with the approximation image processed by UM. (**d**) Enhanced by wavelet transform with the approximation image processed by Lasso regularized deconvolution. (**e**) 1× reference image. (**f**) CMC comparison across enhancement algorithms. The performances of the algorithms with and without SMS (skipped for clear presentation) are identical

Table 4.4. CMC measure and rank 1 performance comparison across enhancement algorithms for 20× images

image (×)	CMC measure (%)	CMC at rank 1 (%)
20× original	58.8	47.3
Eigenface [92]	59.7	48.2
UM	64.3	50.9
LRD	65.3	56.4
wavelet + LRD	66.0	56.4
wavelet + UM	73.6	65.5
wavelet + UM SMS	73.6	65.5
wavelet + Lasso	74.0	63.6
wavelet + Lasso SMS	74.0	63.6
1× reference	74.3	65.5

Table 4.5. CMC measure and rank 1 performance comparison across enhancement algorithms for 10× images

image (×)	CMC measure (%)	CMC at rank 1 (%)
10× original	69.9	61.8
UM	65.8	54.5
LRD	66.1	54.5
wavelet + LRD	65.6	52.7
wavelet + UM	75.7	65.5
wavelet + UM SMS	73.6	63.6
wavelet + Lasso	75.7	63.6
wavelet + Lasso SMS	77.7	69.1
1× reference	74.3	65.5

approach, the Lasso regularized deconvolution method presents a slightly better performance, especially for higher rank recognition. Considering the increased computations required by image deconvolution, the Lasso regularized deconvolution method is well suited for applications placing more emphasis on accuracy, while the UM-based algorithm achieves a better balance between accuracy and computation complexity.

In this work, we also use sharpness measures to predict FRR at different system magnifications and determine whether a postprocessing is necessary. With a sharpness measure selection (SMS) based on the threshold derived from Fig. 4.7, 4 and 15% of the samples from the 20× and 10× original images meet the minimum criterion and hence require no postprocessing. For 20× images, the resulting performance from only processing the images with lower sharpness values than the threshold is identical to the case where all images are processed. As for 10× images, a slight performance improvement is observed from the Lasso regularized deconvolution method, which verifies the suitability of the derived threshold.

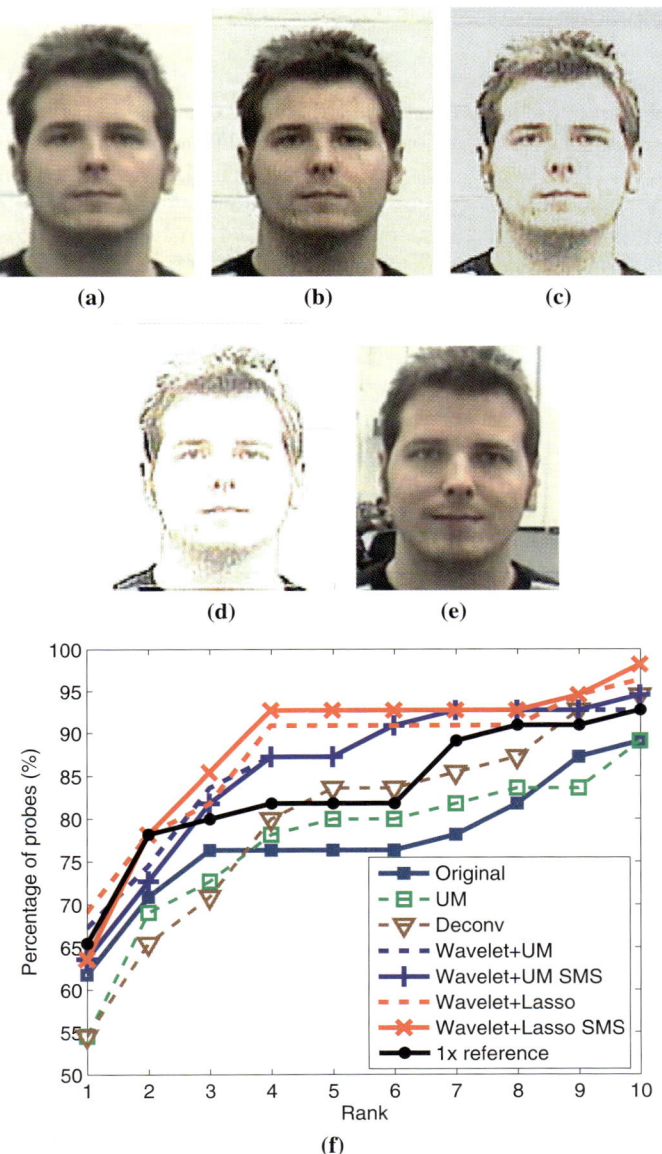

Fig. 4.10. (a) 10× original image. (b) Enhanced by UM. (c) Enhanced by wavelet transform with the approximation image processed by UM. (d) Enhanced by wavelet transform with the approximation image processed by Lasso regularized deconvolution. (e) 1× reference image. (f) CMC comparison across enhancement algorithms

4.6 Conclusions

A unique face database with still images and video sequences, collected from long distances and with high system magnifications, was constructed. This database features various types of degradations encountered in practical long range surveillance applications, with emphasis on magnification blur. Magnification blur was addressed and identified as a major degradation source in face recognition for the first time in this work. A special metric evaluating degradations in face image quality caused by high magnifications was applied and its efficiency in distinguishing low and high magnification images and predicting FRR was illustrated. Image enhancement algorithms were implemented to show that degradations in FRR introduced by magnification blur can be efficiently compensated for by applying the proper deblurring algorithms, such as wavelet-based processing and regularized deconvolution. An improvement of 26% in CMC measure was achieved via assessment and restoration of magnification blur.

In the scope of this chapter, the FRR performances of various enhancement algorithms are studied using high zoom still face images. Our future work will focus on two aspects (1) applications with extreme zoom ($> 30\times$), and (2) applications using video sequences with varying zoom. In addition, to justify the use of high to extreme optical zoom, we will look into super-resolution methods, where low resolution images are registered and interpolated to produce a high resolution image via pure image postprocessing techniques.

5 Core Faces: A Shift-Invariant Principal Component Analysis (PCA) Correlation Filter Bank for Illumination-Tolerant Face Recognition

Marios Savvides, B.V.K. Vijaya Kumar, and Pradeep K. Khosla

5.1 Introduction

Biometric recognition systems [112, 113] are constantly evolving by integrating smarter and more accurate algorithms for identifying people using their physiological characteristics. Among all biometrics, face is of great interest using facial recognition technology because of easy availability of cameras and its nonintrusive nature. However, there are many practical problems faced in face recognition including registration errors, i.e., face images are not centered properly before being given to the classifier, some times resulting in only partial faces being presented to the classifiers. Many recognition algorithms [114–116] are sensitive to such registration errors, even in the scenario where a person provides a perfect frontal pose. Illumination variations is another problem which has been addressed; however these algorithms assume perfectly registered training and test images (the databases used in the experiments were registered by human input). This however, is not the case in a practical face recognition system which will rely on an automated face detector to locate the position of the face [117, 118]. Even the best of face detectors will not provide the registration accuracy of that of a human, and more importantly in scenarios where the face is partially occluded, only partial face image is retrieved. What is the recognition accuracy on a partial face image? How is this registered before it is given to the classifier? Furthermore, what happens when we have both occlusion and illumination variations? Recent research work in using advanced correlation filters for illumination tolerance have shown promising results [119–121]. However, it is not yet clear how these filters capture the variation in given set of training images. Natural question then is how many correlation filters are needed? Correlation filters have many good attributes; such as built-in shift-invariance, i.e., if the test input image is shifted, then the correlation output is also shifted by the same amount and thus peak-based metrics do not change and therefore the classification decision is invariant to such shifts.

In Sect. 5.1.1 we review the popular minimum average correlation energy (MACE) filter background which has had the best success in handling illumination tolerance to get more intuition of how these filters work, and we show how we can improve the current designs to form our hybrid PCA-correlation filter.

5.1.1 Advanced Correlation Filters

Matched filter is the most commonly used correlation filter optimal for detecting targets in white noise [122], however in Biometrics and specifically face recognition, one needs one matched filter for each different appearance of a person's face (pose, illumination, and expression) to be able to successfully do matching. This is both computationally and memory-wise expensive to perform. Here is where advanced correlation filters can overcome some of these difficulties. The synthetic discriminant function (SDF) correlation filter [123] was designed to address this computation and memory issue, by designing a single correlation filter that can provide the same peak response for a set of example training images. While this solves the computation and memory issue by just having a single filter, the filter design only controls the peak response at the origin, thus other points on the correlation points could yield values larger than peak at the origin leading to false detections.

The MACE filter [124] was designed to overcome the limitations of the SDF filter, by not only controlling the peak response at the origin but also in the whole correlation plane; it achieves this by constraining the peak response at the origin for every training image to a specific value (e.g., a correlation peak of 1) and minimizing the average correlation plane energy among all the training images. This energy minimization is equivalent to forcing all the values outside the origin in the correlation plane to go to zero, thus solving the false peak detection problem of the SDF filter. Another filter known as the minimum variance synthetic discriminant function (MVSDF) filter [125] is optimal for detecting a target buried in noise of known power spectral density. Optimal tradeoff SDF (OTSDF) filter [126] optimally trades off noise tolerance for distortion tolerance. Good reference tutorials can be found [127–129] that summarize these filters.

Let $\mathbf{X}$ be a $d \times N$ matrix containing the 2D Fourier transforms of N images lexicographically reordered into each column. Let $\mathbf{h}$ denote our MACE filter in vector format. We can write the peak constraints at the correlation origin as a set of linear equations in vector-matrix format as follows:

$$\mathbf{X}^+\mathbf{h} = \mathbf{u} \quad (5.1)$$

where $\mathbf{c}$ is a vector containing the desired correlation outputs of the N training images. The MACE filter also minimizes the average correlation energy, and this can be written in the Fourier domain for simplicity using Parseval's theorem as shown below.

$$e = \sum_{i=1}^{N} \sum_{x=1}^{M} \sum_{y=1}^{M} c_i(x,y)^2 = \sum_{i=1}^{N} \sum_{u=1}^{d} \sum_{v=1}^{d} |X_i(u,v)|^2 |H(u,v)|^2 = \mathbf{h}^+\mathbf{D}\mathbf{h} \quad (5.2)$$

where $X_i(u,v)$ is the 2D Fourier transform of the ith training image, $H(u,v)$ is the 2D MACE filter and $\mathbf{D}$ is a diagonal matrix containing the average power spectrum of the training images. Minimizing $\mathbf{h}^+\mathbf{D}\mathbf{h}$ subject to the linear constraints $\mathbf{X}^+\mathbf{h} = \mathbf{u}$ yields the following closed form solution for the MACE filter directly in the frequency domain

$$\mathbf{h} = \mathbf{D}^{-1}\mathbf{X}(\mathbf{X}^+\mathbf{D}^{-1}\mathbf{X})^{-1}\mathbf{u} \quad (5.3)$$

Since the MACE filter is designed to yield sharp correlation peaks, i.e., value of 1 at the origin and values close to zero in the rest of the correlation plane we use the peak-to-sidelobe ratio (PSR) defined below in (5.4) to measure this peak sharpness. Figure 5.1 shows how this is computed on a sample correlation output, where the sidelobe region is a 20 × 20 region centered around the peak. A 5 × 5 region around the peak is excluded to denote the correlation value corresponding to the peak region. The larger the PSR, the better match of the target in the scene to the MACE filter.

$$PSR = \frac{peak - mean}{\sigma} \quad (5.4)$$

MACE-type filters in conjunction with the PSR metric provide a recognition algorithm which is tolerant to even unseen illumination variations. One explanation can be seen upon the examination of the MACE formulation in (5.6). We use the MACE example as this provides a clearer mathematical intuitive interpretation of what is happening in the frequency domain. We see that we can write (5.3) in two steps:

$$\mathbf{h} = \mathbf{D}^{-0.5}\mathbf{X}'(\mathbf{X}'^{+}\mathbf{X}')^{-1}\mathbf{u} \quad (5.5)$$

where $\mathbf{X}' = \mathbf{D}^{-0.5}\mathbf{X}$ is a spectral prefiltering step. Thus what is left is a phase-matching process in (5.5). This effect can be shown more clearly in the figure below for unconstrained MACE filters [130].

$$\mathbf{h} = \mathbf{D}^{-1}\mathbf{m} \Rightarrow \mathbf{D}^{-0.5}(\mathbf{D}^{-0.5})\mathbf{m} \quad (5.6)$$

From Fig. 5.2, we see that the UMACE filter is split into two stages (a) a prewhitening step and (b) a phase-matching stage. MACE filters are designed to

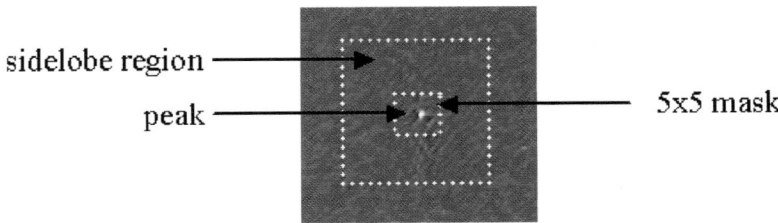

Fig. 5.1. Peak-to-sidelobe ratio (PSR) is a measure of peak sharpness and is used to classify images that belong to a particular MACE filter

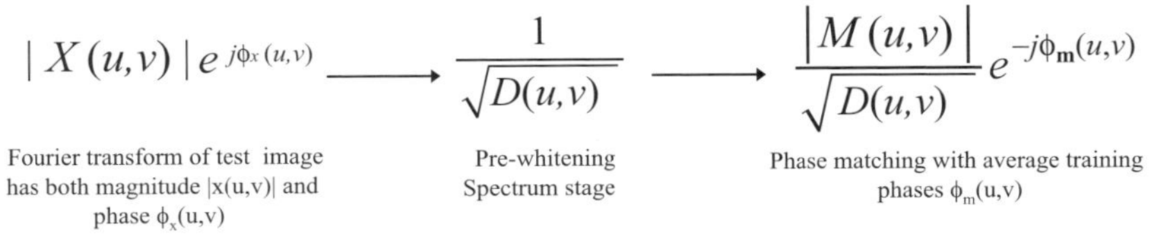

$$|X(u,v)|\,e^{j\phi_x(u,v)} \longrightarrow \frac{1}{\sqrt{D(u,v)}} \longrightarrow \frac{|M(u,v)|}{\sqrt{D(u,v)}} e^{-j\phi_m(u,v)}$$

Fourier transform of test image has both magnitude |x(u,v)| and phase $\phi_x(u,v)$

Pre-whitening Spectrum stage

Phase matching with average training phases $\phi_m(u,v)$

Fig. 5.2. Shows the block diagram of spectral prewhitening being performed followed by phase-matching step

produce sharp peaks, we can think of sharp peaks as delta functions in the ideal case in the spatial domain. A spatial delta function is represented by a constant Fourier transform, thus if we want a delta-function type output in the correlation output, the filter has to somehow (a) flatten the spectrum and (b) cancel the complex phases (u, v) so that we end up with a constant frequency spectrum. We see from Fig. 5.2 that this is exactly what UMACE filter is trying to do. This also explains why illumination tolerance is achieved with these filters, since most images have most dominant concentration of signal energy in the lower frequency spectrum, the prewhitening processing step will emphasize higher frequencies in the test image, where the effects of illumination variation are not so predominant (we assume that illumination variations are mostly in the lower frequency spectrum) followed by the most important step which is the phase matching. If the phase of the Fourier transform matches with the phase of the filter then they will cancel out leaving a constant flat magnitude spectrum (assuming the prewhitening step has flattened the magnitude spectrum).

Hayes [131, 132] and others [133] have done extensive research in image restoration from partial information (such as magnitude retrieval from phase and vice-versa). The conclusion is that most of the intelligibility of the 2D-images is retained in the phase information and not the magnitude of the Fourier transforms. In fact the magnitude information can be retrieved just using the phase information up to a scale factor [134]. We have shown that reduced complexity filters [135] focus on the phase matching part and perform comparably to the full complexity MACE filters, our proposed method we will only focus on the phase information and focus on improved phase matching by setting the magnitude to 1 for all frequencies u, v for both training and test images.

5.2 Eigenphases vs. Eigenfaces

Since we have shown that phase information in the frequency domain is most important we develop in this section a better phase-matching process than correlation filter based on a subspace approach. We use PCA [114] in the Fourier transform domain to get a better representation of the phase spectrum of the training images. This will form a linear subspace that models the frequency phase variations for a particular person's face images. We also show an important fact that just performing PCA in the Fourier domain alone does not achieve any additional benefits using the full frequency spectrum content, in fact it results in the same principal components (or eigenvectors) resulting from PCA on the space domain images (related via an inverse Fourier transform). We will use $\mathbf{C}_s$ and $\mathbf{C}_f$ to denote the covariance matrices of the data in the space domain and frequency domain, respectively.

$$\mathbf{C}_f = \sum_{i=1}^{N} \{\mathbf{T}_{DFT}(\mathbf{x} - \mathbf{m})\}\{\mathbf{T}_{DFT}(\mathbf{x} - \mathbf{m})\}^+ = \mathbf{T}_{\mathbf{DFT}}\mathbf{X}\mathbf{X}^+\mathbf{T}_{\mathbf{DFT}}^{-1} \quad (5.7)$$

where $\mathbf{T_{DFT}}$ is the discrete Fourier transform matrix and $\mathbf{XX^+}$ is the spatial covariance matrix $\mathbf{C}_s$ defined as

$$\mathbf{XX^+} = \sum_{i=1}^{N}(\mathbf{x}-\mathbf{m})(\mathbf{x}-\mathbf{m})^+ = \mathbf{C}_s \tag{5.8}$$

PCA diagonalizes the covariance matrix $\mathbf{C}_f$ using the orthogonal eigenvectors obtained in (5.9)

$$\mathbf{C}_f \mathbf{v}_f = \lambda \mathbf{v}_f \tag{5.9}$$

Substituting $\mathbf{C}_f$ from (5.7) in (5.9) we get

$$\mathbf{T_{DFT}} \mathbf{XX^+} \mathbf{T_{DFT}^{-1}} \mathbf{v}_f = \lambda \mathbf{v}_f \tag{5.10}$$

Premultiplying (5.10) by $\mathbf{T_{DFT}^{-1}}$ we get

$$\mathbf{XX^+} \mathbf{T_{DFT}^{-1}} \mathbf{v}_f = \lambda \mathbf{T_{DFT}^{-1}} \mathbf{v}_f \tag{5.11}$$

We now formulate the space domain PCA and noting that the space domain covariance matrix is $\mathbf{C_s} = \mathbf{XX^+}$

$$\begin{aligned}\mathbf{C}_s \mathbf{v}_s &= \lambda \mathbf{v}_s \\ \mathbf{XX^+} \mathbf{v}_s &= \lambda \mathbf{v}_s\end{aligned} \tag{5.12}$$

Comparing (5.12) with (5.11) we see by inspection that there is a relation between the space and frequency domain eigenvectors related by an inverse Fourier transform as follows:

$$\mathbf{v}_s = \mathbf{T_{DFT}^{-1}} \mathbf{v}_f \tag{5.13}$$

Thus performing PCA directly in the frequency domain alone does not provide any additional advantage by itself, the eigenvectors provided by both approaches are the same and ranked in the same order and only differ by a sign change as shown in Fig. 5.3. However performing eigenanalysis on the phase spectra yields eigenphases as shown in Fig. 5.4 which we will show are tolerant to occlusions and illumination variations in the CMU PIE dataset.

PCA in the frequency domain on only the phase spectrums proves to be very powerful for illumination tolerant face recognition. We show experimental results on the CMU PIE [136] database and compare to other standard illumination tolerant algorithms for comparison such as Fisherfaces [115], 3D linear subspace [116] and traditional PCA [114]. The CMU PIE dataset [136] used consists of 65 people each with 21 different illumination variations captured under no ambient background, this being the hardest illumination set. Figure 5.5 shows 21 sample images from person 1 to denote the type of illumination variations in the dataset. We compared 12 different experiments using a variety of training images as shown in Fig. 5.6. These we split into two types, different training images that contained some illumination

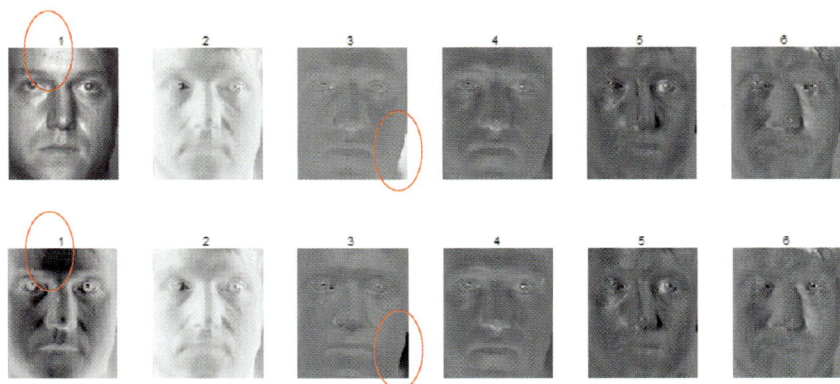

Fig. 5.3. *Top row*: eigenvectors obtained by performing spatial domain PCA. *Bottom row*: inverse Fourier transformed eigenvectors of frequency domain PCA

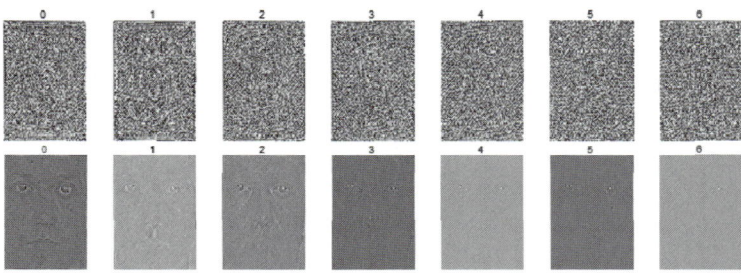

Fig. 5.4. *Top row*: eigenphases (phase angles). *Bottom row*: inverse Fourier transformed eigenphases

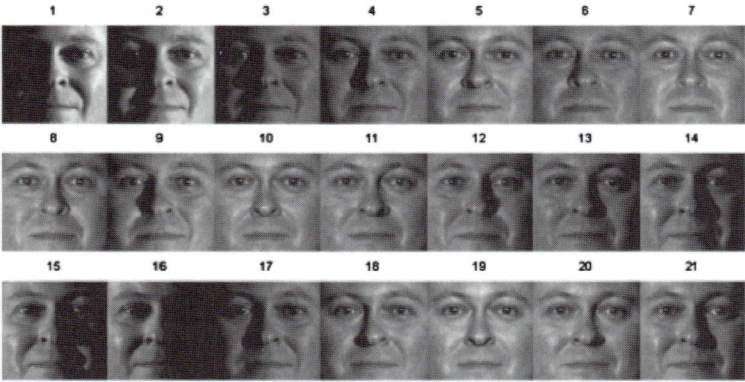

Fig. 5.5. Twenty-one images of Person 1 of the CMU pose illumination expression (PIE) database captured under no ambient lighting

Experiment No. #	Index # of Training Image
1	3, 7, 16
2	1, 10, 16
3	2, 7, 16
4	4, 7, 13
5	1, 2, 7, 16
6	3, 10, 16
7	3, 16, 20
8	5, 6, 7, 8, 9, 10, 18, 19, 20
9	5, 6, 7, 8, 9, 10, 11, 12
10	5, 6, 7, 8, 9, 10,
11	5, 7, 9, 10,
12	7, 10, 19
13	6, 7, 8
14	8, 9, 10
15	18, 19, 20

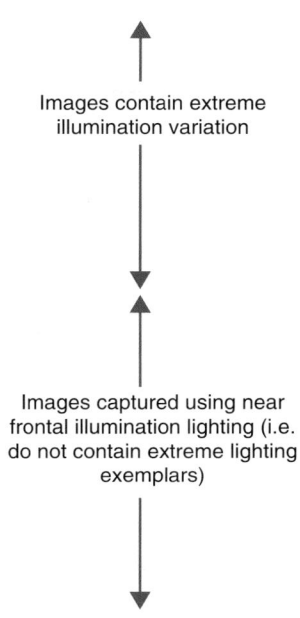

Fig. 5.6. Different experiments using different training scenarios. The indices indicate the type of illumination used for training for each person as shown in Fig. 5.5

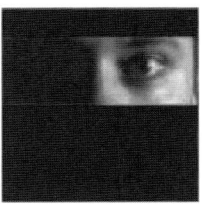

Fig. 5.7. PIE cropped test dataset with only the eye-region visible. This test set has occlusion + illumination variations while the training is performed on whole face images

variation (experiments 1–7) and training images that contained little or no illumination variation (experiments 8–15) where experiment 12 is the hardest using training images 7,10,19 which are frontal neutral lighting. This is the hardest experiment as any classification algorithm must be able to generalize and verify images of that person under illumination variations of any kind (the rest of the 18 illumination variations).

To show that our algorithm can handle illumination and occlusion, we also cropped the test images to only include the eye-region even though the system was trained on whole faces (as shown in Fig. 5.7). Figure 5.8 summarizes the

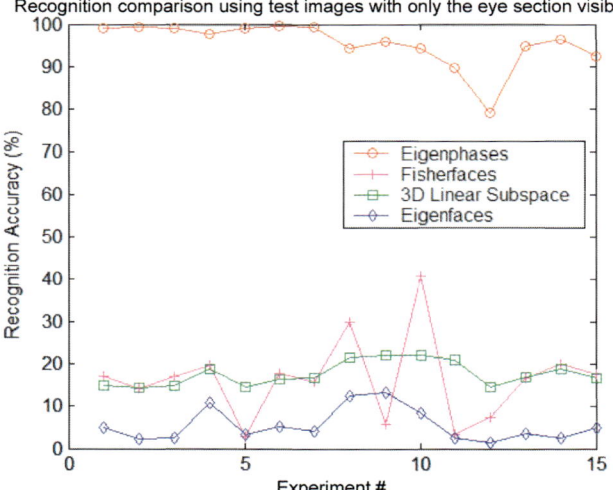

Fig. 5.8. Rank 1 recognition rate for 15 different experiments using the cropped eye-region PIE dataset (but trained on whole face regions)

performance result of each experiment for our eigenphase approach to other methods, clearly showing that while other methods cannot handle occlusion and illumination, even on the hardest experiment 12, eigenphases still perform very well.

While we have demonstrated that phase spectrums capture the discrimination detail in face images and are illumination tolerant, one of the limitations of this approach is the lack of shift-invariance, i.e., if the test image is shifted then performance will degrade. In order to achieve shift-invariance and still have the distortion tolerance to occlusion and illumination we develop CoreFace theory in Sect. 5.3 for a complete shift-invariant hybrid PCA correlation filter.

5.3 CoreFaces

In this section we show how we extend the eigenphase approach to be shift-invariant. The task is to develop a framework to compute the reconstructed phase spectrum at each possible location in the scene to achieve shift-invariance and we show how this will too formulate into a filter bank. Let us assume that we have computed the eigenvectors of the phase-only spectra of the training images which we denote as $\mathbf{v}_i$, these are then placed along the columns of matrix $\mathbf{V}$. We then want to reconstruct the phase-only image at each shift; to do this we need to compute the projection coefficients $\mathbf{p}$ defined as follows:

$$\mathbf{p} = \mathbf{V}^+(\mathbf{x} - \mathbf{m}) \qquad (5.14)$$

Therefore the reconstructed image is computed using the coefficients as follows

$$\mathbf{r} = \mathbf{Vp} + \mathbf{m} \quad (5.15)$$

However we want to compute the phase-only correlation (or inner-product) of this reconstructed image and the test image for all shifts. We write this correlation value c as follows:

$$c = \mathbf{r}^+\mathbf{x} \quad (5.16)$$

Substituting (5.15) in (5.16) we get

$$c = (\mathbf{Vp} + \mathbf{m})^+\mathbf{x} = \mathbf{p}^+\mathbf{V}^+\mathbf{x} + \mathbf{m}^+\mathbf{x} \quad (5.17)$$

$$c = [\mathbf{V}^+(\mathbf{x} - \mathbf{m})]^+\mathbf{V}^+\mathbf{x} + \mathbf{m}^+\mathbf{x}$$
$$= \mathbf{x}^+\mathbf{V}\mathbf{V}^+\mathbf{x} - \mathbf{m}^+\mathbf{V}\mathbf{V}^+\mathbf{x} + \mathbf{m}^+\mathbf{x} \quad (5.18)$$
$$= ||\mathbf{V}^+\mathbf{x}||^2 - (\mathbf{m}^+\mathbf{V})(\mathbf{V}^+\mathbf{x}) + \mathbf{m}^+\mathbf{x}$$

Looking closely at (5.18) we see how we can write the above equation in terms of cross-correlations of the input image and the eigenvectors to compute an correlation output plane c which contains the phase-only cross-correlation of the reconstructed phase-only image and the test image at all possible shifts in the scene image. This is shown in (5.19) where $CORR(\mathbf{a}, \mathbf{b})$ denotes computing the cross-correlation between images $\mathbf{a}$ and $\mathbf{b}$ using FFTs.

$$c(x,y) = \sum_{i=1}^{N} CORR(\mathbf{v}_i, \mathbf{x})^2 - \sum_{i=1}^{N}(\mathbf{m}^+\mathbf{v}_i)CORR(\mathbf{v}_i, \mathbf{x}) + CORR(\mathbf{m}, \mathbf{x})$$
$$(5.19)$$

Thus $c(x,y)$ contains the phase-only correlation at each spatial location (x,y), using this correlation output plane we search for the peak and compute the PSR just as done previously. Figure 5.9 shows the CoreFace output on a face image centered in the scene, giving a very large PSR of 108.45, and the bottom plot depicts the shift-invariance of the CoreFace approach where the same face image was shifted up 40 pixels in the face image where the PSR remains exactly the same but the peak is shifted up by 40 pixels showing the location of the face in the scene. Figure 5.10 shows PSR plot from MACE filter CoreFace trained on person 1 and tested on the whole dataset. The top line depicts the highest authentic PSR plot achieved by CoreFace and the bottom 2 lines depict the maximum impostor PSRs scores from each of 64 people under each of the 21 illumination variations.

Another advantage of the CoreFace method, one can compute (5.19) iteratively starting from the most dominant eigenvectors (i.e., $N = 1 \ldots M$), and computing the PSR after each iteration, if the PSR is above a matching threshold there is no need to continue and compute the correlation of the rest of the eigenvectors. Thereby providing a computationally efficient manner for computing the PSR incrementally if N is very large (Table 5.1).

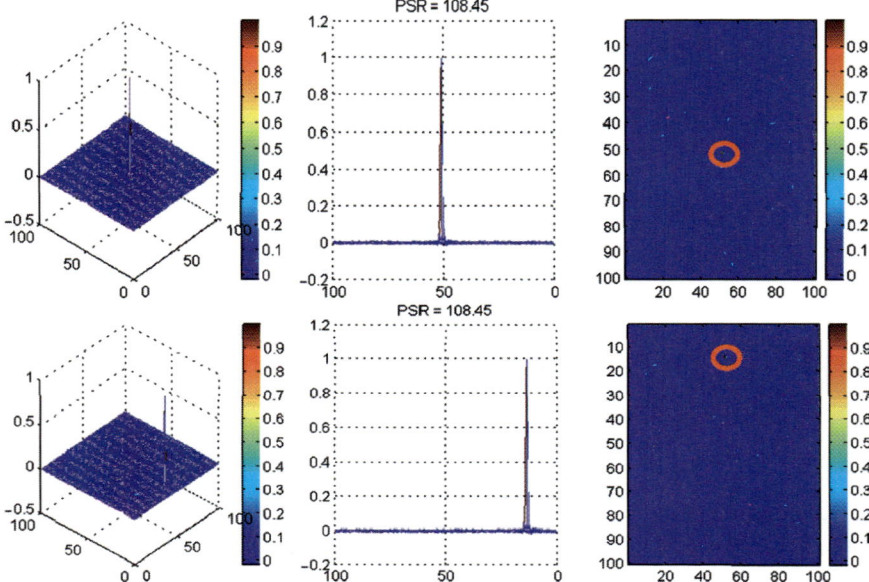

Fig. 5.9. *Top row*: CoreFace output from a face image centered in the scene. *Bottom row*: CoreFace output from the same face image shifted up 40 pixels. The PSR remains exactly the same as the unshifted face and the peak is shifted up by the same amount

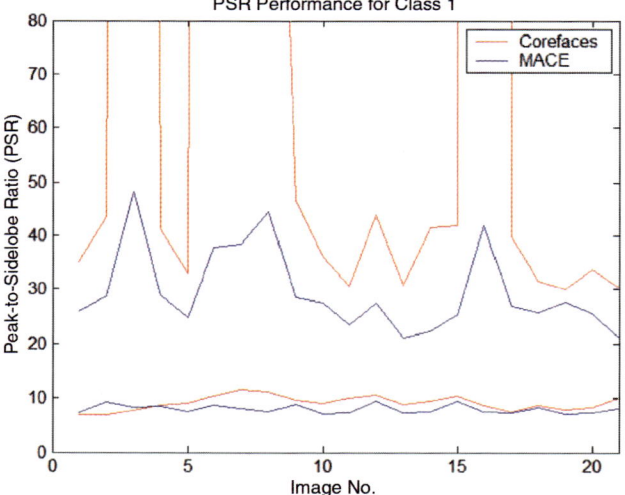

Fig. 5.10. Shows the PSRs from MACE and CoreFaces synthesized from images from class 1 and tested on the whole PIE database. The top plots belong to PSRs from class 1 and the bottom plots belong to the maximum PSRs from the 21 images of the other 64 people (i.e., 1,344 PSRs)

Table 5.1. Face recognition rates on eye-cropped PIE database

method	recognition rate (%)
MACE	95.8
CoreFaces	99.92

5.4 Discussion

This chapter briefly introduced advanced correlation filter designs such as the MACE filter, and showed that these types of filters work by performing a spectral prefiltering operation followed by phase-matching process. With this insight, we show the development of eigenphases which extend traditional linear subspace modeling to spectral phase modeling, showing that we can achieve distortion tolerance to illumination variations and occlusions. We then further extend the eigenphase model to achieve complete shift-invariance by developing the theory behind CoreFaces which is a hybrid PCA correlation filter bank that performs phase matching in a shift-invariant manner in the input scene. Results show the shift-invariance and face recognition improvement of the CoreFace approach over the MACE filters on the CMU PIE dataset.

Part II

Multi-Sensory Face Biometrics

6 Towards Person Authentication by Fusing Visual and Thermal Face Biometrics

Ognjen Arandjelović, Riad Hammoud, and Roberto Cipolla

6.1 Introduction

In this chapter we focus on face appearance-based biometrics. The cheap and readily available hardware used to acquire data, their non-invasiveness and the ease of employing them from a distance and without the awareness of the user, are just some of the reasons why these continue to be of great practical interest.

However, a number of research challenges remain. Specifically, face biometrics have traditionally focused on images acquired in the visible light spectrum and these are greatly affected by such extrinsic factors such as the illumination, camera angle (or, equivalently, head pose) and occlusion. In practice, the effects of changing pose are usually least problematic and can oftentimes be overcome by acquiring data over a time period, e.g., by tracking a face in a surveillance video. Consequently, image sequence or image set matching has recently gained a lot of attention in the literature [137–139] and is the paradigm adopted in this chapter as well. In other words, we assume that the training image set for each individual contains some variability in pose, but is not obtained in scripted conditions or in controlled illumination.

In contrast, illumination is much more difficult to deal with: the illumination setup is in most cases not practical to control and its physics is difficult to accurately model. *Thermal spectrum* imagery is useful in this regard as it is virtually insensitive to illumination changes, as illustrated in Fig. 6.1. On the other hand, it lacks much of the individual, discriminating facial detail contained in visual images. In this sense, the two modalities can be seen as complementing each other. The key idea behind the system presented in this chapter is that robustness to extreme illumination changes can be achieved by *fusing* the two. This paradigm will further prove useful when we consider the difficulty of recognition in the presence of occlusion caused by prescription glasses.

6.1.1 Mono-Sensor Based Techniques

Optical sensors

Among the most used sensors in face biometric systems is the optical imager. This is driven by its availability and low-cost. An optical imager captures the light reflectance of the face surface in the visible spectrum. The visible spectrum provides

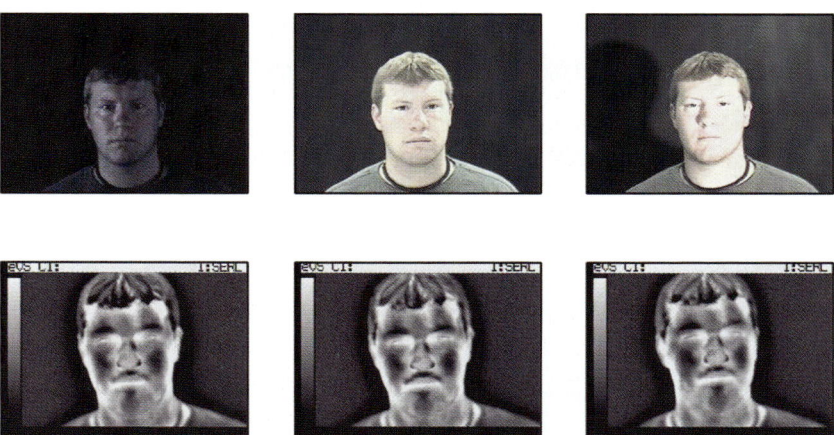

Fig. 6.1. *Sensitivity to lighting conditions*: illumination changes have a dramatic effect on images acquired in the visible light spectrum (*top row*). In contrast, thermal imagery (*bottom row*) shows remarkable invariance

features that depend only on surface reflectance. Thus, it is obvious that the face appearance changes according to the ambient light. In order to overcome the lighting, pose and facial expression changes, a flurry of face recognition algorithms, from the two well-known broad categories, appearance-based and feature-based methods, has been proposed [140]. Appearance-based methods find the global properties of the face pattern and recognize the face as a whole. In contrast, feature-based methods [141–143] explore the statistical and geometrical properties of facial features like eyes and mouth. The face recognition performance depends on the accuracy of facial feature detection. Moreover, local and global lighting changes cause existing face recognition techniques for the visible imagery to perform poorly.

Infrared sensors

Recent studies have shown that face recognition in the thermal spectrum offers a few distinct advantages over the visible spectrum, including invariance to ambient illumination changes [144–147]. This is due to the fact that a thermal infrared sensor measures the heat energy radiation emitted by the face rather than the light reflectance. A thermal sensor generates imaging features that uncover thermal characteristics of the face pattern. Indeed, thermal face recognition algorithms attempt to take advantage of such anatomical information of the human face as unique signatures.

Appearance-based face recognition algorithms applied to thermal IR imaging consistently performed better than when applied to visible imagery, under various lighting conditions and facial expressions [145, 148–150]. Further performance improvements were achieved using decision-based fusion [145]. In contrast to other techniques, Srivastava et al. [151], performed face recognition in the space of

K-Bessel form. First, they decompose each infrared face image using Gabor filters. Then, they represent the face by a few parameters by modelling the marginal density of the Gabor filter coefficients using Bessel functions. This approach has been improved by Buddharaju et al. [152]. Recently, Friedrich et al. [153] showed that IR-based recognition is less sensitive to changes in 3D head pose and facial expression.

6.1.2 Multi-Sensor Based Techniques

As the surface of the face and its temperature have nothing in common, one would state that the extracted cues from both sensors are not redundant and yet complementary. Several attempts have been made in face recognition based on the fusion of different types of data from multiple sensors. Face recognition algorithms based on the fusion of visible and thermal IR images demonstrated higher performance than individual image types [154–157]. Biometric systems that integrate face and speech signals [158], the face and fingerprint information [159], and the face and the ear images [160] improved the accuracy in personal identification.

Recently, Heo et al. [161] proposed two types of visible and thermal fusion technique, the first fuses low-level data while the second fuses matching outputs. Data fusion was implemented by applying pixel-based weighted averaging of coregistered visual and thermal images. Decision fusion was implemented by combining the matching scores of individual recognition modules. To deal with occlusions caused by eyeglasses in thermal imagery, they used a simple ellipse fitting technique to detect the circle-like eyeglass regions in the IR image and replaced them with an average eye template. Using a commercial face recognition system, Faceit [162], they demonstrated improvements in recognition accuracy.

6.2 Method Details

In the sections that follow we explain our system in detail, the main components of which are conceptually depicted in Fig. 6.2.

6.2.1 Matching Image Sets

In this chapter we deal with face recognition from *sets* of images, both in the visual and thermal spectrum. We will show how to achieve illumination invariance using a combination of simple data preprocessing (Sect. 6.2.2), local features (Sect. 6.2.3) and modality fusion (see Sect. 6.2.4). Hence, the requirements for our basic set matching algorithm are those of (a) some pose generalization and (b) robustness to noise. We compare two image sets by modelling the variations within a set using a linear subspace and comparing two subspaces by finding the most similar modes of variation within them.

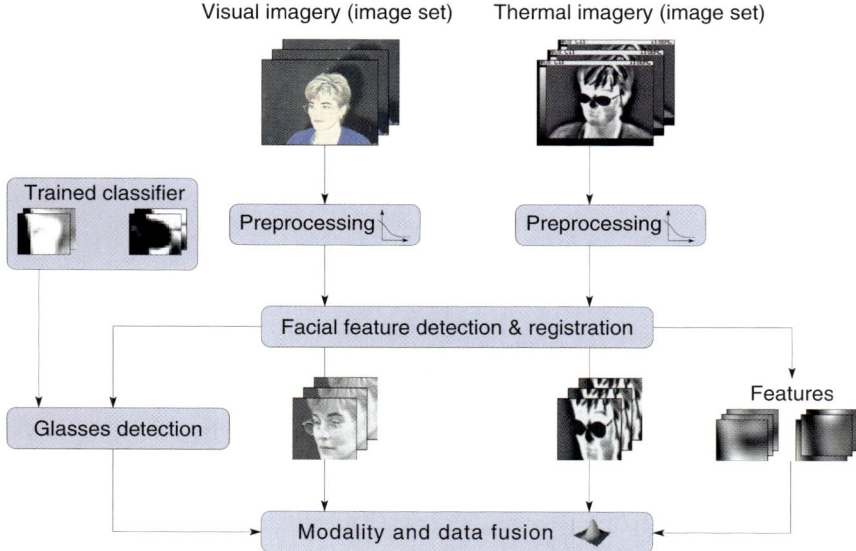

Fig. 6.2. *System overview*: our system consists of three main modules performing (**a**) data preprocessing and registration, (**b**) glasses detection, and (**c**) fusion of holistic and local face representations using visual and thermal modalities

The modelling step is a simple application of principal component analysis (PCA) without mean subtraction. In other words, given a data matrix **d** (each column representing a rasterized image), the subspace is spanned by the eigenvectors of the matrix $\mathbf{C} = \mathbf{dd}^T$ corresponding to the largest eigenvalues; we used 5D subspaces, as sufficiently expressive to on average explain over 90% of data variation within intrinsically low-dimensional face appearance changes in a set.

The similarity of two subspaces U_1 and U_2 is quantified by the cosine of the smallest angle between two vectors confined to them:

$$\rho = \cos\theta = \max_{\mathbf{u} \in U_1} \max_{\mathbf{v} \in U_2} \mathbf{u}^T \mathbf{v}. \tag{6.1}$$

The quantity ρ is also known as the first canonical correlation [163]. It is this implicit "search" over entire subspaces that achieves linear pose interpolation and extrapolation, by finding the most similar appearances described by the two sets [164]. The robustness of canonical correlations to noise is well detailed in [165] (also see [166]).

Further appeal of comparing two subspaces in this manner is contained in its computational efficiency. If $\mathbf{B}_1$ and $\mathbf{B}_2$ are the corresponding orthonormal basis matrices, the computation of ρ can be rapidly performed by finding the largest singular value of the 5×5 matrix $\mathbf{B}_1^T \mathbf{B}_2$ [165].

6.2.2 Data Preprocessing and Feature Extraction

The first stage of our system involves coarse normalization of pose and brightness. We register all faces, both in the visual and thermal domain, to have the salient facial features aligned. Specifically, we align the eyes and the mouth due to the ease of detection of these features (e.g., see [167–171]). The three point correspondences, between the detected and the canonical features' locations, uniquely define an affine transformation which is applied to the original image. Faces are then cropped to 80×80 pixels, as shown in Fig. 6.3.

Coarse brightness normalization is performed by band-pass filtering the images [167, 172]. The aim is to reduce the amount of high-frequency noise as well as extrinsic appearance variations confined to a low-frequency band containing little discriminating information. Most obviously, in visual imagery, the latter are caused by illumination changes, owing to the smoothness of the surface and albedo of faces [173].

We consider the following type of a band-pass filter:

$$\mathbf{I}_F = \mathbf{I} * \mathbf{G}_{\sigma=W_1} - \mathbf{I} * \mathbf{G}_{\sigma=W_2}, \qquad (6.2)$$

which has two parameters – the widths W_1 and W_2 of isotropic Gaussian kernels. These are estimated from a small training corpus of individuals in different illuminations. Figure 6.4 shows the recognition rate across the corpus as the values of the two parameters are varied. The optimal values were found to be 2.3 and 6.2 for visual data; the optimal filter for thermal data was found to be a *low-pass* filter with $W_2 = 2.8$ (i.e., W_1 was found to be very large). Examples are shown in Fig. 6.5. It is important to note from Fig. 6.4 that the recognition rate varied smoothly with changes in kernel widths, showing that the method is not very sensitive to their exact values, which is suggestive of good generalization to unseen data.

The result of filtering visual data is further scaled by a smooth version of the original image:

$$\hat{\mathbf{I}}_F(x, y) = \mathbf{I}_F(x, y) . / (\mathbf{I} * \mathbf{G}_{\sigma=W_2}), \qquad (6.3)$$

where ./ represents element-wise division. The purpose of local scaling is to equalize edge strengths in dark (weak edges) and bright (strong edges) regions of

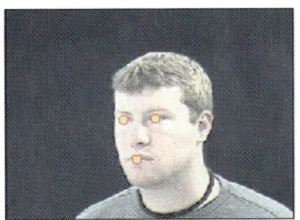  

Fig. 6.3. *Registration*: shown is the original image in the visual spectrum with detected facial features marked by yellow circles (*left*), the result of affine warping the image to the canonical frame (*centre*) and the final registered and cropped facial image

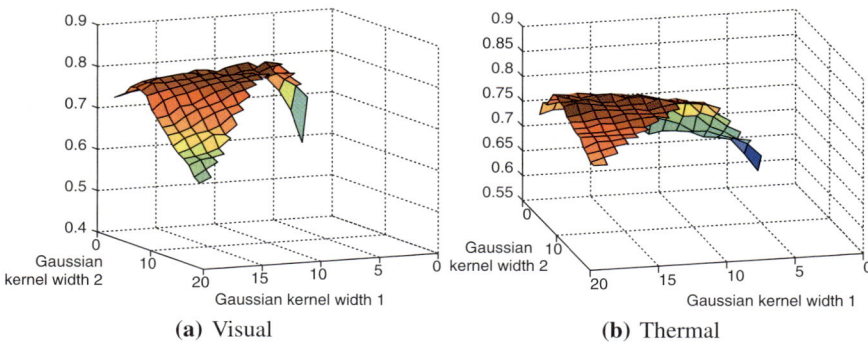

Fig. 6.4. *Band-pass filter*: the optimal combination of the lower and upper band-pass filter thresholds is estimated from a small training corpus. The plots show the recognition rate using a single modality, (**a**) visual and (**b**) thermal, as a function of the widths W_1 and W_2 of the two Gaussian kernels in (6.2). It is interesting to note that the optimal band-pass filter for the visual spectrum passes a rather narrow, mid-frequency band, whereas the optimal filter for the thermal spectrum is in fact a *low-pass* filter

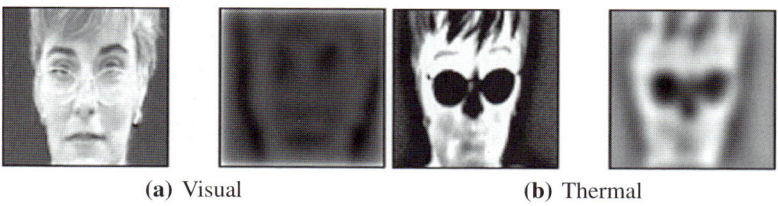

Fig. 6.5. *Preprocessing*: the effects of the optimal band-pass filters on registered and cropped faces in (**a**) visual and (**b**) thermal spectra

the face; this is similar to the self-quotient image (SQI) of Wang et al. [174]. This step further improves the robustness of the representation to illumination changes, see Sect. 6.3.

6.2.3 Single Modality-Based Recognition

We compute the similarity of two individuals using only a single modality (visual or thermal) by combining the holistic face representation described in Sect. 6.2.2 and a representation based on local image patches. These have been shown to benefit recognition in the presence of large pose changes [139].

As before, we use the eyes and the mouth as the most discriminative regions, by extracting rectangular patches centred at the detections, see Fig. 6.6. The overall similarity score is obtained by weighted summation:

$$\rho_{v/t} = \omega_h \cdot \rho_h + \omega_e \cdot \rho_e + (1 - \omega_h - \omega_e) \cdot \rho_m, \qquad (6.4)$$

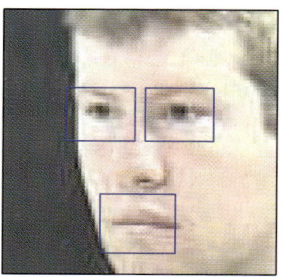

Fig. 6.6. *Features*: in both the visual and the thermal spectrum our algorithm combines the similarities obtained by matching the holistic face appearance and the appearance of three salient local features – the eyes and the mouth

where ρ_m, ρ_e and ρ_h are the scores of separately matching, respectively, the mouth, the eyes and the entire face regions, and ω_h and ω_e the weighting constants.

The optimal values of the weights were estimated from the offline training corpus. For the visual spectrum we obtained $\omega_e = 0.3$, while the mouth region was found not to improve recognition (i.e., $\omega_h = 0.7$). The relative magnitudes of the weights were found to be different in the thermal spectrum, both the eye and the mouth region contributing equally to the overall score: $\omega_e = 0.1$, $\omega_h = 0.8$.

6.2.4 Fusing Modalities

Until now we have focused on deriving a similarity score between two individuals given sets of images in either thermal or visual spectrum. A combination of holistic and local features was employed in the computation of both. However, the greatest power of our system comes from the fusion of the two modalities.

Given ρ_v and ρ_t, the similarity scores corresponding to visual and thermal data, we compute the joint similarity as:

$$\rho_f = \omega_v(\rho_v) \cdot \rho_v + (1 - \omega_v(\rho_v)) \cdot \rho_t. \tag{6.5}$$

Notice that the weighting factors are no longer constants, but *functions*. The key idea is that if the visual spectrum match is very good (i.e., ρ_v is close to 1.0), we can be confident that illumination difference between the two images sets compared is mild and well compensated for by the visual spectrum preprocessing of Sect. 6.2.2. In this case, visual spectrum should be given relatively more weight than when the match is bad and the illumination change is likely more drastic.

The function $\omega_v \equiv \omega_v(\rho_v)$ is estimated in three stages: first (a) we estimate $\hat{p}(\omega_v, \rho_v)$, the probability that ω_v is the optimal weighting given the estimated similarity ρ_v, then (b) compute $\omega(\rho_v)$ in the maximum a posteriori sense and finally (c) make an analytic fit to the obtained marginal distribution. Step (a) is challenging and we describe it next.

Iterative density estimate

The principal difficulty of estimating $\hat{p}(\omega_v, \rho_v)$ is of practical nature: in order to obtain an accurate estimate (i.e., a well-sampled distribution), a prohibitively large training database is needed. Instead, we employ a heuristic alternative. Much like before, the estimation is performed using the offline training corpus.

Our algorithm is based on an iterative incremental update of the density, initialized as uniform over the domain $\omega, \rho \in [0, 1]$. We iteratively simulate matching of an unknown person against a set gallery individuals. In each iteration of the algorithm, these are randomly drawn from the offline training database. Since the ground truth identities of all persons in the offline database is known, for each $\omega = k\Delta\omega$ we can compute the separation, i.e., the difference between the similarities of the test set and the set corresponding to it in identity, and that between the test set and the most similar set that does *not* correspond to it in identity. Density $\hat{p}(\omega, \rho)$ is then incremented at each $(k\Delta\omega, \rho^{p,p})$ proportionally to $\delta(k\Delta\omega)$ after being passed through the sigmoid function. This is similar to the algorithm proposed in [175].

Input: visual data $d_v(person, illumination)$,
thermal data $d_t(person, illumination)$.
Output: density estimate $\hat{p}(\omega, \rho_v)$.

1: Init.
$\hat{p}(\omega, \rho_v) = 0,$

2: Iteration
for all illuminations i, j and persons p

3: Iteration
for all $k = 0, \ldots, 1/\Delta\omega$, $\omega = k\Delta\omega$

4: Separation given ω
$\delta(k\Delta\omega) = \min_{q \neq p}[\omega\rho_v^{p,p} + (1-\omega)\rho_t^{p,p}$
$-\omega\rho_v^{p,q} + (1-\omega)\rho_t^{p,q}]$

5: Update density estimate
$\hat{p}(k\Delta\omega, \rho_v^{p,p}) = \hat{p}(k\Delta\omega, \rho_v^{p,p})$
$+\text{sig}(C \cdot \delta(k\Delta\omega))$

6: Smooth the output
$\hat{p}(\omega, \mu) = \hat{p}(\omega, \mu) * \mathbf{G}_{\sigma=0.05}$

7: Normalize to unit integral
$\hat{p}(\omega, \rho) = \hat{p}(\omega, \rho) / \int_\omega \int_\rho \hat{p}(\omega, \rho) d\rho d\omega$

Fig. 6.7. Offline: optimal fusion training algorithm

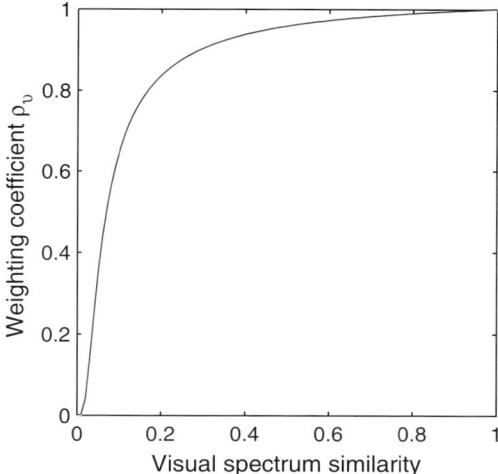

Fig. 6.8. *Modality fusion*: the contribution of visual matching, as a function of the similarity of visual imagery. A low similarity score between image sets in the visual domain is indicative of large illumination changes and consequently our algorithm leant that more weight should be placed on the illumination-invariant thermal spectrum

Figure 6.7 summarizes the proposed offline learning algorithm. An analytic fit to $\hat{p}(\omega_v)$ in the form $(1 + \exp(a))/(1 + \exp(a/\rho_v))$ is shown in Fig. 6.8.

6.2.5 Dealing with Glasses

The appeal of using the thermal spectrum for face recognition stems mainly from its invariance to illumination changes, in sharp contrast to visual spectrum data. The exact opposite is true in the case of prescription glasses, which appear as dark patches in thermal imagery, see Fig. 6.5. The practical importance of this can be seen by noting that in the US in 2000 roughly 96 million people, or 34% of the total population, wore prescription glasses [176].

In our system, the otherwise undesired, gross appearance distortion that glasses cause in thermal imagery is used to help recognition by detecting their presence. If the subject is not wearing glasses, then both holistic and all local patches-based face representations can be used in recognition; otherwise the eye regions in thermal images are ignored.

Glasses detection

We detect the presence of glasses by building representations for the left eye region (due to the symmetry of faces, a detector for only one side is needed) with and without glasses, in the thermal spectrum. The foundations of our classifier are laid in Sect. 6.2.1. Appearance variations of the eye region with and without glasses

Fig. 6.9. *Appearance models*: shown are examples of glasses-on (*top*) and glasses-off (*bottom*) thermal data used to construct the corresponding appearance models for our glasses detector

are represented by two 6D linear subspaces, see Fig. 6.9 for example training data. Patches extracted from a set of thermal imagery of a novel person is then compared with each subspace. The presence of glasses is deduced when the corresponding subspace results in a higher similarity score. We obtain close to flawless performance on our data set (also see Sect. 6.3 for description), as shown in Fig. 6.10.

The presence of glasses severely limits what can be achieved with thermal imagery, the occlusion heavily affecting both the holistic face appearance as well as that of the eye regions. This is the point at which our method heavily relies on decision fusion with visual data, limiting the contribution of the thermal spectrum to matching using mouth appearance only, i.e., setting $\omega_h = \omega_e = 0.0$ in (6.4).

6.3 Empirical Evaluation

We evaluated the described system on the *Dataset 02: IRIS Thermal/Visible Face Database* subset of the *object tracking and classification beyond the visible spectrum (OTCBVS)* database,[1] freely available for download at http://www.cse.ohio-state.edu/OTCBVS-BENCH/. Briefly, this database contains 29 individuals, 11 roughly matching poses in visual and thermal spectra and large illumination variations (some of these are exemplified in Fig. 6.11).

Our algorithm was trained using all images in a single illumination in which all three salient facial features could be detected. This typically resulted in 7–8 images in the visual and 6–7 in the thermal spectrum, see Fig. 6.12, and roughly ±45° yaw range, as measured from the frontal face orientation.

The performance of the algorithm was evaluated both in 1-to-N and 1-to-1 matching scenarios. In the former case, we assumed that test data corresponded to one of people in the training set and recognition was performed by associating it with the closest match. Verification (or 1-to-1 matching, "is this the same person?")

---

[1] IEEE OTCBVS WS Series Bench; DOE University Research Program in Robotics under grant DOE-DE-FG02-86NE37968; DOD/TACOM/NAC/ARC Program under grant R01-1344-18; FAA/NSSA grant R01-1344-48/49; Office of Naval Research under grant #N000143010022.

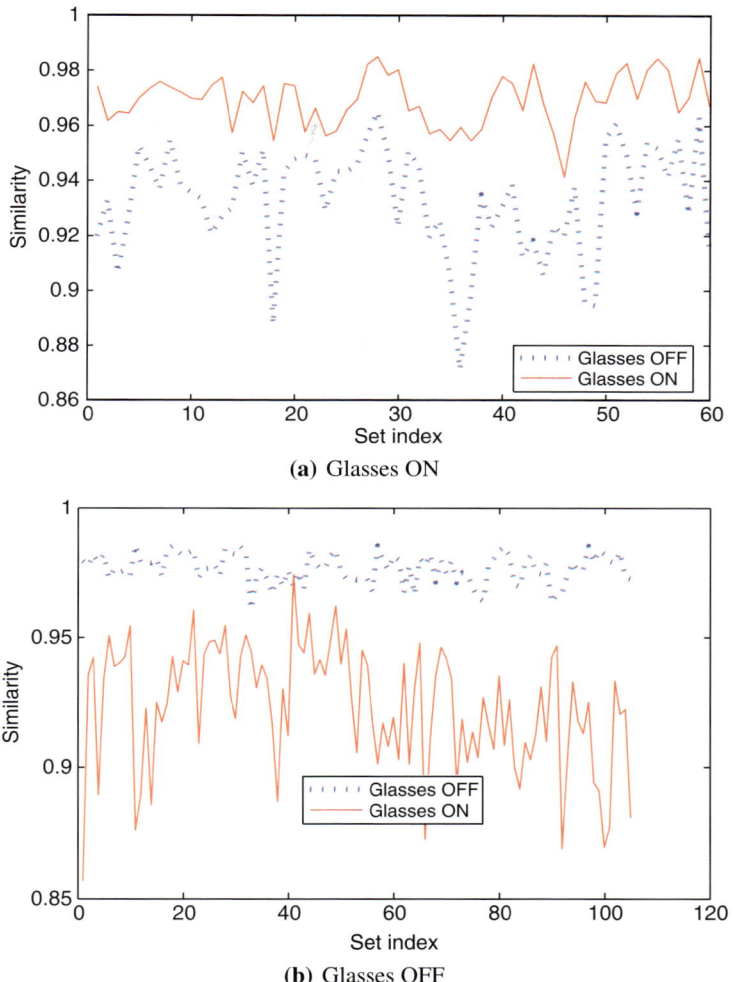

Fig. 6.10. Glasses Detection Results: inter- and intra-class similarities across our data set

performance was quantified by looking at the true positive admittance rate for a threshold that corresponds to 1 admitted intruder in 100.

6.3.1 Results

A summary of 1-to-N matching results is shown in Table 6.1.

Firstly, note the poor performance achieved using both raw visual as well as raw thermal data. The former is suggestive of challenging illumination changes present in the OTCBVS data set. This is further confirmed by significant improvements gained with both band-pass filtering and the SQI which increased the average

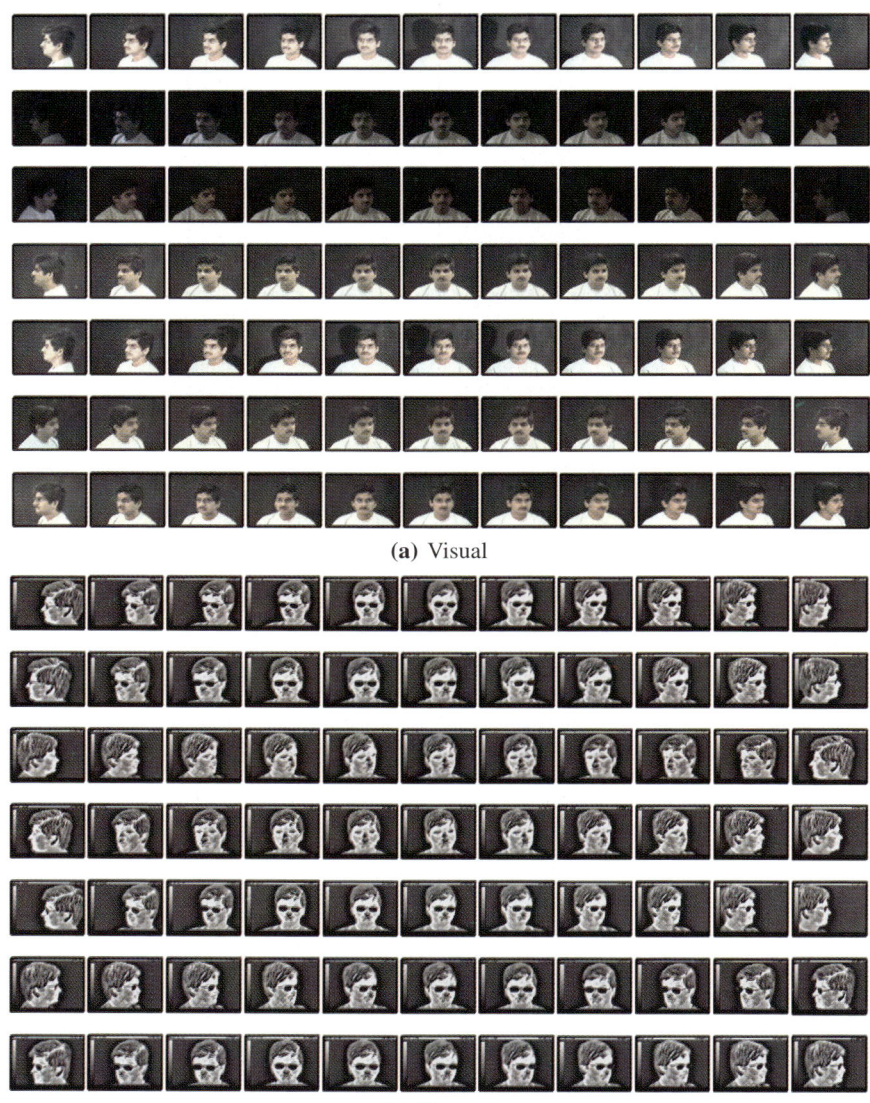

Fig. 6.11. *Example training sets*: each row corresponds to a single training (or test) set of images used for our algorithm in (**a**) the visual and (**b**) the thermal spectrum. Note the extreme changes in illumination, as well as that in some sets the user is wearing glasses and in some not

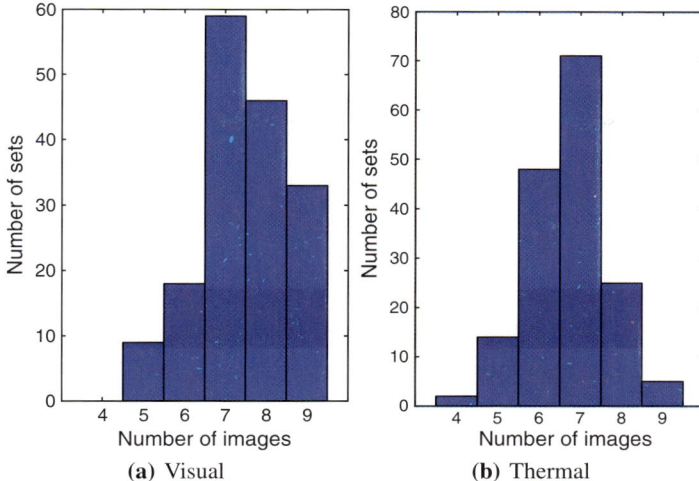

Fig. 6.12. *Training sets*: shown are histograms of the number of images per person used to train our algorithm. Depending on the exact head poses assumed by the user we typically obtained 7–8 visual spectrum images and typically a slightly lower number for the thermal spectrum. The range of yaw angles covered is roughly ±45° measured from the frontal face orientation

Table 6.1. *1-to-N matching (recognition) results*: shown is the average rank 1 recognition rate using different representations across all combinations of illuminations. Note the performance increase with each of the main features of our system: image filtering, combination of holistic and local features, modality fusion and prescription glasses detection

	Representation	Recognition
Visual	Holistic raw data	0.58
	Holistic, band-pass	0.78
	Holistic, SQI filtered	0.85
	Mouth + eyes + holistic Data fusion, SQI filtered	0.87
Thermal	Holistic raw data	0.74
	Holistic raw w/ Glasses detection	0.77
	Holistic, low-pass filtered	0.80
	Mouth + eyes + holistic Data fusion, low-pass filtered	0.82
Proposed thermal + visual fusion	w/o glasses detection	0.90
	w/glasses detection	*0.97*

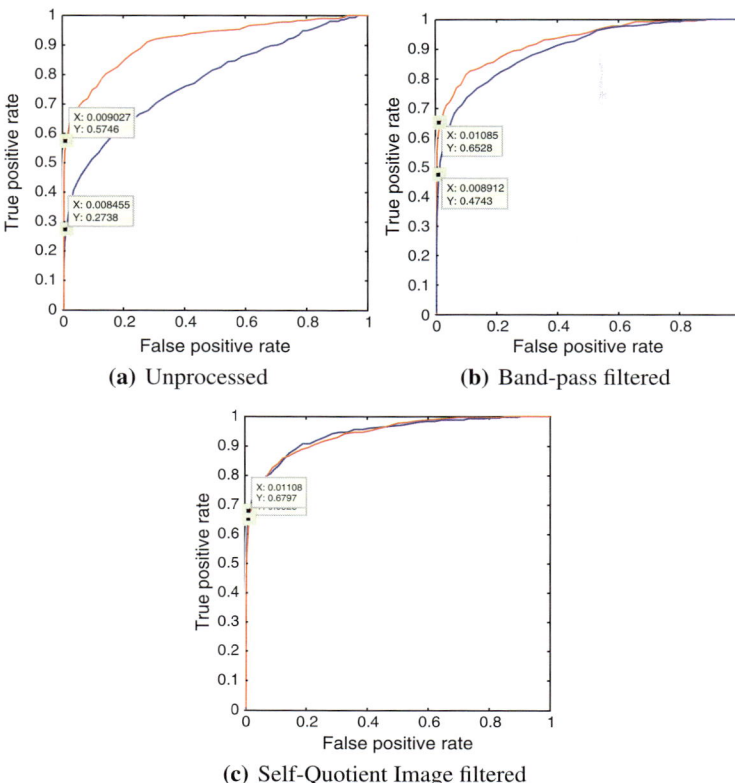

Fig. 6.13. *Holistic representations Receiver–Operator Characteristics*: visual (*blue*) and thermal (*red*) spectra

Table 6.2. *Holistic, 1-to-1 matching (verification)*: a summary of the comparison of different image processing filters for 1 in 100 intruder acceptance rate. Both the simple band-pass filter, and even further its locally scaled variant, greatly improve performance. This is most significant in the visual spectrum, in which image intensity in the low spatial frequency is most affected by illumination changes

Representation	Visual	Thermal
1% intruder acceptance		
Unprocessed/raw	0.2850	0.5803
Band-pass filtered (BP)	0.4933	0.6287
Self-quotient image (SQI)	0.6410	0.6301

recognition rate for, respectively, 35% and 47%. The same is corroborated by the Receiver–Operator Characteristic (ROC) curves in Fig. 6.13 and 1-to-1 matching results in Table 6.2.

Table 6.3. *Isolated local features, 1-to-1 matching (verification)*: a summary of the results for 1 in 100 intruder acceptance rate. Local features in isolation perform very poorly

Representation	Visual (SQI)	Thermal (BP)
	1% intruder acceptance	
Eyes	0.1016	0.2984
Mouth	0.1223	0.3037

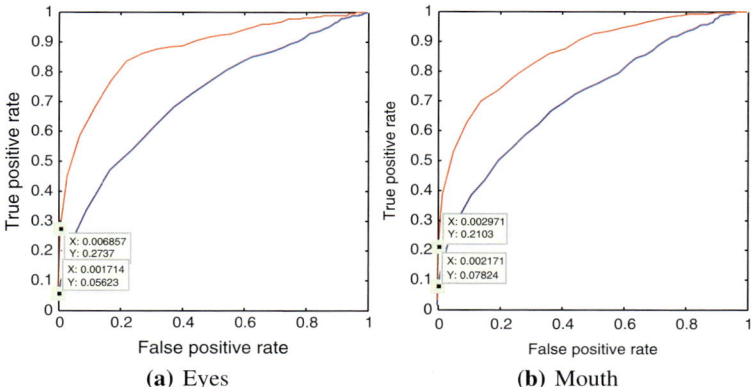

Fig. 6.14. *Isolated local features Receiver–Operator Characteristics*: Visual (*blue*) and thermal (*red*) spectra

On the other hand, the reason for low recognition rate of raw thermal imagery is twofold: it was previously argued that the two main limitations of this modality are the inherently lower discriminative power and occlusions caused by prescription glasses. The addition of the glasses detection module is of little help at this point – some benefit is gained by steering away from misleadingly good matches between any two people wearing glasses, but it is limited in extent as a very discriminative region of the face is lost. Furthermore, the improvement achieved by optimal band-pass filtering in thermal imagery is much more modest than with visual data, increasing performance, respectively, by 35% and 8%. Similar increase was obtained in true admittance rate (42% versus 8%), see Table 6.3.

Neither the eyes or the mouth regions, in either the visual or thermal spectrum, proved very discriminative when used in isolation, see Fig. 6.14. Only 10–12% true positive admittance was achieved, as shown in Table 6.3. However, the proposed fusion of holistic and local appearance offered a consistent and statistically significant improvement. In 1-to-1 matching the true positive admittance rate increased for 4–6%, while the average correct 1-to-N matching improved for roughly 2–3% (Table 6.4).

The greatest power of the method becomes apparent when the two modalities, visual and thermal, are fused. In this case the role of the glasses detection module is much more prominent, drastically decreasing the average error rate from 10% down

Table 6.4. *Holistic and local features, 1-to-1 matching (verification)*: a summary of the results

Representation	Visual (SQI)	Thermal (BP)
	1% intruder acceptance	
Holistic + Eyes	0.6782	0.6499
Holistic + Mouth	0.6410	0.6501
Holistic + Eyes + Mouth	0.6782	0.6558

Table 6.5. *Feature and modality fusion, 1-to-1 matching (verification)*: a summary of the results

Representation	True admission rate
	1% intruder acceptance
Without glasses detection	0.7435
With glasses detection	0.8014

to 3%, see Table 6.1. Similarly, the true admission rate increases to 74% when data is fused without special handling of glasses, and to 80% when glasses are taken into account, see Table 6.5.

6.4 Conclusion

In this chapter we described a system for personal identification based on a face biometric that uses cues from visual and thermal imagery. The two modalities are shown to complement each other, their fusion providing good illumination invariance and discriminative power between individuals. Prescription glasses, a major difficulty in the thermal spectrum, are reliably detected by our method, restricting the matching to non-affected face regions. Finally, we examined how different preprocessing methods affect recognition in the two spectra, as well as holistic and local feature-based face representations. The proposed method was shown to achieve a high recognition rate (97%) using only a small number of training images (5–7) in the presence of large illumination changes.

Our results suggest several possible avenues for improvement. We intend to make further use of the thermal spectrum, by not only detecting the glasses, but also by segmenting them out. This is challenging across large pose variations, such as those contained in our test set. Another research direction we would like to pursue is that of synthetically enriching the training corpus to achieve increased robustness to pose differences between image sets (cf. [177, 178]). Additionally, more advanced set matching methods can be used for better discriminative performance, e.g., [138, 179, 180]. Finally, we note that a research challenge that remains, and which has not been addressed in this chapter, is that of changing facial expression.

7 Multispectral Face Recognition: Fusion of Visual Imagery with Physiological Information

Pradeep Buddharaju and Ioannis Pavlidis

7.1 Introduction

Biometrics has received a lot of attention during the last few years both from the academic and business communities. It has emerged as a preferred alternative to traditional forms of identification, like card IDs, which are not embedded into one's physical characteristics. Research into several biometric modalities including face, fingerprint, iris, and retina recognition has produced varying degrees of success [181]. Face recognition stands as the most appealing modality, since it is the natural mode of identification among humans and is totally unobtrusive. At the same time, however, it is one of the most challenging modalities [182]. Research into face recognition has been biased toward the visual spectrum for a variety of reasons. Among those is the availability and low cost of visual band cameras and the undeniable fact that face recognition is one of the primary activities of the human visual system. Machine recognition of human faces, however, has proven more problematic than the seemingly effortless face recognition performed by humans. The major culprit is light variability, which is prevalent in the visual spectrum due to the reflective nature of incident light in this band. Secondary problems are associated with the difficulty of detecting facial disguises [183].

As a solution to the aforementioned problems, researchers have started investigating the use of thermal infrared for face recognition purposes [184–186]. However, many of these research efforts in thermal face recognition use the thermal infrared band only as a way to see in the dark or reduce the deleterious effect of light variability [187,188]. Methodologically, they do not differ very much from face recognition algorithms in the visual band, which can be classified as appearance-based [189, 190] and feature-based approaches [191, 192].

Recently, attempts have been made to fuse the visual and thermal infrared modalities to increase the performance of face recognition [193–198]. However, almost all these approaches use similar algorithms for extracting features from both visual and thermal infrared images. In this chapter, we present a novel approach to the problem of thermal facial recognition that realizes the full potential of the thermal infrared band. Our goal is to promote a different way of thinking in the area of face recognition in thermal infrared, which can be approached in a distinct manner when compared with other modalities. It consists of a statistical face segmentation and a physiological feature extraction algorithm tailored to thermal phenomenology. The use of vessel structure for human identification has been

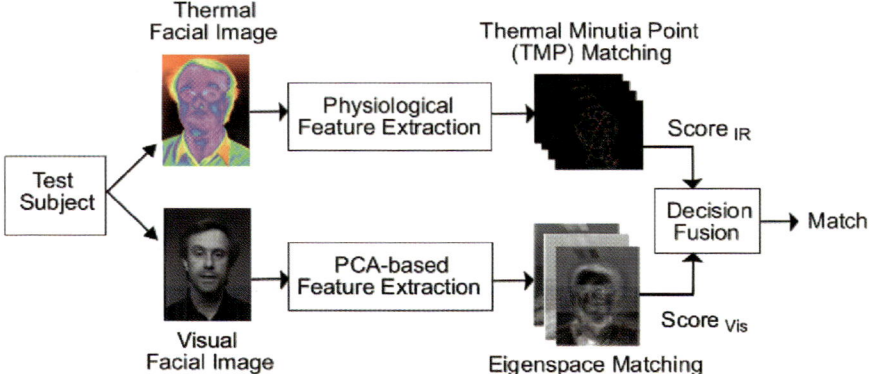

Fig. 7.1. Multispectral face recognition methodology

studied during recent years using traits such as hand vessel patterns [199, 200] and finger vessel patterns [201, 202]. Prokoski et al. anticipated the possibility of extracting the vascular network from thermal facial images and using it as a feature space for face recognition [203]. However, they did not present an algorithmic approach for achieving this. We present a full methodology to extract and match the vascular network from thermal facial imagery [204].

Figure 7.1 depicts the essence of the proposed multispectral face recognition methodology. The goal of face recognition is to match a query face image against a database of facial images to establish the identity of an individual. We collect both thermal and visual facial images of the subject whose identity needs to be tested. We extract the thermal minutia points (TMPs) from the thermal facial image and match them against TMPs of subjects already stored in the database. We then extract the principal components (eiganfaces) from the visual face image and project it to the face space constructed from visual database images. The eigenspace match score is fused with the TMP match score to produce the final match score.

7.2 Physiological Feature Extraction from Thermal Images

A thermal infrared camera with good sensitivity (NEDT $> 0.025°C$) provides the ability to directly image superficial blood vessels on the human face [205]. The pattern of the underlying blood vessels (see Fig. 7.2) is characteristic to each individual, and the extraction of this vascular network can provide the basis for a feature vector. Figure 7.3 outlines the architecture of the feature extraction algorithm.

7.2.1 Face Segmentation

Due to its physiology, a human face consists of "hot" parts that correspond to tissue areas that are rich in vasculature and "cold" parts that correspond to tissue areas

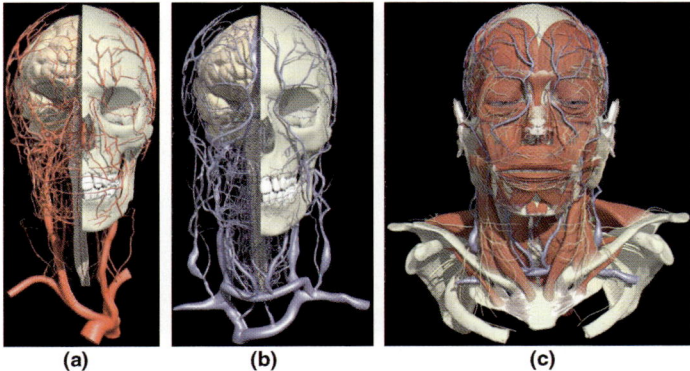

Fig. 7.2. Generic maps of the superficial blood vessels on the face – courtesy of Primal Pictures [206]: (**a**) Overview of arterial network. (**b**) Overview of venous network. (**c**) Arteries and veins together under the facial surface

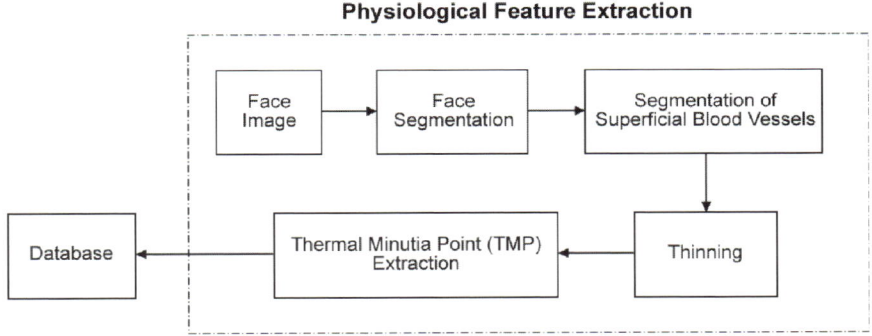

Fig. 7.3. Architecture of physiological feature extraction algorithm

with sparse vasculature. This casts the human face as a bimodal temperature distribution entity, which can be modeled using a mixture of two Normal distributions. Similarly, the background can be described by a bimodal temperature distribution with walls being the "cold" objects and the upper part of the subject's body dressed in cloths being the "hot" object. Figure 7.4b shows the temperature distributions of the facial skin and the background from a typical thermal facial image. We approach the problem of delineating facial tissue from the background using a Bayesian framework [204, 207], because we have a priori knowledge of the bimodal nature of the scene.

We call θ the parameter of interest, which takes two possible values (skin s or background b) with some probability. For each pixel x in the image at time t, we draw our inference of whether it represents skin (i.e., $\theta = s$) or background (i.e., $\theta = b$) based on the posterior distribution $p^{(t)}(\theta|x_t)$ given by:

$$p^{(t)}(\theta|x_t) = \begin{cases} p^{(t)}(s|x_t), & \text{when } \theta = s, \\ p^{(t)}(b|x_t) = 1 - p^{(t)}(s|x_t), & \text{when } \theta = b. \end{cases} \quad (7.1)$$

We develop the statistics only for skin and then the statistics for the background can easily be inferred from (7.1).

According to the Bayes' theorem:

$$p^{(t)}(s|x_t) = \frac{\pi^{(t)}(s)f(x_t|s)}{\pi^{(t)}(s)f(x_t|s) + \pi^{(t)}(b)f(x_t|b)}. \quad (7.2)$$

Here, $\pi^{(t)}(s)$ is the prior skin distribution and $f(x_t|s)$ is the likelihood for pixel x representing skin at time t. In the first frame ($t = 1$) the prior distributions for skin and background are considered equiprobable:

$$\pi^{(1)}(s) = \frac{1}{2} = \pi^{(1)}(b). \quad (7.3)$$

For $t > 1$, the prior skin distribution $\pi^{(t)}(s)$ at time t is equal to the posterior skin distribution at time $t - 1$:

$$\pi^{(t)}(s) = p^{(t-1)}(s|x_{t-1}). \quad (7.4)$$

The likelihood $f(x_t|s)$ of pixel x representing skin at time $t \geq 1$ is given by:

$$f(x_t|s) = \sum_{i=1}^{2} w_{s_i}^{(t)} N(\mu_{s_i}^{(t)}, \sigma_{s_i}^{2(t)}), \quad (7.5)$$

where the mixture parameters w_{s_i} (weight), μ_{s_i} (mean), $\sigma_{s_i}^2$ (variance) : $i = 1, 2$ and $w_{s_2} = 1 - w_{s_1}$ of the bimodal skin distribution can be initialized and updated using

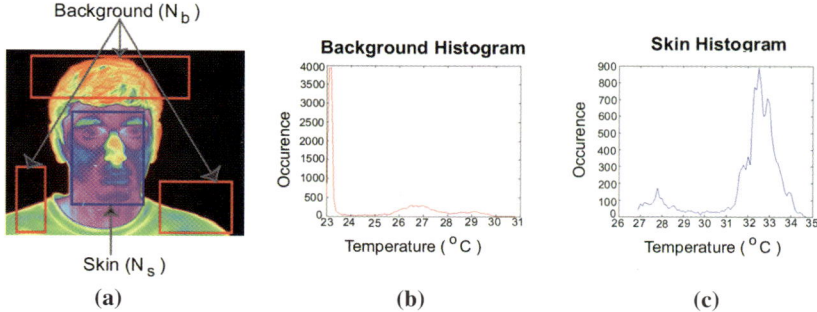

Fig. 7.4. Skin and background: (a) selection of samples for EM algorithm. (b) Corresponding bimodal temperature distribution of background region. (c) Corresponding bimodal temperature distribution of skin region

the EM algorithm. For that, we select N representative facial frames (offline) from a variety of subjects that we call the training set. Then, we manually segment, for each of the N frames, skin (and background) areas, which yields N_s skin (and N_b background) pixels as shown in Fig. 7.4a.

To estimate the mixture parameters for the skin, we initially provide the EM algorithm with some crude estimates of the parameters of interest: $w_{s_0}, \mu_{s_0}, \sigma^2_{s_0}$. Then, we apply the following loop for $k = 0, 1, \ldots$:

$$z_{ij}^{(k)} = \frac{w_{s_i}^{(k)} (\sigma_{s_i}^{(k)})^{-1} \exp\left\{-\frac{1}{2(\sigma_{s_i}^{(k)})^2}(x_j - \mu_{s_i}^{(k)})^2\right\}}{\sum_{t=1}^{2} w_{s_t}^{(k)} (\sigma_{s_t}^{(k)})^{-1} \exp\left\{-\frac{1}{2(\sigma_{s_t}^{(k)})^2}(x_j - \mu_{s_t}^{(k)})^2\right\}},$$

$$w_{s_i}^{(k+1)} = \frac{\sum_{j=1}^{N_s} z_{ij}^{(k)}}{N_s},$$

$$\mu_{s_i}^{(k+1)} = \frac{\sum_{j=1}^{N_s} z_{ij}^{(k)} x_j}{N_s w_{s_i}^{(k+1)}},$$

$$(\sigma_{s_i}^{(k+1)})^2 = \frac{\sum_{j=1}^{N_s} z_{ij}^{(k)} (x_j - \mu_{s_i}^{(k+1)})^2}{N_s w_{s_i}^{(k+1)}},$$

where $i = 1, 2$ and $j = 1, \ldots, N_s$. Then, we set $k = k + 1$ and repeat the loop. The condition for terminating the loop is:

$$|w_{s_i}^{(k+1)} - w_{s_i}^{(k)}| < \epsilon, \, i = 1, 2. \tag{7.6}$$

We apply a similar EM process for determining the initial parameters of the background distributions. Once a data point x_t becomes available, we decide that

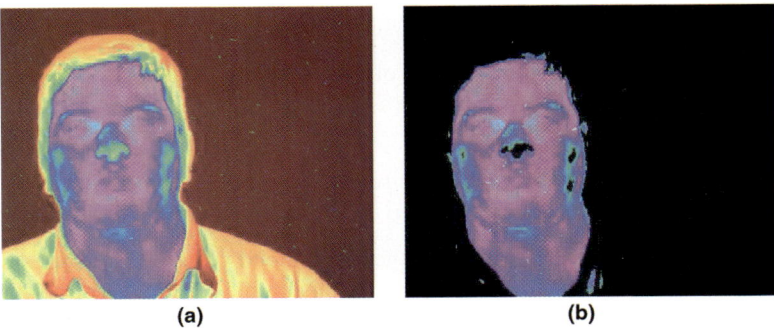

Fig. 7.5. Segmentation of facial skin region: (**a**) original thermal facial image. (**b**) Result of Bayesian segmentation

it represents skin if the posterior distribution for the skin $p^{(t)}(s|x_t) > 0.5$ and that it represents background otherwise. Figure 7.5b depicts the visualization of Bayesian segmentation on the subject shown in Fig. 7.5a. Part of the subject's nose has been erroneously classified as background and a couple of cloth patches from the subject's shirt have been erroneously marked as facial skin. This is due to occasional overlapping between portions of the skin and background distributions. The isolated nature of these mislabeled patches makes them easily correctable through postprocessing. We apply our three-step postprocessing algorithm on the binary segmented image. Using foreground (and background) correction, we find the mislabeled pixels in foreground (and background) and remove them. The specific algorithm that achieves this is the following:

1. Label all the regions in the foreground and background using a simple flood-fill or connected component labeling algorithm [208]. Let the foreground regions be $R_f(i), i = 1, \ldots, N_f$, where N_f represents the number of foreground regions, and let the background regions be $R_b(j), j = 1, \ldots, N_b$, where N_b represents the number of background regions.
2. Compute the number of pixels in each of the foreground and background regions. Find the maximum foreground (R_f^{max}) and background (R_b^{max}) areas:

$$R_f^{max} = \max\{R_f(i), i = 1, \ldots, N_f\},$$

$$R_b^{max} = \max\{R_b(i), i = 1, \ldots, N_b\}.$$

3. Change all foreground regions that satisfy the condition $R_f(i) < R_f^{max}/4$ to background. Similarly, change all background regions that satisfy the condition $R_b(i) < R_b^{max}/4$ to foreground. We found experimentally that outliers tend to have an area smaller than one-fourth of the maximum area, and hence can be corrected with the above conditions. Figure 7.6 shows the result of our postprocessing algorithm.

7.2.2 Segmentation of Superficial Blood Vessels

Once a face is delineated from the rest of the scene, the segmentation of superficial blood vessels from the facial tissue is carried out in the following two steps [205, 207]:

1. The image is processed to reduce noise and enhance edges.
2. Morphological operations are applied to localize the superficial vasculature.

In thermal imagery of human tissue the major blood vessels have weak sigmoid edges, which can be handled effectively using anisotropic diffusion. The anisotropic

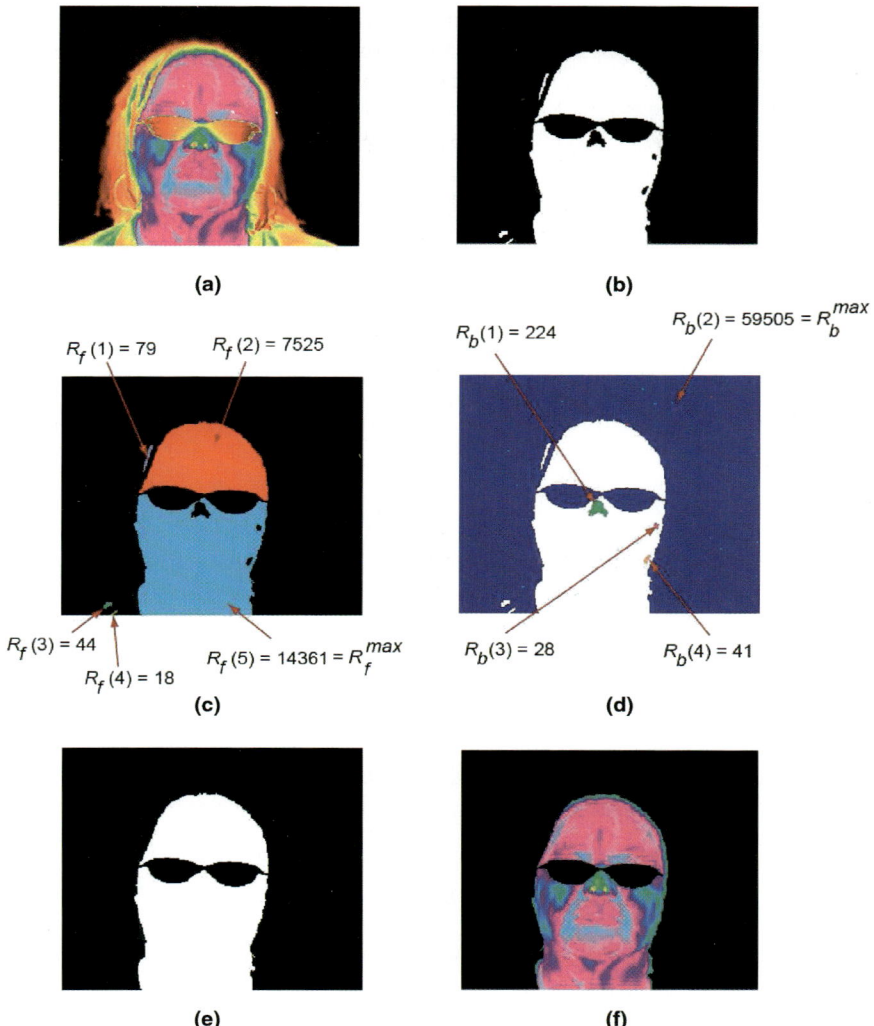

Fig. 7.6. Segmentation of facial skin region: (**a**) original thermal facial image. (**b**) Binary segmented image. (**c**) Foreground regions each represented in different color. (**d**) Background regions each represented in different color. (**e**) Binary mask after foreground and background corrections. (**f**) Final segmentation result after post-processing

diffusion filter is formulated as a process that enhances object boundaries by performing intra-region as opposed to inter-region smoothing. The mathematical equation for the process is:

$$\frac{\partial I(\bar{x}, t)}{\partial t} = \nabla(c(\bar{x}, t)\nabla I(\bar{x}, t)). \tag{7.7}$$

In our case $I(\bar{x}, t)$ is the thermal infrared image, $\bar{x}$ refers to the spatial dimensions, and t to time. $c(\bar{x}, t)$ is called the diffusion function. The discrete version of the anisotropic diffusion filter of (7.7) is as follows:

$$I_{t+1}(x,y) = I_t + \frac{1}{4} * [c_{N,t}(x,y)\nabla I_{N,t}(x,y) \\
+ c_{S,t}(x,y)\nabla I_{S,t}(x,y) + c_{E,t}(x,y)\nabla I_{E,t}(x,y) \\
+ c_{W,t}(x,y)\nabla I_{W,t}(x,y)]. \qquad (7.8)$$

The four diffusion coefficients and four gradients in (7.8) correspond to four directions (i.e., North, South, East and West) with respect to the location (x,y). Each diffusion coefficient and the corresponding gradient are calculated in the same manner. For example, the coefficient along the north direction is calculated as follows:

$$c_{N,t}(x,y) = exp(\frac{-\nabla I_{N,t}^2(x,y)}{k^2}), \qquad (7.9)$$

where $I_{N,t} = I_t(x, y+1) - I_t(x, y)$.

Image morphology is then applied on the diffused image to extract the blood vessels that are at a relatively low contrast compared to that of the surrounding tissue. We employ for this purpose a top-hat segmentation method, which is a combination of erosion and dilation operations. Top-hat segmentation takes two forms. First form is the white top-hat segmentation that enhances the bright objects in the image, while the second one is the black top-hat segmentation that enhances dark objects. In our case, we are interested in the white top-hat segmentation because it helps with enhancing the bright ("hot") ridge like structures corresponding to the blood vessels. In this method the original image is first opened and then this opened image is subtracted from the original image as follows:

$$I_{open} = (I \ominus S) \oplus S, \\
I_{top} = I - I_{open}, \qquad (7.10)$$

where I, I_{open}, I_{top} are the original, opened, and white top-hat segmented images, respectively, S is the structuring element, and $\ominus$, $\oplus$ are morphological erosion and dilation operations, respectively. Figure 7.7a depicts the result of applying

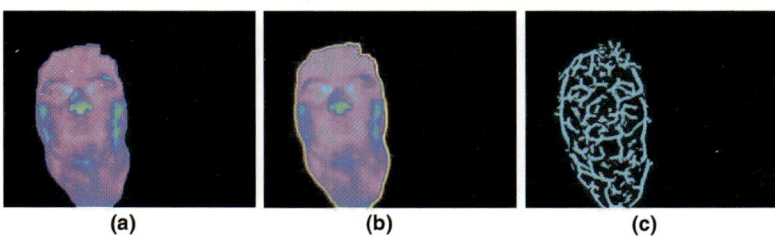

Fig. 7.7. Vascular network extraction: (**a**) original segmented image. (**b**) Anisotropically diffused image. (**c**) Blood vessels extracted using white top-hat segmentation

anisotropic diffusion to the segmented facial tissue shown in Fig. 7.4b, and the Fig. 7.7b shows the corresponding blood vessels extracted using white top-hat segmentation.

7.2.3 Extraction of TMPs

The extracted blood vessels exhibit different contour shapes between subjects. We call the branching points of the blood vessels TMPs. TMPs can be extracted from the blood vessel network in ways similar to those used for fingerprint minutia extraction. A number of methods have been proposed [209] for robust and efficient extraction of minutia from fingerprint images. Most of these approaches describe each minutia point by at least three attributes, including its type, its location in the fingerprint image, and the local vessel orientation. We adopt a similar approach for extracting TMPs from vascular networks, which is outlined in the following steps:

1. The local orientation of the vascular network is estimated.
2. The vascular network is skeletonized.
3. The TMPs are extracted from the thinned vascular network.
4. The spurious TMPs are removed.

Local orientation $\Psi(x, y)$ is the angle formed at (x, y) between the blood vessel and the horizontal axis. Estimating the orientation field at each pixel provides the basis for capturing the overall pattern of the vascular network. We use the approach proposed in [210] for computing the orientation image because it provides pixel-wise accuracy.

Next, the vascular network is thinned to one-pixel thickness [211]. Each pixel in the thinned map contains a value of 1 if it is on the vessel and 0 if it is not. Considering eight-neighbourhood $(N_0, N_1, \ldots, N_7)$ around each pixel, a pixel (x, y) represents a TMP if $(\sum_{i=0}^{7} N_i) > 2$ (see Fig. 7.8).

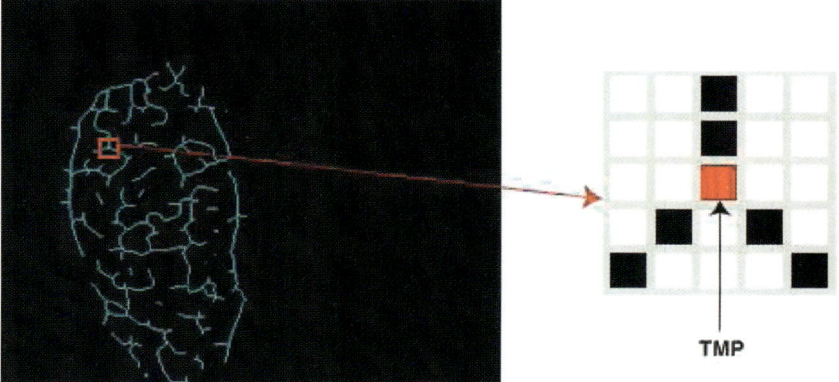

Fig. 7.8. Thermal Minutia Point (TMP) extracted from the thinned vascular network

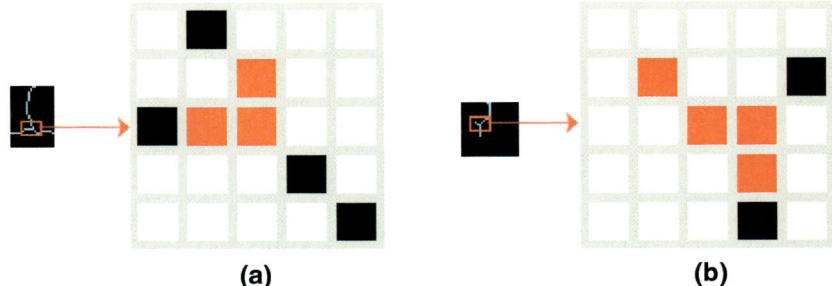

Fig. 7.9. Spurious TMPs: (**a**) clustered TMPs. (**b**) Spike formed due to a short branch

It is desirable that the TMP extraction algorithm does not leave any spurious TMPs since this will adversely affect the matching performance. Removal of clustered TMPs (see Fig. 7.9a) and spikes (see Fig. 7.9b) helps to reduce the number of spurious TMPs in the thinned vascular network.

The vascular network of a typical facial image contains around 50–80 genuine TMPs whose location (x, y) and orientation (Ψ) are stored in the database. Figure 7.10 shows the results of each stage of the feature extraction algorithm on a thermal facial image.

7.2.4 Matching of TMPs

Numerous methods have been proposed for matching fingerprint minutiae, most of which try to simulate the way forensic experts compare fingerprints [209]. Popular techniques are alignment-based point pattern matching, local structure matching, and global structure matching. Local minutiae matching algorithms are fast, simple, and more tolerant to distortions. Global minutiae matching algorithms feature high distinctiveness. A few hybrid approaches [212, 213] have been proposed where the advantages of both local and global methods are exploited. We use a hybrid method [212] to perform TMP matching.

For each TMP $M(x_i, y_i, \Psi_i)$ that is extracted from the vascular network, we consider its N nearest-neighbour TMPs $M(x_n, y_n, \Psi_n)$, $n = 1, \ldots, N$. Then, the TMP $M(x_i, y_i, \Psi_i)$ can be defined by a new feature vector:

$$L_M = \{\{d_1, \varphi_1, \vartheta_1\}, \{d_2, \varphi_2, \vartheta_2\}, \ldots, \{d_N, \varphi_N, \vartheta_N\}, \Psi_i\} \tag{7.11}$$

where

$$\begin{aligned} d_n &= \sqrt{(x_n - x_i)^2 + (y_n - y_i)^2} \\ \varphi_n &= diff(\Psi_n, \Psi_i), \; n = 1, 2, \ldots, N \\ \vartheta_n &= diff\left(\arctan\left(\frac{y_n - y_i}{x_n - x_i}\right), \Psi_i\right) \end{aligned} \tag{7.12}$$

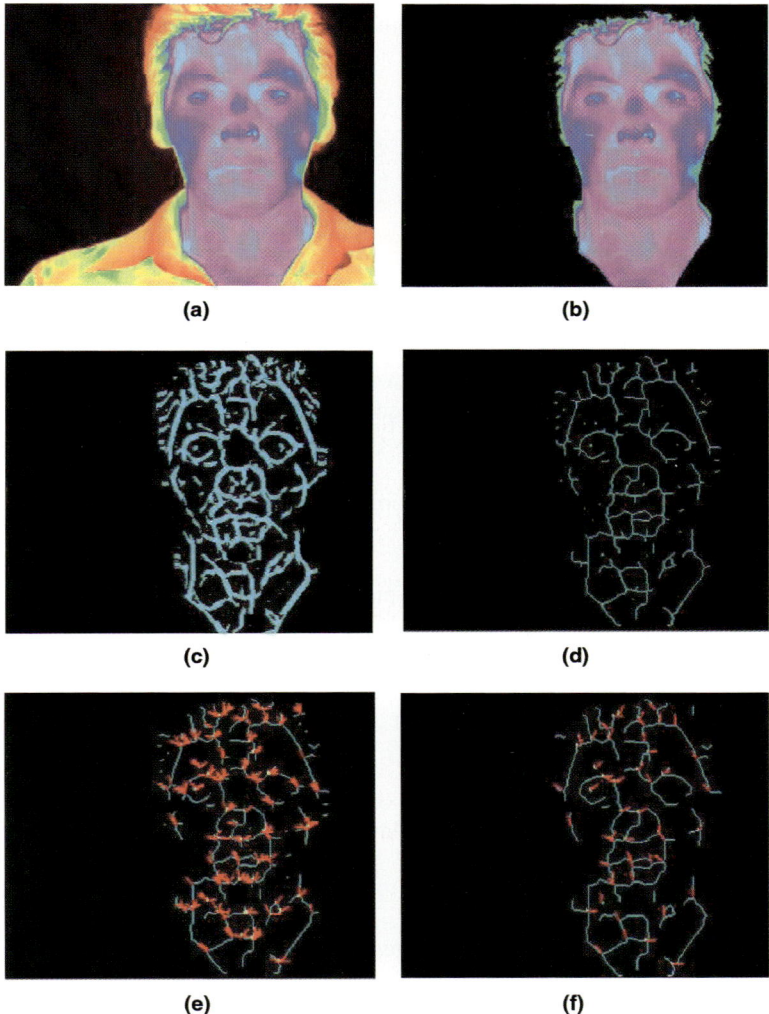

Fig. 7.10. Visualization of the various stages of the feature extraction algorithm: (**a**) a typical thermal facial image. (**b**) Facial tissue delineated from the background. (**c**) Vascular network extracted from thermal facial image. (**d**) Thinned vessel map. (**e**) Extracted TMPs from branching points. (**f**) Spurious TMPs removed

The function $diff()$ calculates the difference of two angles and scales the result within the range $[0, 2\pi)$ [213]. Given a test image $\mathbf{I_t}$, the feature vector of each of its TMP is compared with the feature vector of each TMP of a database image. Two TMPs M and M' are marked to be a matched pair if the absolute difference between corresponding features is less than specific threshold values $\{\delta_d, \delta_\varphi, \delta_\vartheta, \delta_\psi\}$. The threshold values should be chosen in such a way that they accommodate linear deformations and translations. The final matching score between the test image and a database image is given by:

$$Score = \frac{NUM_{match}}{\max(NUM_{test}, NUM_{database})} \quad (7.13)$$

where NUM_{match} represents number of matched TMP pairs, and NUM_{test}, $NUM_{database}$ represent number of TMPs in test and database images, respectively.

7.3 PCA-Based Feature Extraction from Visual Images

Principal component analysis (PCA) is a well known approach for dimensionality reduction of the feature space. It has been successfully applied in face recognition [189, 214]. The main idea is to decompose face images into a small set of feature images called eigenfaces, which can be considered as points in a linear subspace called "face space" or "eigenspace". Recognition is performed by projecting a new face image into this eigenspace and then comparing its position with those of known faces.

Suppose a face image consists of N pixels, so it can be represented by a vector Γ of dimension N. Let $\{\Gamma_i | i = 1, \ldots, M\}$ be the training set of face images. The average face of these M images is given by

$$\Psi = \frac{1}{M} \sum_{i=1}^{M} \Gamma_i. \quad (7.14)$$

Then, each face Γ_i differs from the average face Ψ by Φ_i:

$$\Phi_i = \Gamma_i - \Psi; i = 1, \ldots, M. \quad (7.15)$$

A covariance matrix of the training images can be constructed as follows:

$$C = AA^T, \quad (7.16)$$

where $A = [\Phi_1, \ldots, \Phi_M]$. The top M' eigenvectors $U = [u_1, \ldots, u_{M'}]$ of the covariance matrix A, called eigenfaces, constitute the eiganspace. Figure 7.11 shows the top six eigenfaces extracted from our training set in decreasing order. Given a test image, Γ_{test}, it is projected to the eigenspace and an Ω_{test} vector is obtained as follows:

$$\Omega_{test} = U^T (\Gamma_{test} - \Psi). \quad (7.17)$$

The distances between this vector and the projected vectors from the training images are used as a measure to find the best match in the database. Any standard distance measure such as Euclidean distance, Mahalanobis distance, or MahCosine measure can be used to compare the vectors [197].

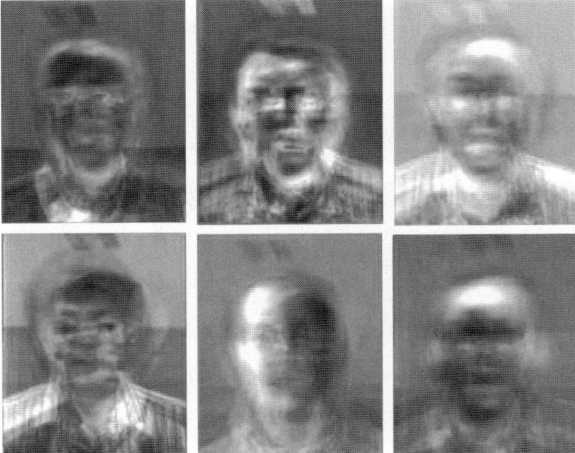

Fig. 7.11. Eigenfaces extracted from our training set that correspond to decreasing order of eigenvalues

7.4 Experimental Results and Discussion

We used the Equinox Corporation's database in our experiments. It is a large database of both infrared (short-, mid-, and long-wave) and visual band images available for public download at *http://www.equinoxsensors.com/products/HID.html*. Image frame sequences were acquired at 10 frames s^{-1} while the subject was reciting the vowel sequence "a,e,i,o,u". The database also consists of subject images wearing glasses and with expressions of happiness, anger and surprise, which were used to account for variation in poses. In order to induce variability in visual band images, three different illumination conditions were used during acquisition – frontal, frontal-left, and frontal-right. For each subject in the database subset we used, images were acquired simultaneously from visual and mid-wave infrared cameras.

We used a total of 4,552 co-registered visual and mid-wave infrared images for our experiments, which comprised of 45 different subjects. For each subject, we used three training images (one per each illumination condition). If the subject was wearing glasses, we included images with and without glasses in the training set. Figure 7.12 shows the training examples of two subjects from the database. For each test image, we applied the physiological face recognition algorithm on the thermal image and the PCA algorithm on its visual counterpart. Finally, we applied decision fusion by combining the scores from the visual and thermal infrared recognition algorithms. Specifically, the fusion was performed by combining the individual scores from each of the algorithms. We found from our experiments that the physiological algorithm on thermal images performed slightly better than the PCA algorithm on visual imagery. The rank 1 recognition rate for thermal images was 97.74%, whereas that of visible images was 96.19%. Since the mismatches in each of these experiments were disjoint, fusion yielded an increase in performance

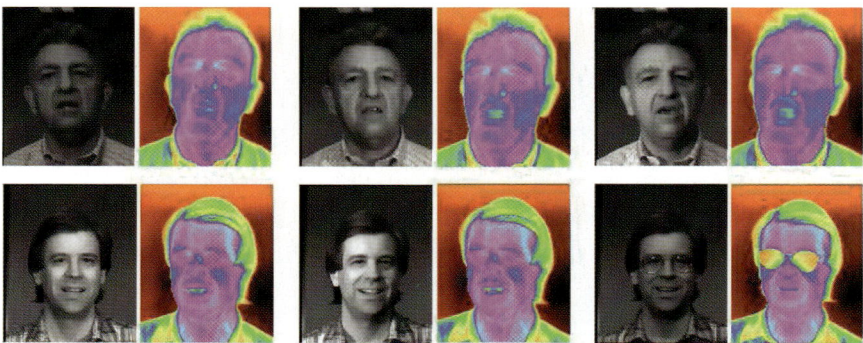

Fig. 7.12. Sample training images of two subjects (one subject per row) in the database

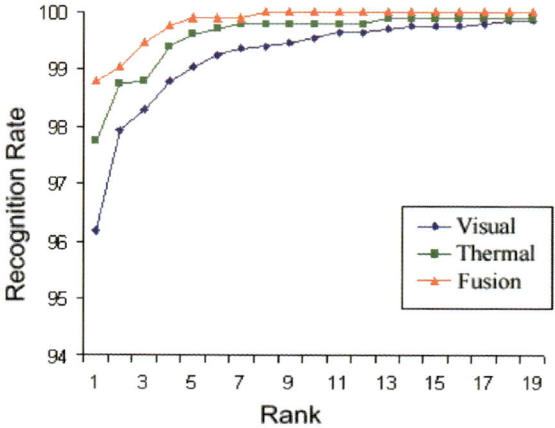

Fig. 7.13. CMC curves of the visual, thermal and fusion face recognition algorithms

with a rank 1 recognition rate of 98.79%. Figure 7.13 shows the CMC curves for the visual, thermal and fusion algorithms.

We noticed that the recognition performance from the physiological algorithm on thermal images can be improved by estimating and eliminating the incorrect TMPs as well as non-linear deformations in the extracted vascular network caused due to large facial expressions and non-linear pose transformations. Figure 7.14 shows an example of non-linear deformations caused in the vascular network between gallery and probe images of the same subject due to pose and facial expression changes. Even though the matching algorithm described in Sect. 7.2.4 works fine with linear transformations in the vascular network, it affords small latitude in the case of non-linear transformations. Our future work is directed towards addressing this issue.

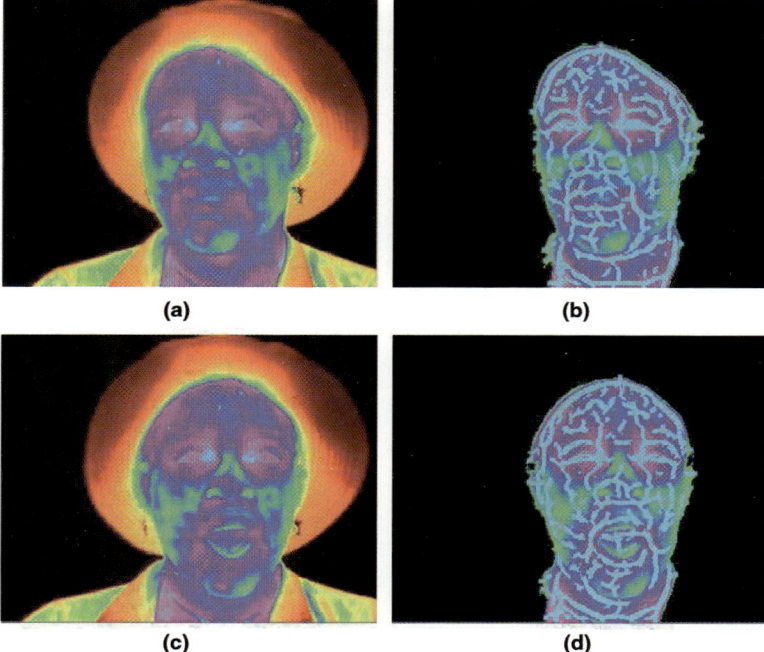

Fig. 7.14. (**a**) Training image and (**b**) corresponding vascular network (overlaid over the segmented image). (**c**) Test image of same subject exhibiting large facial expression and (**d**) corresponding vascular network (overlaid over the segmented image)

There are two major operational limitations in the current physiological feature extraction method:

1. Glasses are opaque in the thermal infrared spectrum and hence block important vascular information around eyes. Also, facial hair curtails the radiation emitted from the covered surface of the skin, and may cause facial segmentation to break down. Figure 7.15 shows examples of failed face segmentation when glasses and facial hair are present.
2. The robustness of the method degrades when there is substantial perspiration. This results in a highly non-linear shift of the thermal map that alters radically the radiation profile of the face. A practical scenario where such a case may arise is when a subject is imaged after a strenuous exercise that lasted several minutes. Another such scenario may arise when a heavily dressed subject is imaged in a very hot environment.

   We have performed an experiment whereby a subject is imaged at the following instances:
   – In a baseline condition (Fig. 7.16, image 1a)
   – After 1 min of rigorous walking (Fig. 7.16, image 2a)

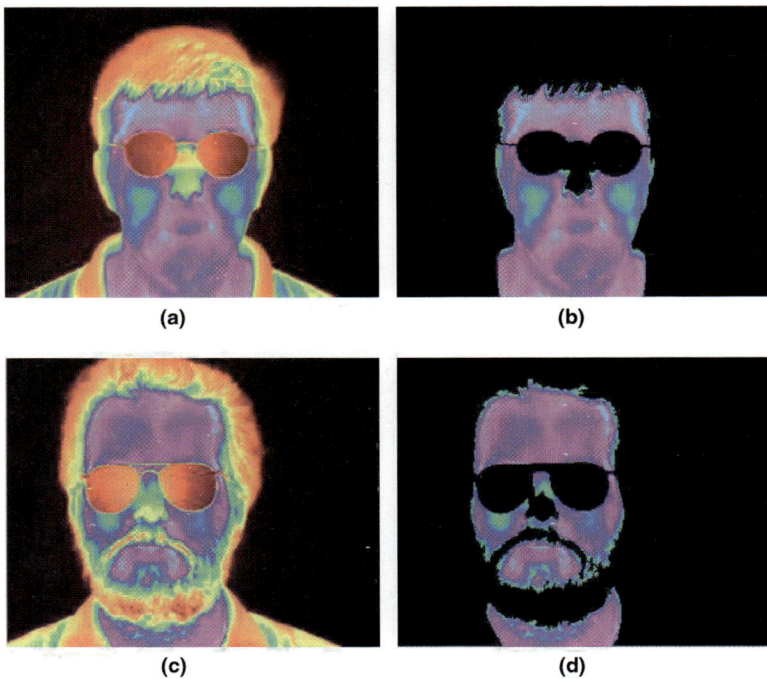

Fig. 7.15. (**a**) Thermal facial image with glasses and (**b**) result of segmentation. (**c**) Thermal facial image with facial hair and glasses and (**d**) result of segmentation

- After 5 min of rigorous walking (Fig. 7.16, image 3a)
- After 5 min of rigorous jogging (Fig. 7.16, image 4a)

Column b of Fig. 7.16 shows the corresponding vessel extraction results. In the case of image 2a, the metabolic rate of the subject shifted to higher gear, but perspiration is still not a major problem. One can find evidence of the higher metabolic rate by looking at the left temporal area, where the region around the rich vasculature has become deeper cyan (hotter) in image 2a with respect to image 1a. This is an example of a positive linear shift (warming up), which the vessel extraction algorithm handles quite well (see image 2b versus image 1b). As the exercise become more strenuous and lasts longer, perspiration increases and introduces a negative non-linear shift (cooling down) in the thermal map. This is especially pronounced in the forehead where most of the perspiration pores are. Due to this, some unwanted noise starts creeping in image 3b, which becomes more dramatic in image 4b. The performance of the vessel extraction algorithm deteriorates but not uniformly. For example, the vessel extraction algorithm continues to perform quite well in the cheeks where perspiration pores

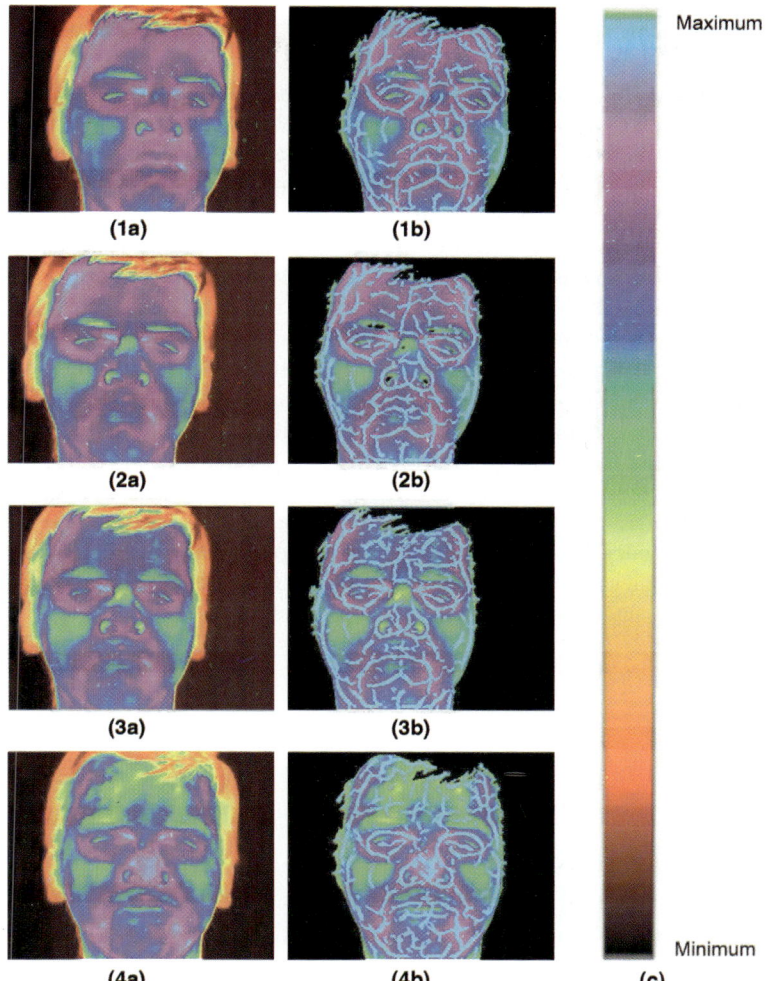

Fig. 7.16. Effect of perspiration on feature extraction. Thermal facial image of a subject (**1a**) at rest, (**2a**) after 1 min of rigorous walking (**3a**) after 5 min of rigorous walking, (**4a**) after 5 min of rigorous jogging, and (**1b,2b,3b,4b**) corresponding vascular network maps, and (**c**) colour map used to visualize temperature values

are sparse and the cooling down effect is not heavily non-linear. In contrast, performance is a lot worse in the forehead area, where some spurious vessel contours are introduced due to severe non-linearity in the thermal map shift.

7.5 Conclusions

We have outlined a novel multi-spectral approach to the problem of face recognition by the fusion of thermal infrared and visual band images. The cornerstone of the approach is the use of unique and time invariant physiological information as feature space for recognition in thermal imagery. The facial tissue is first separated from the background using a Bayesian segmentation method. The vascular network on the surface of the skin is then extracted based on a white top-hat segmentation preceded by anisotropic diffusion. TMPs are extracted from the vascular network and are used as features for matching test to database images. The method although young, performed well on a nontrivial database. We also applied a PCA-based (eigenface) recognition approach on concomitant visual imagery. We have shown that the recognition performance in the thermal domain is slightly better than the visual domain, and that fusion of the two modalities/methodologies is better than either one of them. In a nutshell, this research demonstrated that standard visual face recognition methods can gain in performance if they are combined with physiological information, uniquely extracted in thermal infrared. The most notable outcome besides performance increase is the striking complimentarily of the two modalities/methodolgogies as it is revealed in the experimental results. It is the latter that renders fusion a natural strategy that fits the problem.

8 Feature Selection for Improved Face Recognition in Multisensor Images

Satyanadh Gundimada and Vijayan Asari

8.1 Introduction

This chapter discusses the problems faced by present day face recognition systems in the presence of extreme variations. Even though face recognition technology [215] has progressed from linear subspace methods [216] such as eigenfaces and fisher faces [217–219] to nonlinear methods such as KPCA, KFD [220–223], many of the problems are yet to be addressed completely. In addition to challenges such as expression and pose variations, partial occlusions, the face recognition techniques face a major bottle neck in the form of illumination variation. The chapter addresses the problems of expression variations and partial occlusions by presenting a novel feature selection strategy. The illumination variations are tackled by considering images from multiple sensors.

8.1.1 Sensors and Systems

Recently images from multiple sensors are being utilized for the purpose of obtaining complementary information. Face recognition is one of the applications which can benefit from using multiple sensors. Visual sensors cannot capture enough information for precise face recognition in low to very low illumination conditions. Here, the Long wave infrared image sensor comes into the picture. The infrared sensors capture the amount of heat generated from the objects in the scene and not the light reflected from those objects. Hence this could help in dealing with extreme illumination conditions. Despite its robustness to illumination changes, however, IR imagery has several drawbacks including that it is sensitive to temperature changes in the surrounding environment, variations in the heat patterns of the face and its is opaque to glass. In contrast to IR imagery, visible imagery is more robust to the above factors but as said earlier, very sensitive to illumination changes.

8.1.2 Related Work

While the nature of face imagery in the visible domain is well-studied, particularly with respect to illumination dependence, its thermal counterpart has received less attention. Previous studies have shown that infra red imagery offers a promising alternative to visible imagery for handling variations in face appearance due to illumination changes more successfully. But it is also interesting to observe that, face recognition on thermal images in [224] degrades more sharply than with visible images when probe and gallery are chosen from different sessions. Results in [224] indicate better performances obtained with visible imagery indoors under controlled lighting conditions. But outdoors the thermal image based face recognition system outperformed the visible imagery based one. Also the thermal face recognition results for both indoor and outdoor environments are comparatively less different from each other, thus reiterating that the illumination has little effect on thermal imagery. The conclusion of the studies in [224–226] is that despite the degraded thermal recognition performance, fusion of both visible and thermal modalities yields better overall performance. Most of the studies that were conducted on thermal image face recognition relied on conventional eigenfaces approach. This was particularly relevant for us because, in [224] it can be noticed that while multisession thermal face recognition under controlled indoor illumination was statistically poorer than visible recognition with two standard algorithms, significance was substantially reduced with an algorithm [224] more specifically tuned to thermal imagery. This suggests that previous results reported on thermal face recognition may be incomplete. Hence building of effective algorithms to fuse information from both spectra has the potential to improve face recognition performance. It is possible to realize sensor fusion on different levels: sensor data level fusion, feature vector level fusion, and decision level fusion. In [224] it is consider that fusion on the decision level has more potential applications.

8.1.3 Proposed Methodologies

In addition to multisensor fusion, a novel feature selection strategy is implemented to overcome the above mentioned challenges faced by real time face recognition techniques such as partial occlusions and facial expressions along with illumination variations. Phase congruency based features are used for the purpose. Unlike the edge detectors, which identify the sharp changes in intensity, the phase congruency model detects points of order in the phase spectrum. There is also physiological evidence, indicating that human visual system responds strongly to the points in an image where the phase information is highly ordered. Phase congruency provides a measure that is independent of the overall magnitude of the signal making it invariant to variations in image illumination and/or contrast. Hence phase congruency image maps are used instead of raw intensity images. The facial variations in real world scenario are confined to local regions. Considering additional pixel dependencies across various subregions could help in providing additional information, which in turn could help in improving the classification accuracy. A feature selection policy based on the above discussion in which modular spaces are created with pixels from across various local regions taking into account the locality of such regions is implemented in this chapter. Experiments were conducted on individual and fusion modalities using the proposed face recognition. Both data level and decision level fusion are carried out. Experimental results indicate that the feature selection strategy implemented along with eigen spaces concept resulted in high accuracy compared to raw intensity images for all modalities. It is also observed that recognition accuracy in the case of raw or intensity images is higher for data level fusion compared to all other modalities, where as the application of the proposed face recognition technique provided better accuracy results for decision level fusion.

8.1.4 Organization of the Chapter

The third section explains the need for a proper feature selection strategy and also describes the proposed phase congruency, neighbor hood defined feature selection processes. Section four explains the proposed feature selection strategy. Section five describes the types of fusion techniques that are implemented and section six gives a detailed explanation of the experimental setup and the results obtained. Also a discussion of the results obtained is provided in the same section. Section seven provides the conclusion.

8.2 Phase Congruency Features

Gradient-based operators, which look for points of maximum intensity gradient, will fail to correctly detect and localize a large proportion of features within images. Unlike the edge detectors, which identify the sharp changes in intensity, the phase congruency model detects points of order in the phase spectrum. According to Opeinheim and Lim [227], phase component is more important than the magnitude component in the reconstruction process of an image from its Fourier domain. There is also physiological evidence, indicating that human visual system responds strongly to the points in an image where the phase information is highly ordered. Phase congruency provides a measure that is independent of the overall magnitude of the signal making it invariant to variations in image illumination and/or contrast. Figure 8.1 shows phase congruency image and the corresponding intensity image. The phase congruency technique used in this chapter is based on the one developed by Peter Kovesi [228]. Phase congruency function in terms of the Fourier series expansion of a signal at some location x is given by

$$PC(x) = \frac{\sum_n A_n \cos(\phi_n(x) - \overline{\phi}(x))}{\sum_n A_n} \qquad (8.1)$$

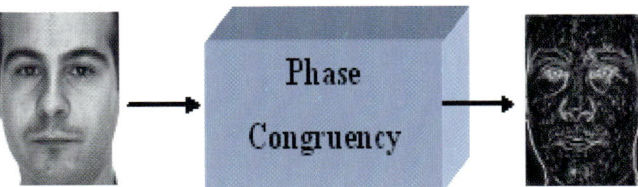

Fig. 8.1. Phase congruency map obtained from the corresponding intensity image

where A_n represents the amplitude of the nth Fourier component, and $\phi_n(x)$ represent the local phase of the Fourier component at position x. $\overline{\phi}(x)$ is the weighted mean of all the frequency components at x. Phase congruency can be approximated to finding where the weighted variance of local phase angles relative to the weighted average local phase, is minimum. An alternative and easier interpretation of phase congruency is proposed in [229]. It is proposed that energy is equal to phase congruency scaled by the sum of the Fourier amplitudes as shown in

$$E(x) = PC(x) \sum_n A_n \qquad (8.2)$$

Hence phase congruency is stated as the ratio of $E(x)$ to the overall path length taken by the local Fourier components in reaching the end point. This makes the phase congruency independent of the overall magnitude of the signal. This provides invariance to variations in image illumination and contrast. $E(x)$ can be expressed as $E(x) = \sqrt{F(x)^2 + H(x)^2}$.

If $I(x)$ is the input signal then $F(x)$ is the signal with its DC component removed and $H(x)$ is the Hilbert transform of $F(x)$ which is a 900 phase shift of $F(x)$. Approximations to the components $F(x)$ and $H(x)$ are obtained by convolving the signal with a quadrature pair of filters. In order to calculate the local frequency and phase information in the signal, logarithmic Gabor functions are used. If $I(x)$ is the signal and M_n^e and M_n^o denote the even symmetric and odd-symmetric wavelets at a scale n. The amplitude and phase of the transform at a given wavelet scale is given by

$$A_n = \sqrt{e_n(x)^2 + o_n(x)^2} \qquad (8.3)$$

$$\phi_n = \tan^{-1}(o_n(x)/e_n(x)) \qquad (8.4)$$

where $e_n(x)$ and $o_n(x)$ are the responses of each quadrature pair of filters. Equation (5) illustrates the response vector.

$$[e_n(x), o_n(x)] = [I(x) * M_n^e, I(x) * M_n^o] \qquad (8.5)$$

$F(x)$ and $H(x)$ can be obtained from

$$F(x) = \sum_n e_n(x) \qquad (8.6)$$

$$H(x) = \sum_n o_n(x) \qquad (8.7)$$

If all the Fourier amplitudes at x are very small then the problem of phase congruency becomes ill conditioned. To overcome the problem a small positive constant ε is added to the denominator. The final phase congruency equation is given by

$$PC(x) = \frac{E(x)}{\varepsilon + \sum_n A_n} \qquad (8.8)$$

One-dimensional analysis is carried out over several orientations, and the results are combined to analyze a two-dimensional signal (image) [228].

8.3 Feature Selection

Facial variations are confined mostly to local regions in reality. Modularizing the images would help in localizing these variations, provided the modules created are sufficiently small. But in doing so, large amount of dependencies among various neighboring pixels are ignored. This can be countered by making the modules larger, but this will result in improper localization of the facial variations. In order to deal with this problem, a novel module creation strategy is implemented in this chapter. It has been proved that, dividing an image of size 64×64 into regions of size 8×8 pixels is appropriate in achieving high classification accuracy [230]. The training phase of the proposed face recognition technique is illustrated in Fig. 8.2. Each image under consideration is divided into small nonoverlapping subregions of size 4×8 pixels. Two of such 4×8 pixel regions from a predefined region (neighborhood)

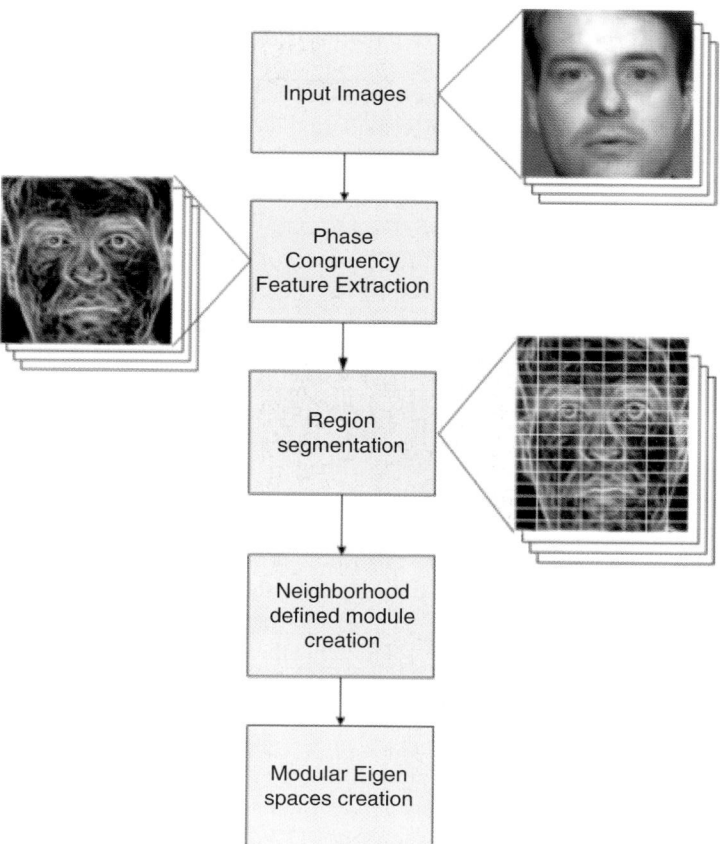

Fig. 8.2. Block diagram illustrating the training phase of the proposed face recognition technique

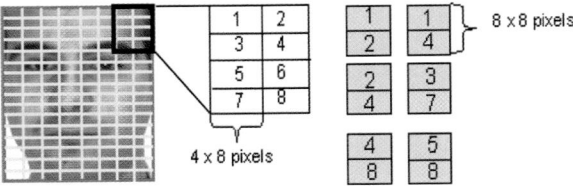

Fig. 8.3. Figure illustrates the creation of modules of size 8×8 by combining blocks of 4×8 pixels from within a neighborhood of 16×16 pixels

within the image can be combined to form a 8×8 pixel region or module. Figure 8.3 illustrates the process of obtaining such regions from each of 16×16 pixels. Twenty-eight such different 8×8 modules can be created (all the combinations). A total of 448 modules are created from a 64×64 face image. By following the above procedure, 448 modules of each image are produced. Eigenspaces are created for each such module. The same procedure of module creation is followed for the probe image, that is, each module is projected onto the corresponding eigenspace and classified according to minimum distance measure. A voting procedure determines the result of the overall classification by considering the individual classification results of all the 448 modules.

8.4 Image Fusion

Image fusion is the process by which two or more images obtained from multiple sensors are combined into a single image, retaining the important features from each of the original images. The most easy and basic image fusion technique is average of the two images. The most popular image fusion algorithm is the one based on the discrete wavelet transform [231]. The wavelet transform of an image provides multistage pyramid decomposition for the image. This decomposition will typically have several stages. There are four frequency bands after each decomposition. These are the low–low, low–high, high–low, and high–high bands. The next stage of the decomposition process operates only on the low–low part of the previous result. This produces a pyramidal hierarchy. We can think of the low–low band as the low pass filtered and subsampled source image. All the other bands which are called high frequency bands contain transform coefficients that reflect the differences between neighboring pixels. The absolute values of the high frequency coefficients represent the intensity of brightness fluctuation of the image at a given scale. The larger values imply more distinct brightness changes which typically correspond to the salient features of objects. Thus, a simple fusion rule is to select the larger absolute value of the two corresponding wavelet coefficients from each of the two source images. There are several other fusion rules [226] that can be implemented. The fusion techniques described above are implemented before any further processing is done. Hence these techniques are called data level or image level fusion. Another type of fusion is to fuse the information that is obtained after certain processing is

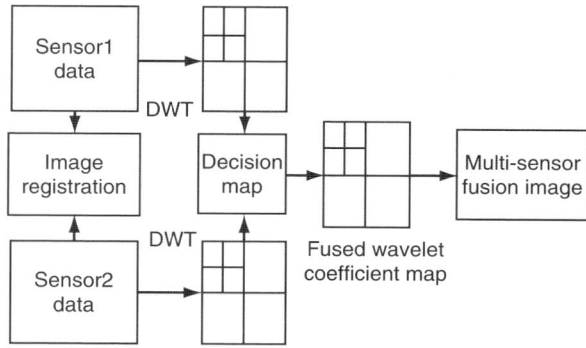

Fig. 8.4. Multisensor image fusion using discrete wavelet transform. The two sensors are visual and thermal sensors, respectively

performed. This type of fusion is called decision level fusion. Both data level and decision level fusion techniques are implemented and evaluated in this paper.

8.4.1 Data Level Fusion

For data level fusion, we implemented a DWT based technique, with a fusion rule that selects coefficients with maximum magnitude. The obtained wavelet feature map is then used to get a fused image by using inverse wavelet transform. The block diagram shown in Fig. 8.4 illustrates the fusion technique.

8.4.2 Decision Level Fusion

Both visual and thermal images are subjected to phase congruency feature extraction and module creation separately. There are a total of 896 modules created, 448 from visual image and 448 from thermal image. Each module is now classified individually after projecting onto the corresponding eigenspace. A voting procedure is now used to classify based on the individual classification results of all the 896 modules. The procedure of concatenating the modules and then classification can be interpreted as decision level fusion. This process is further illustrated in Fig. 8.5.

8.5 Experimental Results

There are two parts in the experimentation. One is to prove that the feature selection process works in the presence of partial occlusions and facial expressions. As the databases with thermal face images with extreme variations are not yet available, the testing of the face recognition techniques is carried out on AR database which has only visual images. From the AR database 40 individuals are chosen randomly to create a test database. Thirteen images of each individual are present in the database. Three images of each individual are used in training the proposed

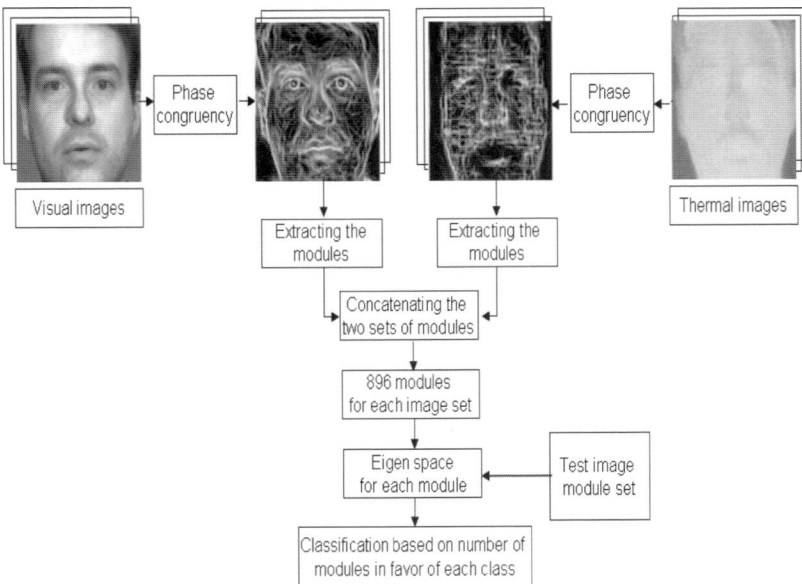

Fig. 8.5. Figure illustrates the concept of decision level fusion. The overall classification is based on the number of modules obtained in favor of each class

Fig. 8.6. The three images of each individual in AR database that are used for training

technique for classification. Figure 8.6 shows the training images of an individual. It can be observed that all the three face images are fairly neutral with little expression variations. The rest of the ten images of each individual in the database are used for testing the proposed technique. The sample test images are shown in Fig. 8.7. The test images are affected due to lighting or (and) expressions or (and) partial occlusions. The graph in Fig. 8.8 illustrates the relationship between percentage of

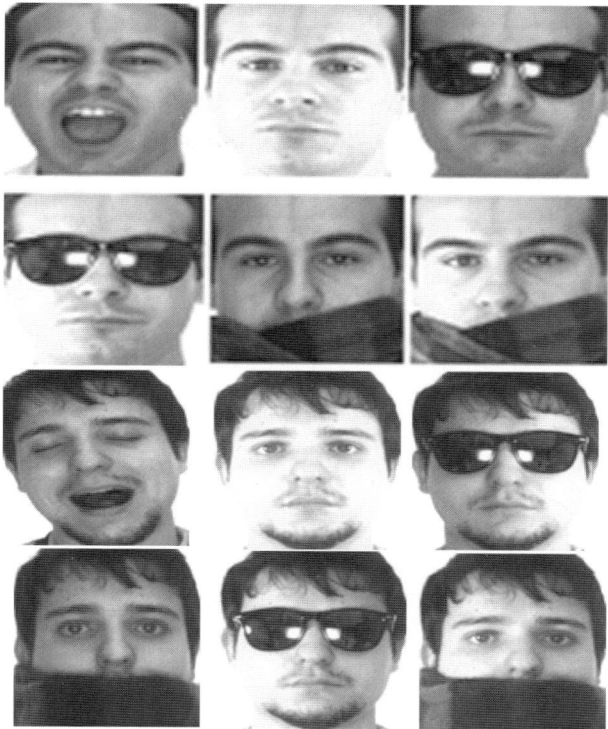

Fig. 8.7. Sample test images of the same person shown in Fig. 8.6

accuracy and the dimensionality of the subspace for various methods such as principal component analysis on holistic faces (PCA), modular PCA (MPCA), principal component analysis on phase congruency features (PPCA), modular subspace approach on phase congruency features (MPPCA) and the proposed method of neighborhood defined module selection on phase congruency features in PCA domain (NPPCA). It can be observed that the use of phase congruency features improves the face recognition accuracy significantly. Also modular subspaces improve the recognition for both intensity and phase congruency features. It can be observed that accuracy has risen by about 10% in the case of NPPCA compared to MPPCA.

The second phase of experimentation is carried out to evaluate the performance of the proposed face recognition technique on fusion of multisensor images, at data level and decision level. Equinox face database which consists of both long wave infra red and visual face images is used for this purpose. A subset of 34 individuals is selected for the experiments. Each individual has a total of 15 images. The thermal and visual images present in this database are preregistered. Instead of cropping the faces manually, a face detection system developed by Viola Jones is used to obtain face images to depict the real-time scenario. This face detection system is used to segment the face part from the background in each image. The corresponding region

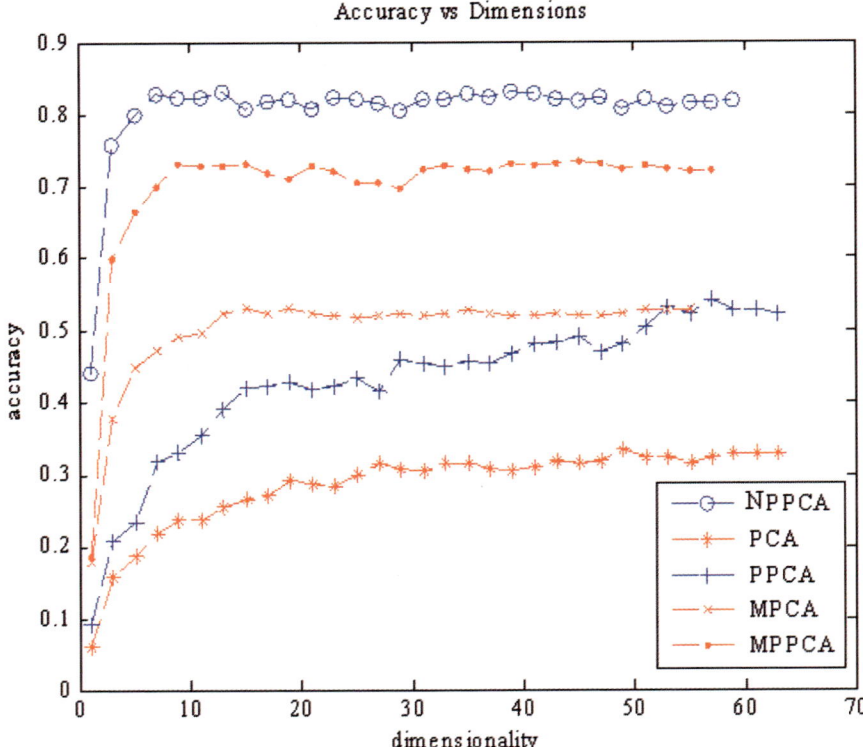

Fig. 8.8. Accuracies of various methods with respect to increase in dimensionality of the subspace are illustrated

in the thermal images is also segmented based on the coordinates of the detected faces in the visual images. Samples images can be seen in Fig. 8.9.

Four face images of each individual are selected randomly and used for training and the rest of the 11 images are used for testing. The experimentation process is divided into two parts. In the first part the proposed neighborhood defined face recognition technique is carried out on raw intensity images. It is observed that the recognition accuracy has improved very much on fused images when compared to either thermal or visual alone. The data level fusion provided better results than the decision level fusion method. In the second case, the proposed face recognition technique is applied on the phase congruency maps. The procedure is carried out on all the three modalities, that is on thermal, visual and fused. It is observed that the recognition accuracy has improved when compared to the corresponding modality in the first stage. Figure 8.10 shows graphs illustrating the improvement in accuracy due to the proposed face recognition technique. A significant result shown that can be observed is that the decision level fusion provided better accuracy by 2% in the case of phase congruency maps.

8 Feature Selection for Improved Face Recognition   119

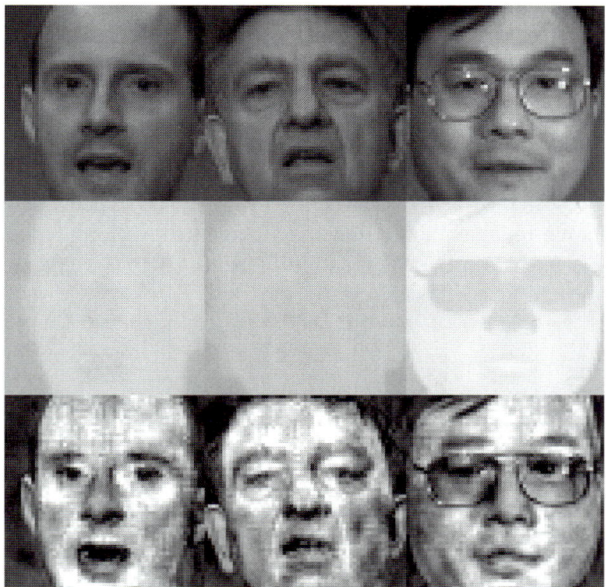

Fig. 8.9. Figure shows visual, thermal and the corresponding fused sample face images in the Equinox face database

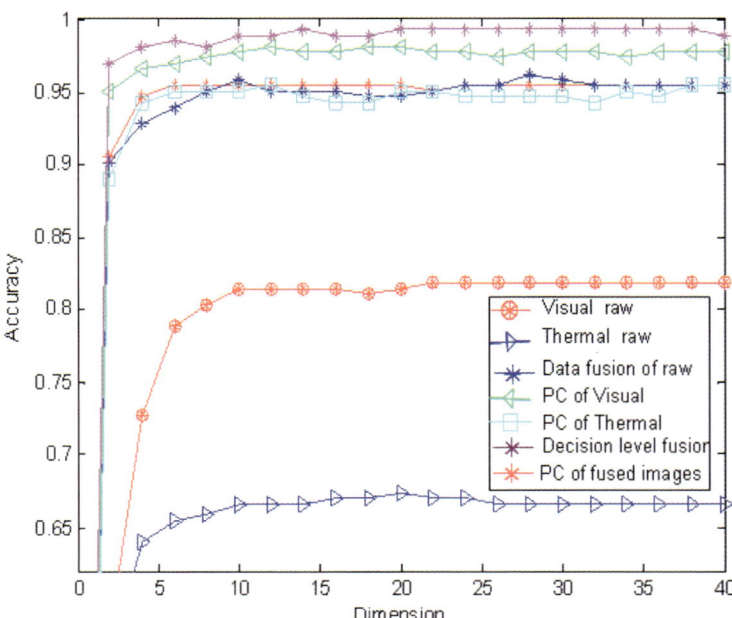

Fig. 8.10. Accuracies of various methods with respect to increase in dimensionality of the subspace

8.6 Conclusion

A novel feature selection policy has been implemented and tested on individual as well as fused modalities. The face recognition technique proposed, indeed provided better accuracy results, in each case. It was observed that, data level fusion is better in the case of raw intensity images where as the decision level fusion outperforms the former in the case of phase congruency images. The experimental results provided in the chapter indicate that multisensor image fusion along with proper feature selection strategy does improve the recognition accuracy to a great extent, especially in the presence of facial variations.

Part III

Multimodal Face Biometrics

9 Multimodal Face and Speaker Identification for Mobile Devices

Timothy J. Hazen, Eugene Weinstein, Bernd Heisele, Alex Park, and Ji Ming

9.1 Introduction

In this chapter we discuss the application of two biometric techniques, face and speaker identification, for use on mobile devices. This research has been spurred by the proliferation of commercially available handheld computers. Because of their mobility and increasing computational power, these devices are fast becoming a pervasive part of our lifestyle. Even formerly specialized devices, such as cellular telephones, now offer a range of capabilities beyond simple voice transmission, such as the ability to take, transmit and display digital images. As these devices become more ubiquitous and their range of applications increases, the need for security also increases. To prevent impostor users from gaining access to sensitive information, stored either locally on a device or on the device's network, security measures must be incorporated into these devices. Face and speaker verification are two techniques that can be used in place of, or in conjunction with, preexisting security measures such as personal identification numbers or passwords.

Handheld devices offer two distinct challenges for standard face and voice identification approaches. First, their mobility ensures that the environmental conditions the devices will experience will be highly variable. Specifically, the audio captured by these devices can contain highly variable background noises producing potentially low signal-to-noise ratios. Similarly, the images captured by the devices can contain highly variable lighting and background conditions. Second, the quality of the video and audio capture devices is also a factor. Typical consumer products are constrained to use audio/visual components that are both small and inexpensive, resulting in a lower quality audio and video than is typically used in laboratory experiments.

To examine these issues we have developed a system that combines two biometric techniques, speaker identification and face identification, for use with a mobile device. We provide a high level overview of our speaker and face identification technologies in Sect. 9.2. Following the description of these technologies, the chapter will focus on the following three research questions:

1. How much improvement in speaker identification performance can be gained by combining the audio and visual biometric information?
2. Can full video information allow for more accurate face identification than single image snapshots?

3. How can speaker identification systems be made more robust to variable environments?

To answer question 1, it has been found that combining speaker and face identification technologies can have a dramatic effect on person identification performance. In one set of experiments, discussed in Sect. 9.3, a 90% reduction in equal error rate in a user verification system was achieved when integrating the face and speaker identification systems.

The answer to question 2 is still largely open for debate. However, in preliminary experiments examining the use of static and dynamic information extracted from video, it was found that dynamic information about lip movement made during the production of speech can be used to complement information from static lip images in order to improve person identification. These results are discussed in Sect. 9.4.

To answer question 3, degradation in speaker identification rates in noisy conditions can be mitigated through the use of noise compensation techniques and/or missing feature theory. Noise compensation involves the adjustment of acoustic models of speech to account for the presence of previously unseen noise conditions in the input signal. Missing feature theory provides a mechanism for ignoring portions of a signal that are so severely corrupted as to become effectively unusable. In Sect. 9.5 we examine the use of two techniques for noise robust speaker identification, the posterior union model (PUM) for handling missing features and universal compensation.

9.2 Person Identification Technologies

9.2.1 Speaker Identification

Speech has long been recognized as a reasonable biometric for person identification. However, speech is a variable signal whose main purpose is not to specify a person's identity but rather to encode a linguistic message. In systems where the linguistic content of the speech is unknown (e.g., for surveillance tasks), text-independent speaker identification systems are generally used. It has been found for many text-independent applications that, even when linguistic knowledge is ignored completely, accurate speaker identification based purely on acoustic information can be performed. The standard approach is to extract wide-band spectral feature vectors from the audio signal (in the form of mel-scale cepstral coefficients or MFCCs [232]) at a fixed interval (typically every 10 ms). The full collection of acoustic features from all utterances in an individual's training set are then pooled together and modeled with a single probability density function constructed from a Gaussian mixture model (GMM). Speaker identification is performed by scoring the MFCC feature vectors against the GMMs of enrolled speakers to generate likelihood scores for these speakers [233].

For the problem of speaker verification (i.e., verifying with a *yes* or *no* decision whether a user is who they claim to be), speaker likelihood scores are typically normalized by a *universal background model* which captures the general distribution of speech over all users. Mathematically, the GMM speaker verification score for a set of acoustic feature vectors $\mathbf{x}_1$ through $\mathbf{x}_N$ for purported user S is modeled probabilistically as follows:

$$\sum_{i=1}^{N} \log \frac{\mathrm{p}(\mathbf{x}_i|S)}{\mathrm{p}(\mathbf{x}_i)} \qquad (9.1)$$

Here, $\mathrm{p}(\mathbf{x}_i)$ represents the GMM for the universal background model.

Text-independent systems have proven to work well for some applications. However, when the linguistic content of the message is known text-dependent speaker recognition systems generally perform better. This is because text-dependent systems can tightly model the characteristics of the specific phonetic content contained in the speech signal. In security applications, where the user is cooperative in the attempt to prove his/her identity, the linguistic content of the speech message is typically prespecified and can be tightly constrained. In this case, a text-dependent system is preferred.

In our work, we have developed a speaker identification system that uses speaker-dependent speech recognition models to perform the speaker identification process [234, 235]. During training, phonetically transcribed enrollment utterances are used to train context-dependent acoustic–phonetic models for each speaker. During testing, a speaker-independent automatic speech recognition system hypothesizes a phonetic transcription for the test utterance. This transcription is then used by the system to score each segment of speech against each speaker-dependent acoustic–phonetic model. Modeling speakers at the phonetic level can be problematic because enrollment data sets are typically too small to build robust speaker-dependent models for every context-dependent phonetic model. To compensate for this difficulty, an adaptive scoring approach can be used in which the specific acoustic–phonetic models for a speaker can be interpolated with the speaker's text-independent GMM model. This improves the robustness of the approach when limited enrollment data are available. Mathematically, the speaker score for our phonetic approach is modeled probabilistically as follows:

$$\sum_{i=1}^{N} \log \left(\lambda_i \frac{\mathrm{p}(\mathbf{x}_i|u_i, S)}{\mathrm{p}(\mathbf{x}_i|u_i)} + (1 - \lambda_i) \frac{\mathrm{p}(\mathbf{x}_i|S)}{\mathrm{p}(\mathbf{x}_i)} \right) \qquad (9.2)$$

Here, a phonetic label u_i is provided from a speech recognition engine for each acoustic feature vector $\mathbf{x}_i$. The interpolation factor λ_i is determined separately for each phonetic unit u_i based on the number of times it appeared in the enrollment data

$$\lambda_i = \frac{\mathrm{count}(u_i)}{\mathrm{count}(u_i) + K} \qquad (9.3)$$

Here K is a predetermined constant (typically 5 in our systems). The interpolation factor prefers the context-dependent model ratio for phone u_i when that phone has been observed often in the enrollment data, but it backs off toward the global GMM approach if u_i is rarely or never seen in the enrollment data.

9.2.2 Face Identification

Identifying people from images of their face is a widely studied problem. In addition to discussion of this topic in other chapters of this book, a thorough review of the literature on this topic is available in [236]. In this chapter, we discuss only the technologies used in our experiments. The primary face identification framework used in our work is largely based on work originally presented in [237].

Face Detection

Before face identification techniques are applied, the face must first be detected and located within a given image. The Viola–Jones face detection algorithm (which is based on a boosted cascade of feature classifiers) is a commonly used approach which we have used as our baseline face detection algorithm [238].

As an alternative, we have also used a fast hierarchical classifier to roughly localize the face in the image [239]. The region around the face is then cropped out from the larger image, histogram equalized, and scaled to a fixed size. Next, a component-based face detector [237] is applied to the extracted region to precisely localize the face and to detect facial components. This method first independently applies component detection classifiers to the face image. Each of these classifiers is trained to detect a particular component, such as a nose, mouth, or left eyebrow. In all, 14 face components are used, and each component classifier is evaluated over a range of positions in the vicinity of the expected location of the desired component. A geometrical configuration classifier is then applied to the combined output of each of the 14 component classifiers from each candidate position. The candidate positions that yield the highest output of the second-level classifier are taken to be the detected component positions. Figure 9.1 illustrates an enrollment image, as well as its selected face region with the positions of the facial components as detected by our system.

 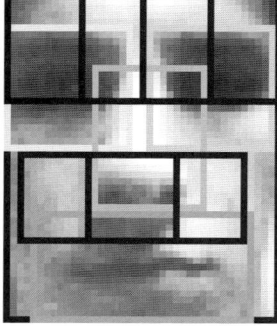

Fig. 9.1. A sample image and its face detection result with the face component regions superimposed

SVM-Based Face Recognition

A common approach to visual feature extraction for face identification is to use an *appearance-based* approach in which the raw image pixels are either used directly or projected into a lower-dimension subspace. Large dimension feature vectors can only be used with classification methods which exhibit robustness to the *curse of dimensionality*, e.g., support vector machines (SVMs) [240]. We have used SVMs as the primary classification technique for face identification in our systems.

In our initial work, presented in [241], we used a full face image compressed to 40×40 gray-scale pixels and histogram normalized to adjust the brightness. Improved results were later obtained by extracting appearance-based features from 10 of the 14 component regions found during the face detection process [242]. The ten selected components are similarly scaled in size and normalized, and then combined into a single feature vector which serves as input to the face recognition component.

For face recognition, a one-vs.-all SVM classification scheme is used, where one classifier is trained to distinguish each person in the database from all the others. In the training process, the feature vectors corresponding to a person's training images are used as positive examples for the classifier, and the feature vectors extracted from images of all other users are used as negative examples. The SVM-training process finds the optimal hyperplane in the feature space that separates the positive and negative data points. Since the training data may not be separable, a mapping function corresponding to a second-order polynomial SVM kernel function is applied to the data before training.

The runtime verification process consists of computing the output score for the purported user's SVM classifier [240]. The scores are zero-centered. In other words, a score of zero means the data point lies directly on the decision hyperplane, and positive and negative scores mean the data point lies on the positive and negative example side of the decision hyperplane, respectively. The absolute value of the SVM output is a multiple of the distance from the decision hyperplane, and could be normalized to produce the distance. Thus, a highly positive score represents a large degree of certainty that the data point belongs to the person the SVM was trained for, and a highly negative score represents the opposite.

GMM-Based Face Recognition

In our work on audio–visual speech recognition, we have used appearance-based visual features extracted from the raw images of the mouth region [243]. We have since adapted this approach to person identification using GMMs (identical in nature to those used in the speaker identification field). Because probabilistic classifiers, such as the GMM, typically require lower-dimension feature spaces to avoid problems of sparse-training data, a dimensionality reducing transform is often required. In experiments discussed in Sect. 9.4, we present results on GMM-based person identification using visual information derived only from the lip region of the face.

9.2.3 Multimodal Fusion

In our work, a simple linear weighted summation is employed for the classifier fusion where the weights for each classifier are trained discriminatively (on held-out development data) to minimize classification error. For the combination of face and speech classifiers, only one fusion parameter (the ratio of the weights of the classifiers) needs to be learned. A simple brute force sampling of different ratios can be used in this case. More complicated techniques (such as gradient descent training) could be used in situations where more than two scores must be fused.

9.3 Multimodal Person ID on a Handheld Device

9.3.1 Overview

Our initial experiments in multimodal person identification were performed using iPAQ handheld computers. A login scenario that combined face and speaker identification techniques to perform the multibiometric user verification process was devised. When "logging on" to the handheld device, users snapped a frontal view of their face, spoke their name, and then recited a prompted lock-combination phrase consisting of three randomly selected two digit numbers (e.g., "25–86–42"). The system recognized the spoken name to obtain the "claimed identity." It then performed face verification on the face image and speaker verification on the prompted lock-combination phrase. Users were "accepted" or "rejected" based on the combined scores of the two biometric techniques.

Speech data were collected utilizing the built-in electret condenser microphone of the iPAQ. Face data were collected using a 640×480 CCD camera located on a custom-built expansion sleeve for the iPAQ. The iPAQ handheld computer, combined with the custom sleeve, was the handheld device platform used for pervasive computing research in MIT's Project Oxygen [244]. An image of the iPAQ with the expansion sleeve is shown in Fig. 9.2. Because of the computation and memory limitations, the images and audio were captured by the handheld device, but then transmitted over a wireless network to servers which perform the operations of face detection, face identification, speech recognition, and speaker identification. In the future we expect the computational and memory components of handheld devices to improve such that our systems can be deployed directly on these small handhelds.

9.3.2 Data Collection

For our set of "enrolled" users, we collected face and voice data from 35 different people. Each person performed eight short enrollment sessions, four to collect image data and four to collect voice data. For each voice session, each user recited 16 prompted lock-combination phrases. For each image collection session, users captured 25 frontal face images in a variety of rooms in our laboratory with different lighting conditions. No specific constraints were placed on the distribution of the

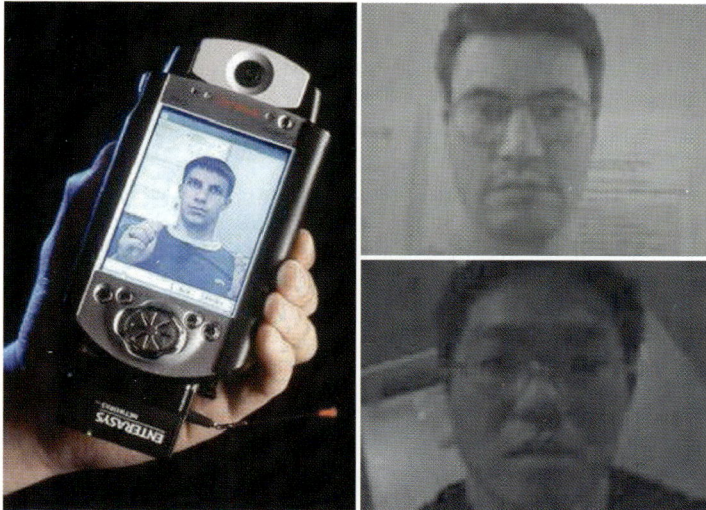

Fig. 9.2. The iPAQ handheld computer used in our study, along with two sample images collected in the iPAQ

locations and lighting conditions; users were allowed to self-select the locales and lighting conditions of their images. To illustrate the quality of the images, Fig. 9.2 shows two sample images captured during the data collection.

During image collection, the Viola–Jones face detector [238] was applied to each captured image to verify that the image indeed contained a valid face. This face detector occasionally rejected images when it failed to locate the face in the image with sufficiently high confidence. When this occurred the user was instructed to capture a new image. Due to a conservative face detection confidence threshold, no false positives (i.e., images with incorrectly detected faces) were observed from this face detector during the data collection.

Each voice and image session was typically collected on a different day, with the time span between sessions often spanning several days and occasionally a week or more. In total this yielded 100 images and 64 speech samples per enrolled user for training. An additional set of four enrollment sessions of audio data (i.e., 64 additional utterances) from 17 of the training speakers was available for development evaluations and multimodal weight fusion training.

A separate set of evaluation data was collected to perform user verification experiments. For this evaluation set, we collected 16 sample login sessions from 25 of the 35 enrolled users. This yielded 400 unique utterance/face evaluation pairs from enrolled users. We also collected 10 impostor login sessions from 20 people not in the set of enrolled users for an additional 200 utterance/face evaluation pairs from unenrolled people. Each utterance/face pair collected from out-of-set impostors was used to generate an impostor example for each of the 35 enrolled users yielding 7,000 impostor examples.

9.3.3 Training

The face and speaker systems were trained on the enrollment data for the 35 enrolled users. To train the fusion weights, one of the four face enrollment sessions was held-out and a development face identification system was trained on the remaining three face sessions. Face identification scores from this held-out set were pairwise combined with speaker identification scores generated for utterances from the existing speaker identification development set. The true in-set examples and in-set impostor examples were provided to the weight training algorithm to generate the multimodal fusion weights.

9.3.4 Face Detection Issues

The performance of a face identification system is affected by the quality of the images it is provided. If the system tightly controls the user and rejects images in which the head is tilted or rotated, the face is contorted in any unusual fashion, etc., then the variance of the data will be reduced and improved performance should be expected. In our work we initially collected facial images within a system running the Viola–Jones face detector. In our evaluations we have used a component-based face detection algorithm which is more conservative in its detection decisions. As a result, a sizable number of images in the training and evaluation data sets were rejected by the component-based face detection algorithm.

To detail the effect of the face detection algorithm upon the face identification results, two experiments were conducted: one where the conservative face detection decisions were used, and a second experiment where the face detection algorithm was forced to output a detected face even if the image's detection score fell below the standard acceptance threshold. These two experiments allow us to examine the tradeoff between the added gain in accuracy enabled by stricter control in the input facial images, and the potential added inconvenience of having users retake snapshot images until the face detection algorithm accepts one.

9.3.5 Experimental Results

Table 9.1 shows our user verification results for three systems (face ID only, speaker ID only, and our full multimodal system) under two different face detection conditions. The results are reported using the equal error rate (EER) metric. The EER is the point in the detection-error tradeoff (DET) curve where the likelihood of a false

Table 9.1. User verification results expressed as equal error rates (%), over three systems (face only, speaker only, and multimodal fusion), using two different face detection scenarios

system	forced face detection (%)	conservative face detection (%)
face	4.87	2.57
speaker	1.66	1.63
fused	0.66	0.15

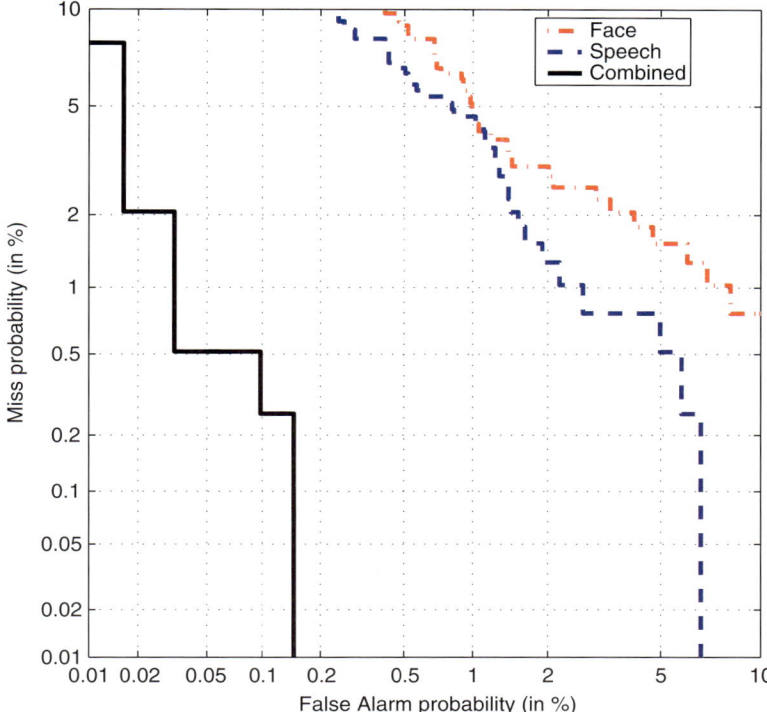

Fig. 9.3. DET curves for face and speech systems run independently and in combination when tested using impostors unknown to the system and when using the conservative face detection threshold

acceptance of an impostor (i.e., a *false alarm*) is equal to the likelihood of false rejection of the true user (i.e., a *miss*).

In the table's forced face detection results, all evaluation sessions are used. However, for the conservative face detection results, 12% of the images were rejected. In this case, the system's verification results were computed using only the 88% of the data that passed the more conservative face detection threshold. Though the speaker identification system is unaffected by the face identification method that is used, the speaker identification EERs are different in the two columns because rejection of an image causes the companion spoken phrase from an evaluation login session to also be discarded, thus altering the speaker identification results slightly.

In examining the results, one can see that the face identification system using conservative face detection thresholding shows a nearly 50% improvement in EER performance over the system using forced face detection. Of course, the improved performance does come with a cost: in a deployed system, a user would face the added inconvenience of providing a new snapshot image whenever the face detector rejects an image.

Next, the results show that the speaker identification system is performing better than the face identification component, though the performance is of the same order of magnitude. When the two systems are used in combination, significant improvements are obtained over the use of either modality by itself. When using conservative face detection, the addition of the face identification system to the speaker identification system produced a 90% relative reduction in the EER from 1.63 to 0.15%. DET curves for the three systems (when using conservative face detection) are shown in Fig. 9.3. These results demonstrate that highly accurate biometric authentication can be obtained via the multibiometric approach of combining speaker identification and face identification technology.

9.4 The Use of Dynamic Lip-Motion Information

When performing face identification, an interesting question to ask is whether full motion video provides any substantive advantage over the use of individual still images. Video has been shown to be useful for face identification by providing a collection of temporally related images. Increased robustness can be obtained using video because face detection results can be interpolated over multiple frames, and results from frames with poor images can be discounted or ignored when considered jointly with other better scoring frames [245]. However, one might wonder if the actual dynamic motion of facial features themselves, such as the motion of the lips when someone is talking, can be used to identify a person. Or even more importantly, can the dynamic lip information provide any significant improvement over using only the static information available from the individual frames extracted from the video?

To examine this issue we have performed experiments using the AV-TIMIT video corpus [246]. This corpus was originally collected for use in audio–visual speech recognition experiments. It contains read sentences recorded in a quiet room using a high-quality digital camera for the video and a far-field array microphone for the audio. The first ten utterances recorded for each user were used to train the face and speaker identification system and five additional utterances were used for our evaluation. In total the corpus contains recordings from 221 different people (yielding $221 \times 5 = 1{,}105$ evaluation utterances). Because the AV-TIMIT corpus being used was recorded in quiet office conditions and the training data come from the same session as the evaluation data, this person identification task does not represent realistic conditions. To make the task more challenging, our face identification system only uses the lip region portion of the video. Despite the unrealistic conditions of the task, we can still use this corpus to compare different visual features and to examine the effect of fusing audio and visual information.

From each individual frame, the image of the lips is represented using the top components from a principal component analysis (PCA) rotation applied to a discrete-cosine transform of the image from the lip region (a.k.a. *eigenlips* [247]). We refer to these feature vectors as the *static PCA* features. The first-order time difference between PCA vectors in sequential image frames is used to represent the dynamic changes in lip images. We refer to these feature vectors as the *dynamic PCA* features.

Because statistical classifiers require a tradeoff between the increased specificity from larger feature space dimensionality and the susceptibility of large dimension classifiers to overtraining, we have evaluated the system using several different feature vector dimensionalities. We have also constructed feature vectors using static PCA features only, using PCA difference features only, and using a combination of the static and dynamic features. To perform person identification in this system, the individual feature vectors are modeled using a single GMM per speaker. Table 9.2 shows the closed-set person recognition performance on the 221 person AV-TIMIT corpus using eight different feature vectors configurations used in our experiments. The results show that static lip information is more useful than dynamic lip information, but that improvements in person identification can be achieved by combining the static and dynamic information.

Table 9.3 shows the individual results of the audio-only and visual-only person identification system for closed-set person recognition. The table also shows the combined audio–visual result when linearly combining the audio and visual scores. In this case, the optimal weighting of 0.95 for the audio stream and 0.05 for the visual stream yields an error rate of 0.27% (3 errors out of 1,105 trials). When ratio of the audio weight to the visual weight is varied between 0.8/0.2 and 0.98/0.02 the person identification is never worse than 0.54% (6 errors out of 1,105 trials).

These results on AV-TIMIT demonstrate, once again, the power of combining audio and visual information for person identification. In on-going research, our

Table 9.2. Person identification results from visual lip images using static PCA features, dynamic PCA features, and a fusion of the static and dynamic features

lip image feature vector	person ID error rate (%)
48-dimension static PCA features	6.0
96-dimension static PCA features	3.6
192-dimension static PCA features	4.1
48-dimension dynamic PCA features	6.6
96-dimension dynamic PCA features	7.7
192-dimension dynamic PCA features	17.8
48-dimension static PCA features + 48-dimension dynamic PCA features	2.1
96-dimension static PCA features + 96-dimension dynamic PCA features	2.1

Table 9.3. Person identification results for audio-only, visual-only, and audio–visual systems using audio and lip-image information from the AV-TIMIT corpus

system	person ID error rate (%)
audio only	1.2
visual only	2.1
audio–visual	0.27–0.54

group is currently moving beyond systems using the high-quality, single-session AV-TIMIT video, and toward the creation of a system that can handle video collected using commercial-off-the-shelf web cameras and handheld devices.

9.5 Noise Robust Speaker Identification

As discussed earlier, one of the great challenges of performing speaker or face identification in mobile applications is the possibility of severe variations in the feature measurements due to the environmental conditions (i.e., background noise, lighting conditions, etc.). One technique for addressing this problem is the application of missing feature theory. The basic premise of missing feature theory is that some features of the observation space may be so corrupted that they become useless for the task of person identification and should be ignored. For speaker identification this could involve either temporal corruption (e.g., a brief impulsive noise such as a door slam) or spectral corruption (e.g., a noise in a narrow spectral band such as a police siren). Comparable analogies could also be drawn for face identification (e.g., sudden severe shadows, occlusions of portions of the face, etc.). One could also view the problem of multimodal fusion within the missing feature theory framework, where either of the audio or visual feature streams could be unreliable at any point in time and ought to be ignored in deference to the more reliable information stream.

In some situations, the corruption may not be so severe that it completely masks all usable information within a feature. In this case, a means of accounting for the corrupting noise in the observation of a feature is more desirable than completely ignoring the feature. In an ideal situation, models for biometric features could be trained from enrollment data collected under all of the corrupting conditions the user may encounter. Unfortunately, this is not feasible for most mobile applications and methods for compensating for unseen conditions must be employed.

In our work we have investigated the problem of robust speaker identification in noisy environments. In particular we have examined a missing feature approach called the *posterior union model*, and a noise compensation technique called *universal compensation*. Though we have not yet extended this work beyond speaker identification experiments, we believe these ideas can be extended to the problems of face identification and the fusion of multimodal information.

9.5.1 The Posterior Union Model

The basic premise behind missing feature theory (as we apply it to speaker identification) is that improved performance can be achieved by utilizing only information about features that can be reliably extracted from the input signal. Thus, if an input signal can be represented as a collection of independent features $X = \{x_1, x_2, \ldots, x_N\}$, then there exists some optimal subset of uncorrupted features $X_{sub} \subseteq X$ that can be used as the basis for the speaker identification decision. This problem can be expressed probabilistically as

$$[S', X'_{sub}] = \arg \max_{S, X_{sub}} P(S|X_{sub}) \qquad (9.4)$$

where S represents a specific speaker and X_{sub} represents a specific subset of features from X. The expression seeks to find the most likely speaker S' by jointly maximizing the posterior probability over all speakers and all possible feature subsets X_{sub} in X. Here X'_{sub} is the optimal feature subset found for the most likely speaker S'. Using Bayes' rule the expression is rewritten as

$$[S', X'_{sub}] = \arg \max_{S, X_{sub}} \frac{p(X_{sub}|S)P(S)}{p(X_{sub})} \qquad (9.5)$$

where $P(S)$ is generally given a uniform distribution and $p(X_{sub})$ is a normalizing term that is independent of the speaker S.

The PUM generalizes the problem by removing the constraint that an exact set of optimal features, X'_{sub}, be found. Instead, for a given number of features M, PUM makes the following assumption:

$$p(X'_{sub}|S, M) \approx \sum_{X_{sub} \subseteq X_N^M} p(X_{sub}|S) \qquad (9.6)$$

Here, X_N^M is the collection of all combinations of sets of M features chosen from the full N features in X. The approximation assumes that the sum of $p(X_{sub}|S)$ over all X_{sub} drawn from X_N^M is dominated by the optimal subset of M features. This reduces the problem to finding the optimal number of reliable features M, but not the exact subset. In practice, individual features are rarely completely reliable or completely unreliable, but somewhere in between. Thus, the use of the union model allows a softer probabilistic decision than forcing features to either be used or discarded. Details of the PUM implementation can be found in [248].

9.5.2 Universal Compensation

If we consider that individual features may be only partially corrupted, then the missing feature approach should be amended to account for partially corrupted features. The universal compensation technique provides just such a mechanism. Instead of decomposing the features in X into reliable features that are used and unreliable features that are ignored, the features can be decomposed into subsets containing variable degrees of corruption. In this formulation we can use the expression

$$p(X|S) = \sum_{l=0}^{L} p(X_l|S, \Phi_l) P(\Phi_l|S) \qquad (9.7)$$

where Φ_l represents a level of corruption and X_l represents the specific set of features in X which are corrupted at level Φ_l. In this case the PUM can be extended such that it considers the optimal number of features corrupted at each corruption level Φ_l and not just those that are completely clean or completely corrupted. Details of this formulation are found in [249].

In practice for speaker identification tasks, the universal compensation technique is applied by taking clean audio training data and adding noise at variable signal-to-noise ratios to simulate the different corruption levels Φ_0 to Φ_L. We have primarily added white noise to the clean training to simulate the corruption, but different types of noises could be used depending on the expected environments. Models for each speaker at each corruption level are trained. During evaluation on unseen data the posterior union model is used to select the number of features from the full set that optimally match each corruption level.

9.5.3 Experimental Results

To demonstrate the effectiveness of the posterior union model and universal compensation techniques, we conducted experiments on a handheld device database collected at MIT. The database was designed to study speaker verification in realistic noisy conditions with limited enrollment data [250]. The database contains 48 enrolled speakers (26 male, 22 female) and 40 impostors (23 male, 17 female), each reciting short ice cream flavor phrases.

In our primary experiments, users enrolled into the system by speaking four examples of a specific phrase into the handheld device. The enrollment session was conducted in a quiet office environment using an external ear-piece microphone. For each enrolled user, speaker identification models were trained from the four enrollment examples. Low-pass filtered white noise was added to each example at nine different signal-to-noise ratios between 4 and 20 dB (increasing 2 dB every step). This gives a total of ten corruption levels (including the no corruption condition) for the training phase. To evaluate the system, the same enrolled users and the 40 previously unseen impostors recited new evaluation phrases using the same handheld device. However, the evaluation data were instead collected outdoors next to a noisy street intersection using the internal microphone of the device.

The speaker identification system uses phrase-dependent hidden Markov models to represent each speaker in the enrollment set. The features used to represent the acoustic information are modeled with subband spectral components derived from decorrelated log filter-bank amplitudes collected from 20-ms wide time windows sampled every 10 ms. In total two energy and time difference values were used to represent the features within ten different spectral subbands. The PUM is thus tasked with selecting the optimal number of subbands corrupted at each of the ten different noise corruption levels. Details of the system used in this experiment can be found in [249].

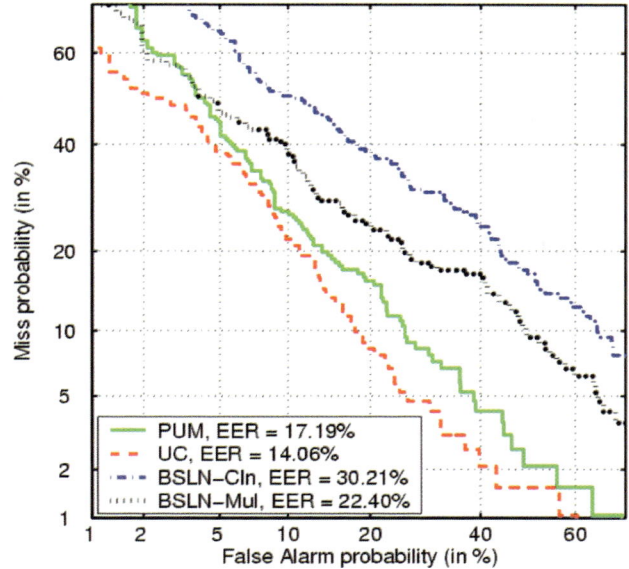

Fig. 9.4. DET curves comparing four limited enrollment speaker verification systems trained in a clean environment and tested in a mismatched noisy environment

For our experiments, we implemented four different systems all based on the same set of acoustic features:

- BSLN-Cln: a baseline system trained only on the clean office data and tested using the full set of acoustic features.
- BSLN-Mul: the baseline system trained on the full set of clean and artificially corrupted data pooled together to train a single multicondition model for each speaker.
- PUM: a system trained only on the clean office data but allowed to select the optimal number of reliable subband components using the PUM approach.
- UC: a system trained on the clean and artificially corrupted data using the PUM approach to optimally select number of subbands matching each corruption level.

The experimental results are shown in Fig. 9.4. The figure shows that a baseline speaker verification system trained in a quiet environment performs quite poorly when it is then used in a noisy environment (next to a noisy street intersection in this case). However, by artificially adding various levels of white noise to the training material, the EER of the system is reduced from 30.2 to 22.4%. If the posterior union model is used in conjunction with the baseline system, the EER is reduced from 30.2 to 17.2%. Finally, if the PUM is combined with the system trained using varying levels of artificially added noise (i.e., the universal compensation approach), the EER is further reduced to 14.1%. These results show that techniques do exist to improve the robustness of speaker identification even in noisy environments that are mismatched with the systems training conditions.

9.6 Summary

In this chapter, we have shown the power of combining face and speaker identification techniques for improved person identification. In Sect. 9.3, we demonstrated that a multibiometric approach can reduce the EER of a user verification system on a handheld device by up to 90% when combining audio and visual information. In Sect. 9.4, we showed that dynamic information captured from a person's lip movements can be used to discriminate between people, and can provide additional benefits beyond the use of static facial features. In Sect. 9.5, we addressed the problem of robust speaker identification for handheld devices and showed the benefits of the PUM and the universal compensation techniques for handling corrupted audio data. In future work we plan to extend the use of the PUM to different facial feature vectors extracted from images as well as to the multimodal fusion of different audio and visual features.

10 Quo Vadis: 3D Face and Ear Recognition?

I. Kakadiaris, G. Passalis, G. Toderici, N. Murtuza, and T. Theoharis

10.1 Introduction

In recent years, among the many biometric modalities, the face has received the most interest. Not only is face recognition one of the most widely accepted modalities, but also advances in processing power have allowed the development of more complex algorithms while still providing a relatively rapid response to queries. Ear recognition, on the other hand, is a relatively new field. However, it holds a lot of promise as the complexity of the human ear is even higher than that of the face, and ears are subject to fewer deformations. In addition, the problem of recognizing an ear is not too different from that of recognizing a face. Therefore, part of the large body of research available in 3D face recognition can be inherited to 3D ear recognition. Both face and ear recognition require no contact with the subject, thus being more easily accepted by the public, compared to other biometrics such as fingerprints.

Face recognition has been traditionally performed using 2D (visible spectrum) images, while hybrid approaches have used infrared spectrum images, and even 3D geometry. Infrared face recognition has not been widely adopted due to the high cost of the infrared cameras necessary to acquire the data. In contrast, the cost of 3D scanners has dropped dramatically, so it has become feasible to deploy them in the field, and therefore the interest in developing algorithms that use 3D data has increased.

The main reason for using information from 3D data as a biometric (both for faces and ears) is that the data acquired by 3D acquisition devices are invariant to pose and lighting conditions, these being the major challenges that face recognition algorithms must cope with. Moreover, image-based face recognition algorithms are more susceptible to impostors. Indeed, if an individual is able to take an image of a subject allowed to enter a facility, they may use a printout of that image in order to break in. To avoid this, the face recognition algorithm must be coupled with liveness test algorithms. Attempting such an attack on a system based on 3D data would be much more difficult, since the attackers would need to obtain an accurate 3D model (sculpture) of the person whom they would like to impersonate.

The challenges of 3D face and ear recognition which concern our approach are the following:

- *Accuracy gain*: a significant gain in accuracy with respect to 2D face and ear recognition systems must justify the introduction of 3D recognition systems.
- *Efficiency*: 3D capture devices generate substantially more information than 2D cameras. Using this large volume of information is expensive in terms of computation time and storage requirements. Therefore, the algorithms we develop need to be efficient both in time and space.
- *Automation*: the systems must be completely automated. It would be unreasonable to have a person assigned to the face or ear recognition system in order to, for example, have them select landmarks on the meshes.
- *Capture devices*: most high-resolution capture devices were developed for medical and other low-volume applications. A deployable 3D face or ear recognition system must be able to process several persons a minute if it is to be used in high-traffic areas.
- *Testing databases*: there are only few large databases of 3D faces and even fewer databases of 3D ears which are widely accepted.
- *Robustness*: the system must perform robustly and reliably under a variety of conditions (e.g., lighting, pose variation, facial feature variation).

Section 10.2 provides an overview of some recent methods used for face and ear recognition. Section 10.3 presents a general deformable model-based framework for matching 3D shapes, and then Sects. 10.4 and 10.5 present its application in 3D face and ear recognition, respectively.

10.2 Related Work

10.2.1 Face Recognition

Face recognition has received unprecedented attention in recent years. At the IEEE Computer Vision and Pattern Recognition conference in 2006, over 80 papers were published with applications to the subject. However, many authors still report results on small databases, or on databases which are not available for others to use. Therefore, many of the results reported are not directly comparable.

To create a common basis for comparing algorithms, and to determine whether 2D face recognition may be used in highly secure locations, the FERET database [251] was used. During the Face Recognition Vendor Test (FRVT) 2002 [252], which followed, it has been shown that the algorithms at the time were not capable of offering the degree of reliability needed. More recently, the Face Recognition Grand Challenge (FRGC) dataset was made available for researchers by NIST [253]. It contains both 2D and 3D range images. The *FRGC* has released data on two occasions. The first dataset released as part of *FRGC* in the spring of 2004, namely *FRGC v1*, included 943 range images, all having only neutral expressions. In spring 2005, the *FRGC v2* database was released, and it included over 4,000 range images, of which over 40% exhibit nonneutral facial expressions. The largest face recognition database to date, which contains both 2D and 3D data, will be used in the FRVT 2006 [254]. Several researchers have now reported results using the *FRGC v1 and v2* databases.

Lu and Jain [255] use a generic 3D model to create user-specific deformable models in the neutral position. In the identification phase, the distance returned by the iterated closest point (ICP) algorithm is used for matching all user-specific models to the new dataset. They report 92.1% rank 1 identification on a subset of *FRGC v2*. Wang et al. [256] use a global optimization procedure to generate a conformal map of the geometry, which then is processed using image processing techniques. The results are encouraging, but are generated using only 100 3D scans.

Russ et al. [257] use principal components analysis (PCA) on range images generated after the data have been aligned to a generic 3D model. The results were presented on subsets of *FRGC v2*, and indicate that this improvement outperforms the pure PCA approach, but still suffers from facial expressions. Lin et al. [258] compute summation invariant images from the raw 3D data. Using the baseline PCA approach included in the *FRGC v2*, but on cropped images, they report verification rates between 80.82 and 83.13%.

Husken et al. [259] present a multimodal approach that uses hierarchical graph matching (HGM). They used the same approach on both the 2D images and on the range images. However, the graph is created using the 2D image and transferred to the range image directly. The verification rate at 0.001 false accept rate (FAR) of the HGM solely on range images is lower than on 2D images on the *FRGC v2* database. Range images alone yield 86.9% verification rate, but the fusion of the two modalities results in 96.8% verification. Maurer et al. [260] also present a multimodal approach tested on the *FRGC v2* database, and report a 87% verification rate at 0.01 FAR.

Bronstein et al. [261] represent the geometry by isometrically embedding it into a low-dimensional Euclidean space by using multidimensional scaling. The recognition is accomplished using canonical forms with promising results on a proprietary database. The authors also propose an approach using Generalized MDS [262]. However, the results are reported only on a subset of 180 images from *FRGC v2*.

10.2.2 Ear Recognition

The usefulness of the human ear as a biometric was first recognized in the 1980s. The most famous early study was presented by Iannarelli [263] in 1989. He collected about 10,000 ear images and manually sectioned each ear image into equal sized triangles. Then, by measuring the distances between specific points on the ear, he indicated that ear data could be used as a biometric for identification purposes.

One of the first computer vision attempts at using the ear for recognition purposes was performed by Burge and Burger [264]. They used a Canny edge detector to extract the edges from a 2D image, estimated curves of the ear, and used a Voronoi neighborhood graph of the curves for matching. Chang et al. [265] use PCA for matching 2D ear images and achieve a 71.6% rank 1 recognition rate.

Along with the development of 3D scanners, 3D ear databases have been created and algorithms developed for using such data. Yan et al. [266–269] use an ICP-based approach which registers 3D ear datasets to compute a distance score based on the mean-squared distance between the registered meshes. In the initial implementation [268], they segmented the ear data manually from the mesh, and achieved a 98.8% rank 1 recognition rate. In their recent publication [269], they use a slightly larger database (415 subjects), and they improved the algorithm by making the ear extraction completely automated, achieving a rank 1 identification rate of 97.6%.

10.3 Methods

10.3.1 Generic 3D-Driven Recognition System

We have developed a system which is capable of using 3D data as input, along with a suitable model to output metadata information. The metadata information is then used for recognition. The model is user-defined, for instance, a generic face or a generic ear. This model needs to be constructed only once, and it can handle objects belonging to the same class. Once the data are acquired, the model is fitted to the data and used to generate a geometry image and a normal map, which are transformed into the wavelet domain. Only a small part of the wavelet coefficients are stored as metadata. In other words, the raw data are transformed from the 3D space to a regular grid representation of lower dimensionality that allows the application of state-of-the-art wavelet analysis algorithms.

Our recognition procedure can be divided into two distinct phases, enrollment and recognition:

- *Enrollment*. Raw data acquired by the scanner are converted to metadata and stored in a database. The following steps describe the conversion from raw data to metadata (Fig. 10.1):
    1. *Acquisition*: the sensor acquires raw data which are converted into a polygonal representation (a 3D mesh). A preprocessing step takes place to alleviate scanner-specific issues.
    2. *Alignment (Sect. 10.3.4)*: the data are aligned to the model into a unified coordinate system using a multistage alignment method.
    3. *Deformable model fitting (Sect. 10.3.5)*: the model is fitted to the data.
    4. *Metadata generation (Sect. 10.3.6)*: generate a geometry image and a normal map from the fitted model. Convert these images into the wavelet domain using the Haar decomposition, and if greater accuracy is desired, then also transform them using the steerable Pyramid transform. Store only the most significant coefficients.
- *Recognition*. We repeat the procedure used for enrollment, except that instead of storing the metadata, we employ them for comparison with the metadata in the database.

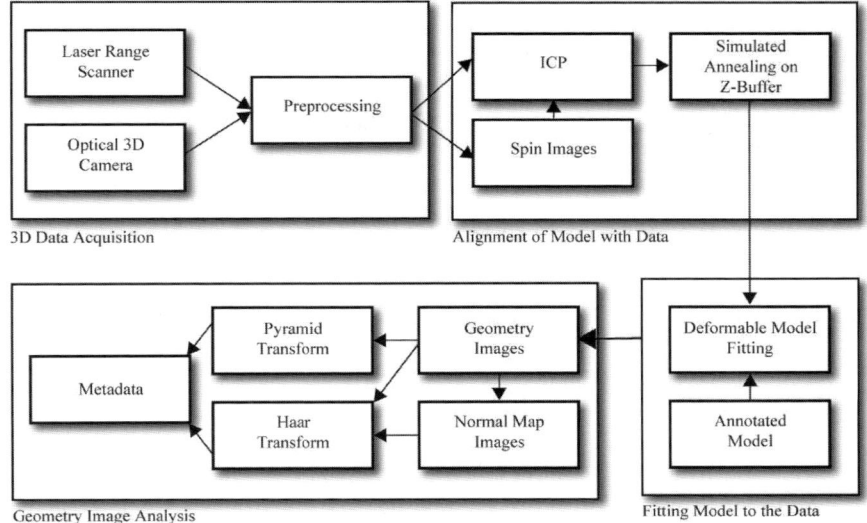

Fig. 10.1. Enrollment phase of the proposed integrated 3D face recognition system

10.3.2 Data Preprocessing

The purpose of preprocessing the data generated by the sensors prior to applying our algorithm is the elimination of any sensor-specific issues, and the unification of data into a common format which is accepted by our program. The preprocessing consists of the following filters, applied in the given order:

1. *Median cut*: this filter removes spikes from data. Spikes are common if the data are acquired using a laser range scanner. Data acquired from stereo scanners may also need spike removal but to a lesser extent.
2. *Hole filling*: both laser scanners and stereo systems are prone to producing holes in the meshes they generate.
3. *Smoothing*: all high-resolution scanners produce noisy data in real-life conditions.
4. *Subsampling*: our deformable model fitting (Sect. 10.3.5) effectively resamples the data, making the resolution of the input data less relevant. If the number of polygons is high enough, we subsample the data for efficiency purposes, without losing performance in the recognition phase. Subsampling further reduces the noise in the geometry.

The current generation of scanners outputs either a range image or a 3D mesh. We implemented the above filters to operate on the native representations of the scanners (Fig. 10.2), and on one-neighborhoods for both representations. Our experiments indicate that the filters perform better if the native data representation is used for filtering, and then the output is converted to a common polygonal format.

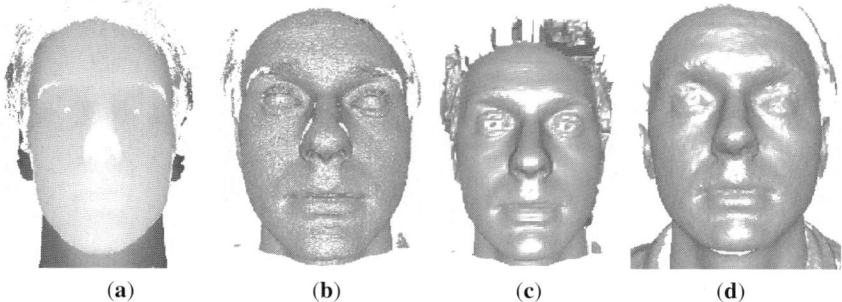

Fig. 10.2. Sensor-dependent preprocessing. Laser range scanner: (**a**) input depth image, (**b**) raw polygonal data (200,000 triangles), and (**c**) processed data (16 K). Stereo camera: (**d**) raw data (34,000 triangles)

10.3.3 Annotated Model

Our approach uses a parameterized annotated model which depends on application. For face recognition, we use an annotated face model (AFM), while for ear recognition we use an annotated ear model (AEM). This model is central to our work, since we use it in the alignment, fitting, and metadata generation; we created the models using 3D modeling tools and we ensured that they are anthropometrically correct using Farkas' data [270]. Using information from facial physiology and from the anthropometry of the ear, we annotated the models into regions corresponding to the same anatomical features. Finally, we applied a continuous global UV parameterization on the model. The injective property of our parameterization allows us to map all vertices of the model from $\mathbb{R}^3$ to $\mathbb{R}^2$, and vice versa. Therefore, we define the model both as polygonal data in $\mathbb{R}^3$ and in $\mathbb{R}^2$ as a *geometry image* [271–273]. A geometry image is a regular sampling of the model represented as a 2D image with three channels, each channel corresponding to the x, y, and z components. Since local neighborhoods on the mesh are preserved (i.e., neighboring vertices are preserved even in the geometry image), we can reconstruct an approximated version of the original mesh from the geometry image. The approximation level depends on the resolution of the image (the sampling of the mesh).

10.3.4 Alignment

Our work on face recognition has indicated that the alignment of the annotated model (AM) to the data is a key part of any geometric approach [274]. If a misalignment error occurs, it cannot be corrected in the later stages of this or similar approaches. We present a novel multistage alignment method that offers robust and accurate alignment even when relatively large deformations occur in the input data.

Before fitting, we align each preprocessed dataset with the annotated model. The alignment stage computes a rigid transform, combining rotation and translation, which brings the data as close as possible to the model. We propose a multistage alignment algorithm which propagates the alignment variables from one stage to the next. The first algorithm is more resilient to local minima, while the next two algorithms provide greater degrees of accuracy:

1. *Spin images*: first, we establish a plausible initial correspondence between the data and the model. If we do not expect large rotations and translations in the database, this step may be omitted. To compute the rough initial alignment, we employ the spin images algorithm developed by Johnson [275]. We compute correspondences on a subset of the vertices in the data. Next, they are grouped into geometrically consistent groups. The transformations they yield are checked to determine whether they rotate the data by an acute angle, based on the assumption that the input data are neither flipped on the z-axis nor inverted on the y-axis. This is a reasonable assumption, since the meshes computed by scanners respect the fact that the y-axis is vertical, while the z-axis is the depth. If the scanner does not follow these assumptions, the data can be transformed in the preprocessing stage. These checks are essential due to the bilateral symmetry property of the human face (if applied for face recognition).
2. *Iterative closest point (ICP)*: we use the ICP algorithm [276] as the main step of the alignment process. To obtain better alignment, we extended the original approach in a number of ways. We exploit the model annotation by assigning a different weight for each annotated region, and we compute a weighted least squares solution for the rigid transformation. Additionally, points belonging to the surface boundaries are rejected [277]. This additional constraint ensures that no residual error is introduced into ICP by the nonoverlapping parts of the two surfaces. If the final transform is not satisfactory, we have the option of running a trimmed ICP algorithm [278].
3. *Simulated annealing on z-buffers*: this is the refinement step, which ensures that the data are well aligned to the model. We apply the global optimization algorithm known as the enhanced simulated annealing (ESA) [279] on the difference of the z-buffers of the model and the data.

10.3.5 Deformable Model Fitting

The purpose of fitting the model to the data is to capture the geometric information of the desired object. To fit the model to the data we utilize the elastically adapted deformable model framework of Metaxas and Kakadiaris [280].

The analytical formulation adaptive is given by:

$$M_q \frac{d^2q}{dt^2} + D_q \frac{dq}{dt} + K_q q = f_q,$$

where M_q is the mass matrix, D_q is the damping matrix, K_q is the stiffness matrix, and f_q are the external forces. The external forces drive the deformation. The stiffness matrix defines the resistance against the deformation, while the mass and damping matrices control the velocity and the acceleration of the vertices. The method

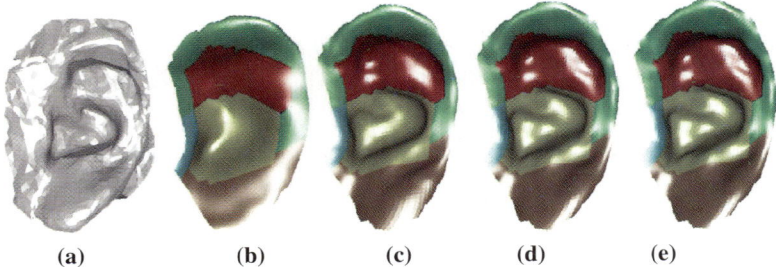

Fig. 10.3. Fitting process of the AEM: (**a**) the ear data, (**b**) subdivided AEM (SAEM), (**c**) fitted SAEM after 8 iterations, (**d**) 32 iterations, and (**e**) 64 iterations

used to solve this equation is based on the finite element method approximation. Figure 10.3 depicts the fitting process of our AEM at various iterations.

Mandal et al. [281,282] use subdivision surfaces to represent the geometry of the deformable model. Subdivision surfaces [283] offer increased flexibility and computational efficiency when compared to parametric surfaces. We use our annotated model as the base for the subdivision surface. The subdivision is performed using Loop's scheme [284]. The analytical formulation remains unchanged, but the FEM implementation is integrated with the subdivision surface computation. Therefore, we can solve the equation at a specific resolution (limit surface) while simultaneously applying the computed forces back to the control mesh (the low-resolution annotated model). For an annotated model with n vertices and m triangles we have $3n$ degrees of freedom in the control mesh and $4^l m$ finite elements (where l is the level of the limit surface).

The polygonal data act as attractors for the vertices of the subdivision surface, thus driving the deformation. At each iteration, we need to compute several nearest neighbor searches, which implemented naively have a complexity of $O(k)$, where k is the number of triangles of the data. Therefore, to make the system operate at speeds reasonable for deployment, we employ a space partitioning technique [285] which lowers the average cost of a search from $O(k)$ to $O(log\,k)$. When the process concludes after a given number of iterations, the annotated model has converged to the polygonal data.

10.3.6 Geometry Image Representation

The fitted annotated model from the previous step retains its native properties, thus allowing us to create a geometry image and a normal map from it (Fig. 10.4c,d). Since the model is UV parameterized, then for each (u, v) pair there is one point $(x, y, z) \in \mathbb{R}^3$ belonging to the model. We sample the UV space ($[0..1] \times [0..1]$) at regular intervals. For each sampled (u, v) value, we use the corresponding (x, y, z) as a "pixel value." The result is called a *geometry image*. The resolution of the sampling is correlated to the resolution of the subdivided surface. We compute the

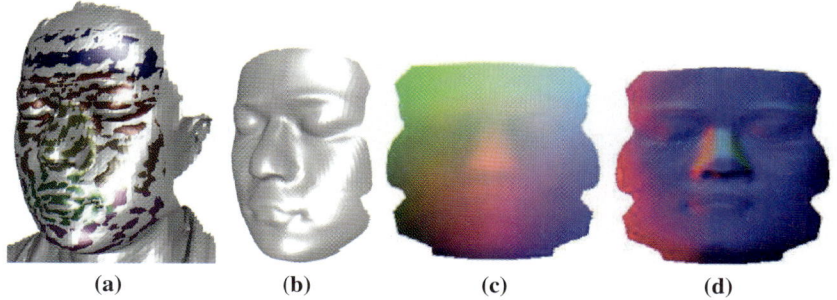

Fig. 10.4. Full face model after fitting: (**a**) Fitted model overlayed on the face data, (**b**) fitted model geometry, (**c**) corresponding geometry image, and (**d**) corresponding normal map

normal map directly from the geometry image. In contrast to the geometry image, the normal map contains 3D normal vectors to the surface as its pixel values.

We treat each channel of the geometry image and normal map independently. Each channel is analyzed using a transform. The coefficients are stored as metadata. We use two different transforms, thus we have two sets of coefficients: the Haar and the Pyramid transform. The Pyramid transform is a more computationally intensive transform, and therefore we may choose not to use it if the system needs to be tuned for speed. Currently, we apply the Haar transform on both the normal map and the geometry image, while we apply only the Pyramid transform to the geometry image.

 – *Haar wavelets.* The choice of Haar wavelets was based on their properties. The transform is conceptually simple, and computationally efficient. The amount of memory needed to compute the transform is equal to the input image size, and the transform is exactly reversible, thus allowing us to compare the wavelet coefficients directly. The Haar wavelet transform is performed by applying a low-pass and a high-pass filter on a 1D input, then repeating the process on the two resulting outputs. Since we are working with images, there will be four outputs for each level of the Haar wavelet (low–low, low–high, high–high, high–low). We compute a level 4 decomposition, meaning that we apply the filters four times, which on the geometry images we have, yields 256 16×16 wavelet packets. Each packet contains a different amount of energy from the initial image. It is possible to ignore most of the packets without losing significant information (Fig. 10.5). For a full discussion, please see [286, 287].
 – *Pyramid transform.* The second transform decomposes the images using the complex version [288] of the steerable Pyramid transform [289], a linear multiscaled, multiorientation image decomposition algorithm. The resultant representation is translation-invariant and rotation-invariant. This feature is desirable to address possible positional and rotational displacements caused by facial expressions. Our algorithm applies a three-scale, ten-orientation complex

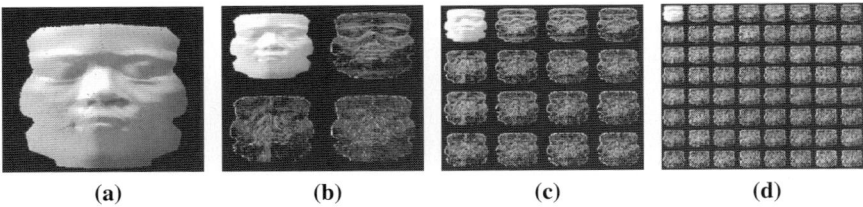

Fig. 10.5. Haar wavelet analysis for the normal map image from Fig. 10.4: (**a**) zero level, (**b**) first level, (**c**) second level, and (**d**) third level. Note that the real numbers were mapped to a gamma corrected gray-scale for visualization purposes

steerable Pyramid transform to decompose each channel of the deformation image. Only the low-pass orientation subbands at the furthest scale are stored. This enables us to compare two images directly and robustly using multiple orientations.

10.3.7 Distance Metrics

In the authentication phase, the comparison between two datasets is performed using the metadata information. We keep metadata as the coefficients of the geometry image, and the normal map of each dataset. Additionally, we may have two coefficient types for each: the Haar coefficients and, optionally, the Pyramid coefficients. To compare the metadata, we need to define a distance metric for each type of coefficient:

Haar metric: in the case of Haar wavelets, the metric we use is weighted L^1.

Pyramid metric: We use a modified version of the complex version of the structural similarity index (CW-SSIM) [290], which we will discuss below.

Fusion: when both types of coefficients are used, we fuse the distances given by the Haar and the Pyramid metrics. We will present the various ways of fusing scores after we explain the Pyramid metric.

The Pyramid Metric

To quantify the distance between the two compressed deformation images of the probe and gallery, we need to assign a numerical score to each annotated region F_k of the model. Note that F_k may be distorted in different ways by facial expressions in the case of the face model. To address this, we employ the CW-SSIM index algorithm. CW-SSIM is a translational insensitive image similarity measure inspired by the structural similarity (SSIM) index algorithm [291]. A window of size 3 is placed in the X direction (first channel). The window then traverses across the input image one step at a time. In each step, we extract all wavelet coefficients associated with F_k. This results in two sets of coefficients $p_w = \{p_{w,i} | i = 1, \ldots, N\}$

and $g_w = \{g_{w,i} | i = 1, \ldots, N\}$, drawn from the probe image and the gallery image, respectively. The distance measure between the two sets is a variation of the CW-SSIM index originally proposed by Wang and Simoncelli [290]

$$\tilde{S}(p_w, g_w) = 1 - \left(\frac{2 \sum_{i=1}^{N} |p_{w,i}||g_{w,i}| + K}{\sum_{i=1}^{N} |p_{w,i}|^2 + \sum_{i=1}^{N} |g_{w,i}|^2 + K} \right) \cdot \left(\frac{2 | \sum_{i=1}^{N} p_{w,i} g_{w,i}^*| + K}{2 \sum_{i=1}^{N} |p_{w,i} g_{w,i}^*| + K} \right)^r.$$

The first component of the subtrahend measures the equivalence of the two coefficient sets. If $p_w = g_w$, then distance 0 is achieved. The second component reflects the consistency of phase changes, which is insensitive to the translational changes caused by facial expressions. The exponent r emphasizes the second component. Experimentally, we found $r = 7$ optimal for most cases, but r should be increased if a strong facial expression between P and G is known or detected. K is a small positive number to insure stable behaviors when tiny numbers are input.

As the sliding window moves, the local $\tilde{S}(p_w, g_w)$ at each step w is computed and stored. The weighted sum of the local scores from all windows gives the distance score of F_k in the X direction

$$e_x(F_k) = \sum_{w=1}^{N} (b_w \cdot \tilde{S}(p_w, g_w))$$

where b_w is a predefined weight depending on which subband the local window lies on. The scores for the Y and Z directions are computed similarly. By summing the scores in the X, Y, and Z directions, the total distance score of F_k is computed. The discrete sum of the scores for all F_ks is the overall distance between the probe image P and the gallery image G

$$d^{CW-SSIM}(P, G) = \sum_{k=1}^{N} (e_x(F_k) + e_y(F_k) + e_z(F_k)).$$

Score Fusion

Since we use both CW-SSIM and the L^1 metric from our previous work [274], we compute the distance between a probe and gallery by the following fusion method:

1. *Score computation*: compute the distances between the probe and the gallery using $L^1 Haar$, and store this into a vector H. Next, compute the distances between the probe and the gallery using the CW-SSIM method, and store this into a vector CW.
2. *Normalization*: compute the mean (μ), and standard deviation (σ) of both H and CW. Then let $H_n = \frac{H - \mu_H}{\sigma_H}$ and $CW_n = \frac{CW - \mu_{CW}}{\sigma_{CW}}$.
3. *Fusion*: compute final distance: $Distance = w_1 * H_n + w_2 * CW_n$, where w_1 and w_2 are the weights for the two scores, and they are proportional to the descriptive power of the two metrics.

10.4 3D Face Recognition

10.4.1 Databases

We report our results on 3D face recognition using two databases. The first is the well-known *FRGC v2* database [253]. The second database is a collection of 3D faces acquired at the University of Houston (UH).

To demonstrate the sensor-invariant nature of the proposed algorithm, we combine the *UH* database with *FRGC v2*.

The *FRGC v2* database contains 4,007 3D scans of 466 persons. The scanner used during the acquisition was a Minolta 910 laser scanner. It produces range images with resolution of 640×480 if the subject occupies the full field of view. The scans were acquired in a controlled environment, and they contain various facial expressions (e.g., happiness, surprise). The subjects are 57% male and 43% female, with the following age distribution: 18–22 (65%) years old, 23–27 (18%), and 28 (17%) years or over. The database contains annotation information such as gender and type of facial expression. Figure 10.6 depicts examples of the various expressions present in the database.

The *UH* database contains 884 3D facial datasets acquired using the 3dMD™ Qlonerator optical system (with one-pod and two-pod setups) over a period of 1 year. Compared to *FRGC v2*, the *UH* database is more challenging as the subjects were encouraged to assume various facial expressions and were allowed to wear accessories (e.g., glasses or hats).

We extend the *FRGC v2* database with the *UH* database to report results on a mixed-scanner database. The extended database contains a total of 4,891 datasets, of which 82% were acquired using a laser scanner, and the rest were acquired using an optical scanner.

10.4.2 Results

We report our results using two different scenarios: identification and verification. In the identification scenario, we divide the database into two distinct sets: a gallery and a probe set. Each probe has exactly one corresponding dataset in the gallery set. Therefore, each subject is represented by only one gallery dataset, increasing the experiment's difficulty. Since there are many ways of dividing the database into a probe and a gallery set, we chose to use the first dataset of each subject as gallery, and the rest as probes. The performance is measured by using a cumulative matching characteristic (CMC) curve and the rank 1 recognition rate is reported.

In the verification scenario, we measure the verification rate at 0.001 FAR. The verification rate is defined as the fraction of datasets that are positive (e.g., claiming to be who they really are) which are classified as positive. The FAR is defined as the fraction of datasets that are negative (e.g., pretending to be somebody else), but are classified as positive. The results are plotted using a receiver operating characteristic (ROC) curve which plots verification rate as a function of FAR. The *FRGC v2*

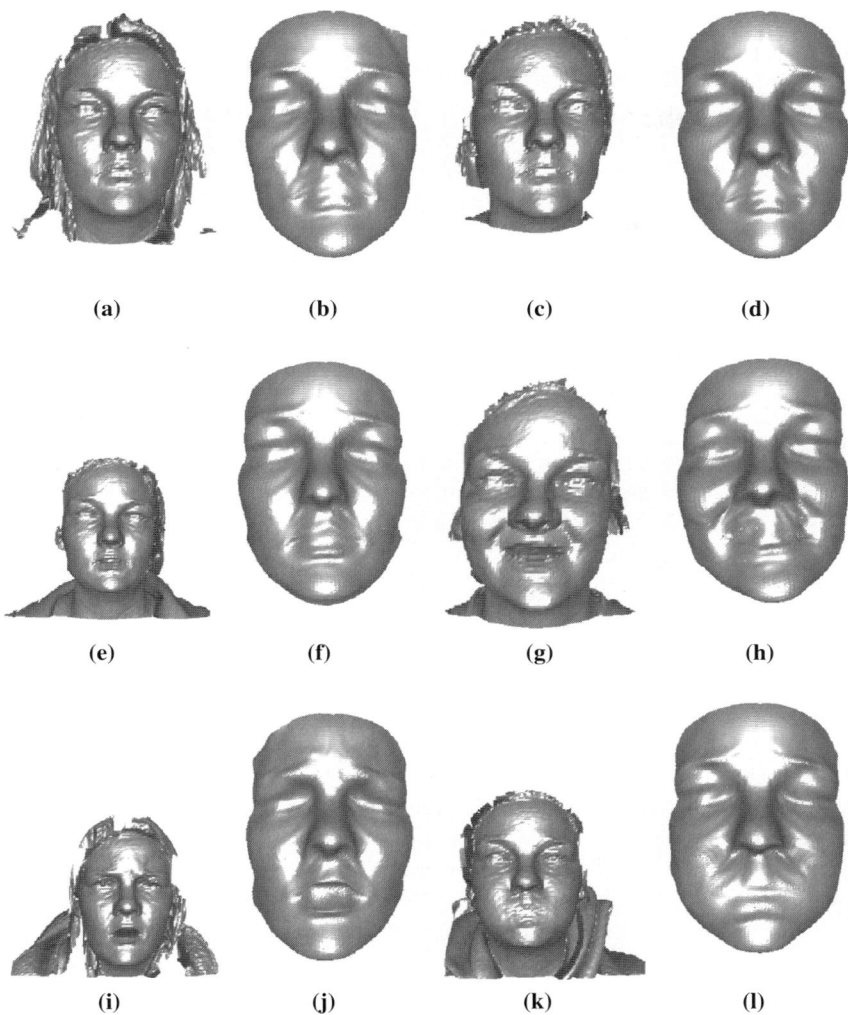

Fig. 10.6. The subjects in the *FRGC v2* database were asked to exhibit various facial expressions. Notice that the subjects clothing and parts of the hair are also present in the input, and that they vary between sessions. Raw data and the fitted model: **(a)**–**(f)** blank stare, **(g)**,**(h)** smile, **(i)**,**(j)** surprise, and **(k,l)** puffy cheeks

database defines three possible selections of datasets (referred to as ROC I, ROC II, and ROC III). In ROC I all the data are within semesters, in ROC II they are within 1 year, while in ROC III the samples are between semesters. These experiments are of increasing difficulty.

Table 10.1. Verification rates of our system at 0.001 FAR on the *FRGC v2* database

	ROC I (%)	ROC II (%)	ROC III (%)
fusion	97.3	97.2	97.0
Haar	97.1	96.8	96.7
Pyramid	95.2	94.7	94.1

Experiment 1: Transforms

The purpose of this experiment is to evaluate the performance of the two transforms, and to provide a reference score on the *FRGC v2* database. Using the fusion of the scores provided by our two wavelet transforms, we obtained a verification rate of 97.3% for ROC I at 0.001 FAR. The weighted sum method gave the best results. We present the verification results at 0.001 FAR in Table 10.1, and the full ROC curves in Fig. 10.7.

Although the Pyramid transform and the CW-SSIM are far more computationally expensive than the Haar transform and the L^1 metric, they achieve poorer results. However, when the Pyramid scores are fused with the Haar scores, they improve the overall performance. To the best of our knowledge, this is the highest performance reported on the *FRGC v2* database for the 3D modality.

Experiment 2: Facial Expressions

Facial expressions have traditionally decreased the performance of face recognition systems. In this experiment, we evaluate the impact of facial expressions on the performance of our system. All datasets in *FRGC v2* are annotated, and one of the categories recorded is the facial expression. We chose to divide the database into two distinct sets: the first set contains nonneutral facial expressions only, while the second set contains datasets that were annotated as having a neutral facial expression.

Figure 10.8 depicts the ROC curves for the two subsets and the entire *FRGC v2* database. We compare the performance of the two subsets to the performance on the entire set at 0.001 FAR in Table 10.2. The average decrease of 1.56% in verification between the full database and the subset containing only facial expressions is very modest when compared to most other systems, given the fact that this subset contains the most challenging datasets from the entire database and is fully automatic. The small decrease in performance can be attributed to the use of the deformable model framework and the annotated face model.

Experiment 3: Multiple Sensors

The purpose of this experiment is to evaluate the performance of our system using data from multiple sensors. Verification experiments depend heavily on the pairs chosen for evaluation. In the absence of any good way of designing such pairing we

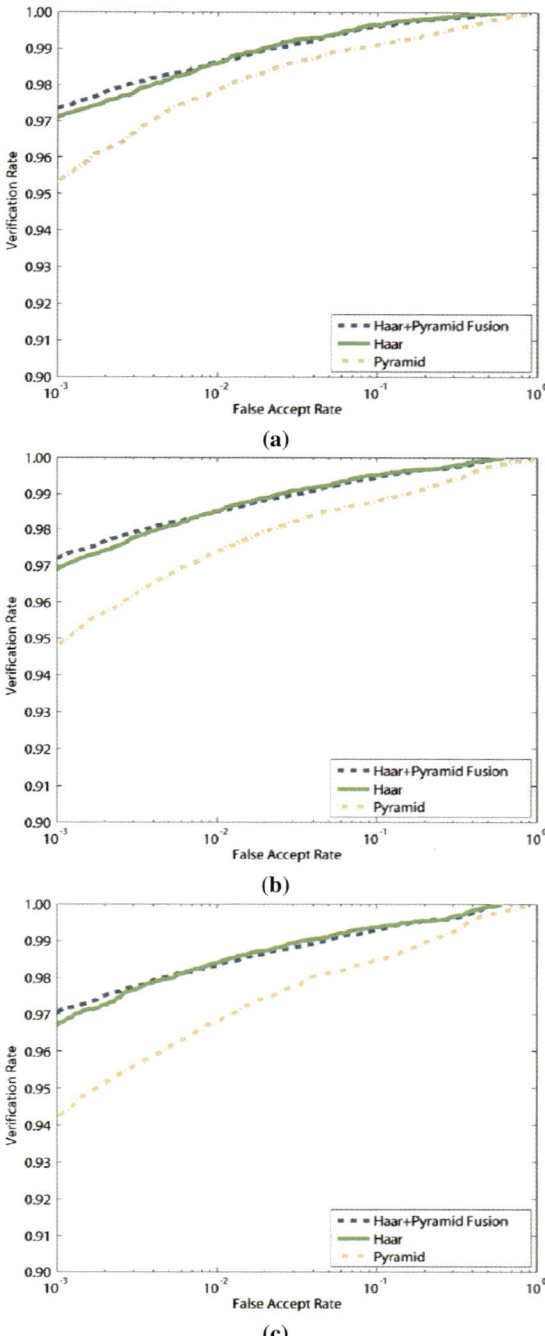

Fig. 10.7. Performance of our system using different transforms (Haar and Pyramid) as well as their fusion on the *FRGC v2* database. Results reported using: (**a**) ROC I, (**b**) ROC II, and (**c**) ROC III

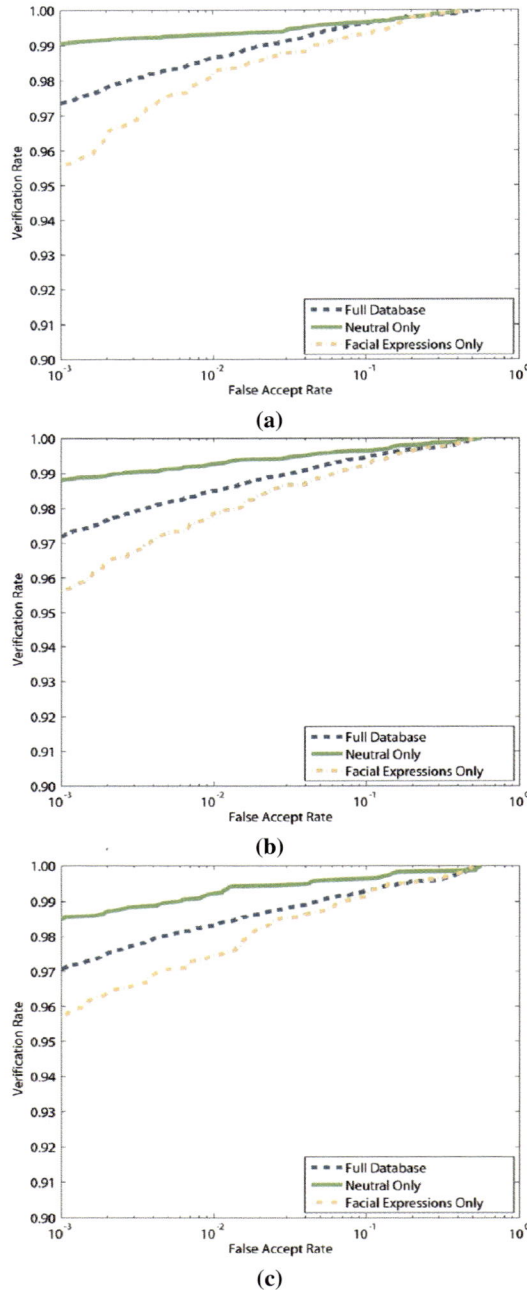

Fig. 10.8. We have divided the *FRGC v2* database into two subsets: the first containing only nonneutral facial expressions and the second one only neutral expressions. Comparison of performance vs. the full database using: (**a**) ROC I, (**b**) ROC II, and (**c**) ROC III

Table 10.2. Performance of our system at 0.001 FAR on the full *FRGC v2* database, on a subset containing only nonneutral facial expressions, and on a subset containing only neutral expressions

	ROC I (%)	ROC II (%)	ROC III (%)
full database	97.3	97.2	97.0
nonneutral expressions	95.6	95.6	95.6
neutral expressions	99.0	98.7	98.5

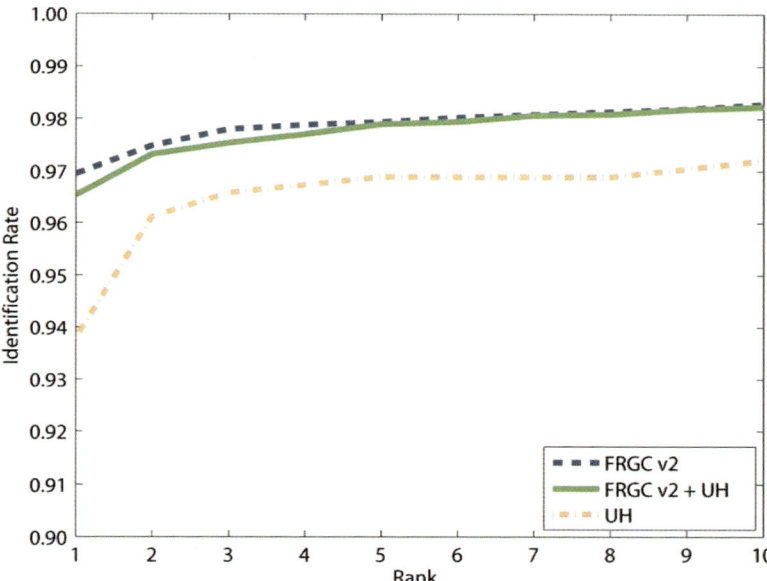

Fig. 10.9. System performance for identification experiment on different databases: *FRGC v2* database with 466 gallery and 3,541 probes (laser scanner), *UH* database with 240 gallery and 644 probes (optical scanner), and *FRGC v2 + UH* database with 706 gallery and 4,185 probes (both scanners)

have opted for an identification experiment, which we consider more representative and more easily duplicated.

On the *FRGC v2* set the rank 1 identification rate was 97.0%, while for the *UH* set, the system achieved 93.8%. Figure 10.9 depicts the full CMC curve. The combined experiment yielded a rank 1 recognition rate of 96.5%, which represents a drop in performance of only 0.5% when compared to the original *FRGC v2* experiment, demonstrating the system's robustness when data from multiple sensors are included in the same database.

10.4.3 Discussion

The results indicate that our algorithm is robust with respect to facial expressions. A few cases still remain in which the distance between the gallery and the probe of the same subject may be larger than the distance between the gallery and a probe belonging to another subject. This may happen if exaggerated facial expressions occur, as they tend to change even the most stable landmarks on the face.

We also have to note that the *FRGC v2* database contains eight datasets which do not properly represent a human face. The reason for this is that during the laser scanning process the subjects moved, and therefore the resulting mesh does not accurately represent their face. Therefore, an algorithm which uses the 3D shape as a feature cannot possibly achieve 100% verification rate on this database.

10.4.4 3D Face Recognition Hardware Prototype

We designed and built a prototype field-deployable 3D face recognition system (Fig. 10.10). It consists of a 3dMD™ optical scanner using a one-pod configuration, which is connected to a PC. A webcam captures a continuous video stream which is used to detect whether a person is facing the 3D camera. When the subject is facing the camera and remains relatively still for more than 2 s, the system triggers the optical scanner and the 3D data of the individual's face are captured. The system can either enroll the subject into the database, or perform a scenario-specific task. In an identification scenario, the system will display the closest five datasets to the operator. In a verification scenario, the system will determine whether the subject is who they claim to be, based on a preset distance threshold.

The system's characteristics are:

- *Automation*. We used only fully automatic methods for triggering the capture and processing the 3D data. To determine whether a subject is facing the camera, the OpenCV [292] implementation of a face detector is employed.
- *Space efficiency*. The raw 3D data captured by the scanner are usually of the order of 2 or 3 MB, depending on how close the subject is to the camera.
- *Time efficiency*. Regardless of the scenario (enrollment, identification, and verification), the most time-consuming step of our algorithm is metadata generation, which takes on average 15 s, with unoptimized code. In the verification and identification scenarios we need to match the computed metadata to entries from the database. The current prototype, using an Opteron 250 (2.4 GHz), can process over 4,000 comparisons per second. If the database size is larger, adding multiple PCs on the backend will linearly reduce the time needed to compute the scores.

Chen et al. [293] claim that the usage of 3D cameras is impractical because of the fact that they need to be calibrated and because they rely on a structured light pattern that is deemed intrusive. In our experience, they are not problematic. The calibration only needs to be performed once every two weeks, and the structured light pattern is barely visible. Moreover, the system allows the acquisition to occur

Fig. 10.10. Prototype system using a 3dMD™ optical scanner with a one-pod configuration

in a very natural way, since the subject is not required to assume any specific pose, except to center their facial image in the screen facing them.

10.5 3D Ear Recognition

We have adapted our generic 3D object recognition algorithm to work on ear datasets. There are several differences between the baseline approach, which is used for face recognition, and ear recognition. We discuss the differences in the following section.

10.5.1 Ear-Specific Issues

Ear data are generally more difficult to acquire than face data. There are several reasons for this:

- *Stereo scanner-specific issues*. The ear capture system used three Qlonerator pods from 3dMD™. Each pod must have a good view of the ear. This is difficult to achieve in practice, as the subject must be very cooperative.

- *Laser scanner-specific issues.* The laser scanner which was used to acquire the *UND* database is not able to capture the full ear. Areas behind the ear lobe and all the data that are not directly visible to the sensor are missing. Therefore, the ear structure is not preserved in its entirety.
- *Amount of data.* Both the laser scanner and the stereo scanner generate a very large amount of data, because the scanners are not able to differentiate between the ear itself and the head. Therefore, in order for our alignment algorithm to work properly, we segmented the ear manually.
- *Intersubject variability.* There is a large variation in ear shape and size among the subjects in the databases on which we have tested our algorithm.

10.5.2 Annotated Ear Model

We constructed an AEM by using the average ear measurements provided by Farkas. Figure 10.11 depicts our ear model, along with an example mesh, while Fig. 10.3 depicts the ear model being fitted to the data.

After analyzing our ear dataset, we noticed that the deviation from the mean measurements of the ear was substantial. Farkas based his studies of facial measurements of Caucasian Americans and did not study other races (Fig. 10.12). Some ear datasets in our collection were out of range of the measurements made by Farkas. Therefore, when building the ear model we located specific features that are common in all ears.

The first region of interest is the curvature of the *Concha* (Fig. 10.13). The *Concha* has a concave shape that is found in all ears. The second is the *Scapha region* which has a distinct curvature that can be used in modeling. The last region is the *Helix* whose shape is distinct from all regions of the ear and can help in the alignment phase of our algorithm. The *Lobule* is not used for modeling due to it being the most elastic portion of the ear.

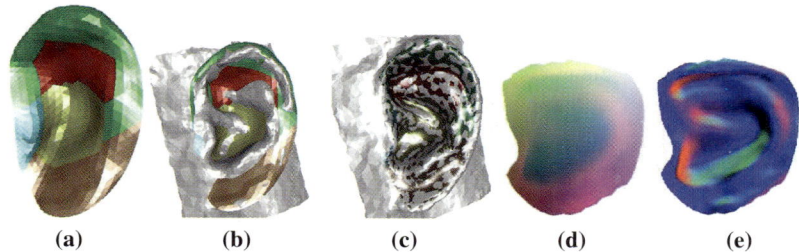

Fig. 10.11. (**a**) The ear model, (**b**) the aligned data overlaid with the ear model, (**c**) the ear model fitted onto the data, (**d**) the corresponding geometry image, and (**e**) the corresponding normal map

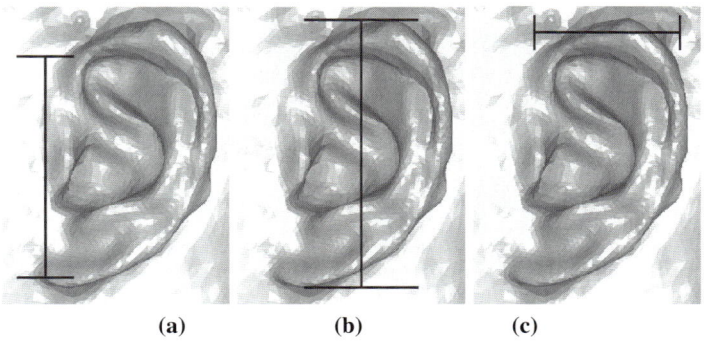

Fig. 10.12. Ear measurement norms: (**a**) width of the ear (mean 36.9 mm, std dev 2.4 mm), (**b**) length of the ear (mean 62.7 mm, std dev 3.6 mm), and (**c**) morphological width of the ear (mean 50.2 mm, std dev 3.9 mm) [270]

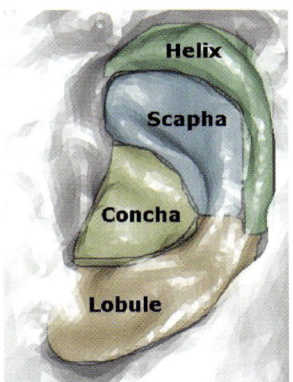

Fig. 10.13. The outer ear anatomy

10.5.3 Ear-Specific Algorithm

Since human ears vary in shape and size quite significantly, we address this issue by not only estimating the rigid alignment of the data to the model, but also by computing the proper scale. Computing the scale factor is performed independently of the original alignment.

The scale factor is computed by minimizing the ICP alignment error between the scaled ear model and the data. The minimization procedure finds the best scale factors on the X, Y, and Z directions. To speed up the process, we choose to limit the search space for the scale factors to a set of known values. The minimization algorithm is ESA. Figure 10.14 depicts several steps from the minimization process.

The distance metrics used are also slightly different than those used for face recognition. Instead of making use of the UV parameterization of the model, and

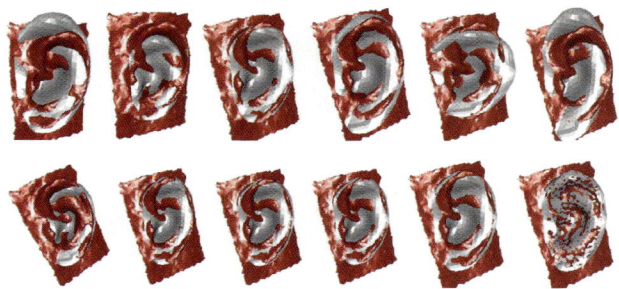

Fig. 10.14. Optimization steps while minimizing the alignment error between the scaled ear model and ear data

Fig. 10.15. Renderings of good quality ear data

computing the distance between meshes using the geometry images, we use ICP. We convert the geometry images back to meshes, then compute the ICP registration between the gallery and the probe, and use the L^2 distance returned by the alignment algorithm as a distance score.

10.5.4 3D Ear Databases

In the Fall of 2005, we acquired data from the ears of 461 subjects for a total of 1,419 datasets (average of three sessions per subject). Three 3dMD^{TM} Qlonerator [229] sensors were used to capture the back, the inside, and the front of the ear. To the best of our knowledge, this is the first dataset of 3D polygonal ear data captured which contains a 180° view of the ear (Fig. 10.17). Previous studies [266, 294, 295] on 3D ear recognition have used range images of profile views of the head.

The *UH* ear dataset consists of 201 good quality ear datasets (Fig. 10.15) out of the 1,419 captured. The rest of the ear datasets were unusable due to badly reconstructed geometry (Fig. 10.16). The number of unique subjects present in the final database is 110. To date and to the best of our knowledge, this is the only ear dataset captured with a 3D optical sensor. When compared to ear datasets acquired using range cameras, there is much more information and curvature with respect to the outer and inner area of the ear (Fig. 10.17).

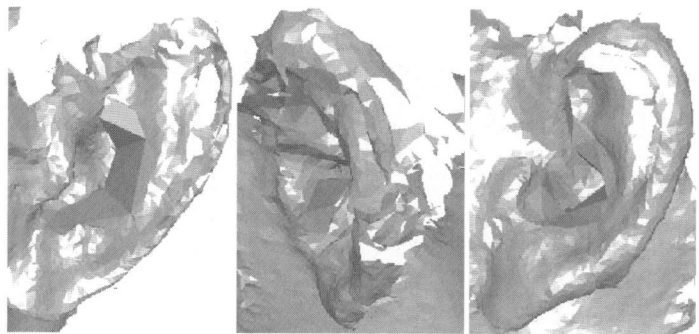

Fig. 10.16. Ear data with noise, holes, and inner ear data missing

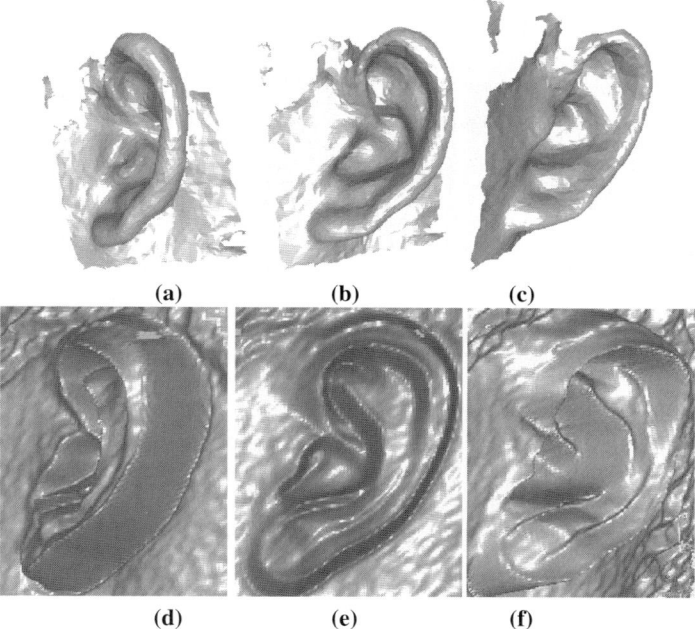

Fig. 10.17. Comparison of ear data quality from back, side, front views. (**a–c**) Data obtained using the multiple 3dMD™ Sensor setup. (**d–f**) Range images converted to a 3D mesh

10.5.5 Results

There was no significant difference in performance between the use of normal maps and geometry images in the case of ear datasets. In fact, the scores had a very high correlation, meaning that regardless of the fusion strategy, it is not possible to improve the score by combining the scores obtained using normal maps with those computed using only geometry images.

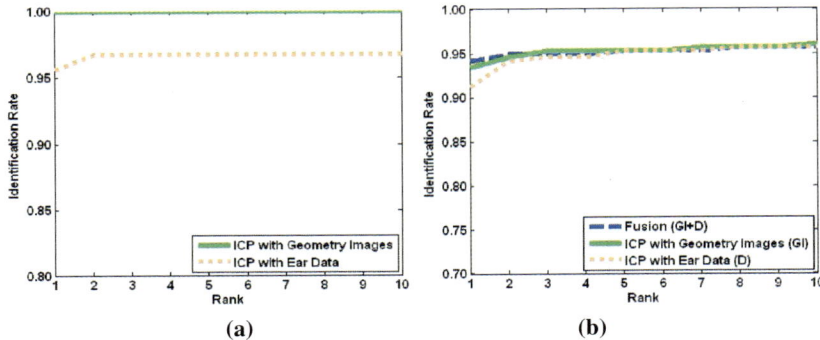

Fig. 10.18. (a) CMC curves of the *UH* ear database and (b) CMC curves of our subset from the *UND* ear database

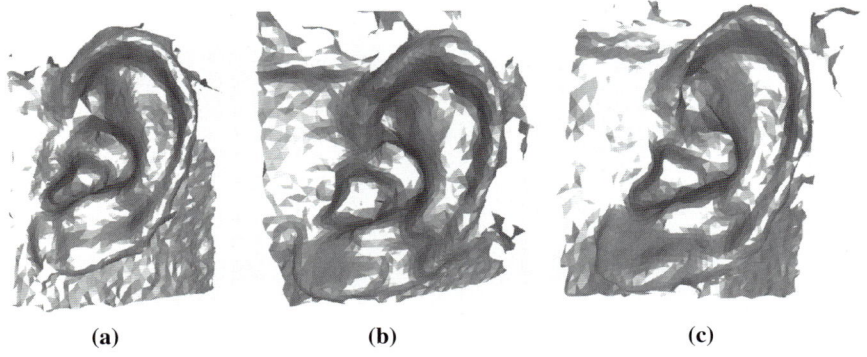

Fig. 10.19. Failure caused by using the L^2 metric after performing ICP on raw data (*UH* database): (a) impostor, (b,c) images belonging to the same subject. The distance between (a) and (b) was smaller than the distance between (b) and (c), therefore causing a failure

Figure 10.18 depicts the CMC curves for both the *UH* and the subset of *UND* ear databases. On the *UH* database, L^2 after aligning geometry images with ICP yielded 100.0% rank 1 recognition, while L^2 with ear data yielded 95.6%. As expected, we had a slightly lower recognition rate when using the L^2 metric on raw data. The reasons are that the data contain noise and the manually segmented ear data are not perfect. The geometry images created by our deformable model approach, on the other hand, are much more resilient to such problems.

The results on the *UND* subset exhibit the same trend: using raw data gives lower performance (91.3%) than when using geometry images (93.47%). Most failures are due to lack of inner ear information. By not having a good sampling of inner ear data, the automatic scaling process cannot find a good alignment between ear model and ear data. By having a bad alignment, the geometry image that is created after fitting will be an incorrect representation of the ear's geometry. A failure arises when

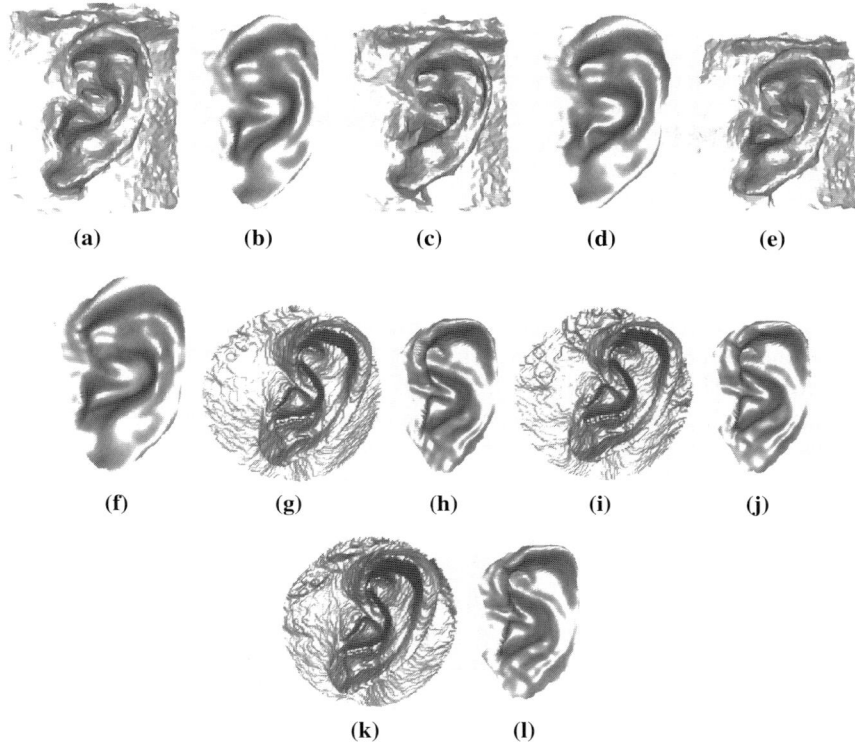

Fig. 10.20. Variation of the ear datasets within subject. Most of the variation comes from the noise introduced by the capture process. Ear data and the corresponding fitted model belonging to a subject from three sessions: (**a–f**) *UH* and (**g–l**) *UND* database

a bad geometry image is compared to a good geometry image of the same subject. Fusing the two methods of ICP surface matching improves the rank 1 recognition rate to 94.2% (Fig. 10.18b).

10.5.6 Discussion

The performance of 3D ear recognition is directly related to the quality of the input data. The *UND* database contains many ear datasets that do not have the 3D description of the full ear lobe, which causes both our approach and ICP to fail occasionally (Fig. 10.17). The datasets in the *UH* database have more data available, which allows our method to perform better than ICP (Fig. 10.19). Figure 10.20 depicts the variation in raw data, and fitted model for a typical subject chosen from the *UH* and another subject chosen from the *UND* database.

10.6 Conclusion

We have presented a general algorithmic solution for 3D object recognition, and how to adapt it for 3D face and 3D ear recognition. By utilizing a deformable model we map the 3D geometry information onto a 2D regular grid, thus combining the rich information of the 3D data with the computational efficiency of 2D data. A multistage fully automatic alignment algorithm and the advanced wavelet analysis resulted in state-of-the-art performance on the publicly available *FRGC v2* face database. We have also presented encouraging results on our own 3D ear database and on a subset of the publicly available *UND* ear database.

The current trend in biometrics is to achieve higher accuracy and robustness by using multiple biometric modalities. It is becoming increasingly accepted that no biometric modality can provide 100% verification rate at very low FARs in large databases. Any hope for achieving such an accuracy figure can only result from the fusion of multiple modalities. We have observed the same phenomenon in 3D face recognition at a smaller scale: our verification rates increased significantly by fusing position data and normal data as well as fusing Haar and Pyramid wavelet transforms of the above two types of data. We expect that the fusion of 3D face and 3D ear biometric data from same individual will provide good fusion results. To test this hypothesis a large dataset of 3D face and 3D ear data needs to be acquired. An additional advantage of such a fusion is that the cost of the 3D capture device can be amortized over these two modalities.

11 Human Recognition at a Distance in Video by Integrating Face Profile and Gait

Xiaoli Zhou, Bir Bhanu, and Ju Han

11.1 Introduction

It has been found to be difficult to recognize a person from arbitrary views in reality, especially when one is walking at a distance in real-world outdoor conditions. For optimal performance, the system should use as much information as possible from the observations. A fusion system, which combines face and gait cues from video sequences, is a potential approach to accomplish the task of human recognition.

The general solution to analyze face and gait video data from arbitrary views is to estimate 3D models. However, the problem of building reliable 3D models for nonrigid face with flexible neck and the articulated human body from low-resolution video data remains a hard one. In recent years, integrated face and gait recognition approaches without resorting to 3D models have achieved some progress. In [296], Kale et al. present a gait recognition algorithm and a face recognition algorithm based on sequential importance sampling. The fusion of frontal face and gait cues is performed in the single camera scenario. In [297, 298], Shakhnarovich et al. compute an image-based visual hull from a set of views of four monocular cameras. It is then used to render virtual canonical views for tracking and recognition. The gait recognition scheme is based on silhouette extent analysis. Eigenfaces are used for recognizing frontal face rendered by the visual hull. They discuss the issues of crossmodal correlation and score transformations for different modalities and present the fusion of face and gait.

Most current gait recognition algorithms rely on the availability of the side view of the subject since human gait or the style of walking is best exposed when one presents a side view to the camera. For face recognition, on the other hand, it is preferred to have frontal views. These conflicting requirements are easily satisfied by an individual classifier for face or gait, but pose some challenges when one attempts to integrate face and gait biometrics in real-world applications. In Kale's and Shakhnarovich's fusion systems [296–298], both use the side view of gait and the frontal view of face. In Kale's work [296], the subjects are walking in a single camera scenario. For face recognition, only the final segment of the database presents a nearly frontal view of face and it is used as the probe. The galley consists of static faces for the corresponding subjects. Therefore, they perform still-to-video face recognition. In Shakhnarovich's work [297, 298], four cameras must be used to get both the canonical view of gait and the frontal view of face simultaneously.

In this chapter, an innovative system is proposed, aiming at recognizing non-cooperating individuals at a distance in a single camera scenario. Information from two biometric sources, face profile and gait, is combined. We use face profile instead of frontal face in the system since a side view of face is more likely to be seen than a frontal view of a face when one exposes the best side view of gait to the camera. It is very natural to integrate information of the side face view and the side gait view. However, it is difficult to get reliable information of a face profile directly from a low-resolution video frame for recognition tasks because of limited resolution. To overcome this problem, we use resolution enhancement algorithms for face profile analysis. We first reconstruct a high-resolution face profile image from multiple adjacent low-resolution video frames. The high-resolution face profile image fuses both the spatial and temporal information present in a video sequence. The approach relies on the fact that the temporally adjacent frames in a video sequence, in which one is walking with a side view to the camera, contain slightly different, but unique, information for face profile [299]. Then, we extract face profile features from the high-resolution face profile images. Finally, a dynamic time warping (DTW) method [300] is used to match face profiles based on absolute values of curvature. For gait, we use gait energy image (GEI), a spatiotemporal compact representation, to characterize human walking properties [301]. Recognition is carried out based on the direct GEI matching. Face profile cues and gait cues are integrated by three schemes. The first two are *Sum* rule and *Product* rule [302]. The last one is an indexing-verification scheme, which consolidates the accept/reject decisions of multiple classifiers [303].

This chapter is organized as follows. Section 11.2 presents the overall technical approach. It explains the construction of a high-resolution face profile image and describes the generation of GEI. It presents the details of face profile recognition and gait recognition. It provides a description of the fusion of face profile and gait, and the classification methods. In Sect. 11.3, a number of dynamic video sequences are tested. Experimental results are compared and discussed. Finally, Sect. 11.4 concludes this chapter.

11.2 Technical Approach

The overall technical approach is shown in Fig. 11.1. A simple background subtraction method [304] is used for human body segmentation from video data. For each video sequence in the gallery, we construct a high-resolution face profile image from low-resolution face profile images, and a GEI from the binary silhouette image sequences. Then, we extract face profile features from each high-resolution profile image to form face feature gallery. During the testing procedure, each testing video is processed to generate both the high-resolution face profile image and the GEI. The face profile features are extracted from the high-resolution face profile image and compared with face profile features in the gallery using DTW. The GEI is directly

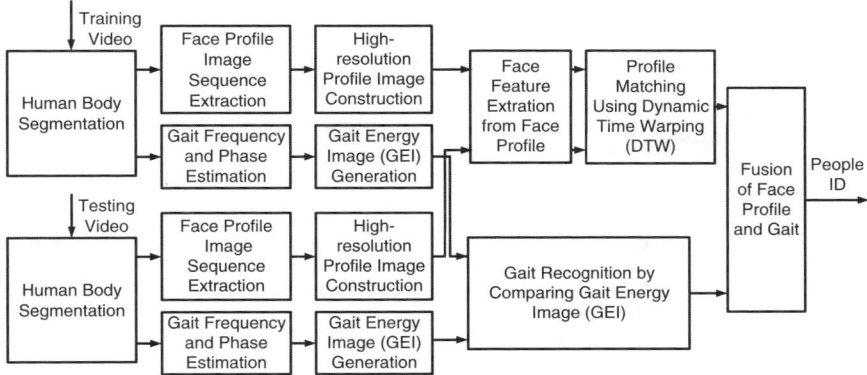

Fig. 11.1. Technical approach for integrating face profile and gait in video

compared with the GEI templates in the gallery. Finally, different fusion strategies are used to combine the results of the face profile classifier and the gait classifier to improve recognition performance.

11.2.1 High-Resolution Image Construction for Face Profile

Multiframe resolution enhancement seeks to construct a single high-resolution image from multiple low-resolution images. These images must be of the same object, taken from slightly different angles, but not so much as to change the overall appearance of the object in the image. The idea of super-resolution was first introduced for multiframe image restoration of band-limited signals in 1984 [305]. In the last two decades, different mathematical approaches have been developed. All of them seek to address the question of how to combine irredundant image information present in multiple images.

In this chapter, the original low-resolution face profile images are first localized and extracted from the segmented human body obtained from multiple video frames. A human body is divided into two parts according to the proportion of its parts [306]: from the top of the head to the bottom of the chin, and then from the bottom of the chin to the bottom of the foot. Human head is defined as the part from the top of the head to the bottom of the chin. Considering the height of hair and the length of neck, we obtain the original low-resolution face profile images by cutting the upper 16% of the segmented human body. Before multiple low-resolution face images are fused to construct a high-resolution face image using the resolution enhancement method, they are aligned by affine transformation and motion estimates are computed to determine pixel displacements between them. Then, an iterative method [307] is used to construct a high-resolution face profile image from aligned low-resolution face profile images.

The Imaging Model

The imaging process, yielding the observed face profile image sequence f_k, is modeled by [307]

$$f_k(m,n) = \sigma_k(h(T_k(F(x,y)))+\eta_k(x,y)) \qquad (11.1)$$

where

1. f_k is the sensed image of the tracked face profile in the kth frame.
2. F is a high-resolution image of the tracked face profile in a desired reconstructed view. Finding F is the objective of the super-resolution algorithm.
3. T_k is the 2D geometric transformation from F to f_k, determined by the 2D motion parameters of the tracked face profile in the image plane. T_k is assumed to be invertible and does not include the decrease in the sampling rate between F and f_k.
4. h is a blurring operator, determined by the point spread function (PSF) of the sensor. We use a circular averaging filter with radius 2 as PSF.
5. η_k is an additive noise term.
6. σ_k is a downsampling operator which digitizes and decimates the image into pixels and quantizes the resulting pixel values.

The receptive field (in F) of a detector whose output is the pixel $f_k(m,n)$ is uniquely defined by its center (x,y) and its shape. The shape is determined by the region of the blurring operator h, and by the inverse geometric transformation T_k^{-1}. Similarly, the center (x,y) is obtained by $T_k^{-1}(m,n)$. The resolution enhancement algorithm aims to construct a higher resolution image $\hat{F}$, which approximates F as accurately as possible, and surpasses the visual quality of the observed images in $\{f_k\}$.

The Super-Resolution Algorithm

The algorithm for creating higher resolution images is iterative. Starting with an initial guess $F^{(0)}$ for the high-resolution face profile image, the imaging process is simulated to obtain a set of low-resolution face profile images $\{f_k^{(0)}\}_{k=1}^K$ corresponding to the observed input images $\{f_k\}_{k=1}^K$. If $F^{(0)}$ were the correct high-resolution face profile image, then the simulated images $\{f_k^{(0)}\}_{k=1}^K$ should be identical to the observed low-resolution face profile image $\{f_k\}_{k=1}^K$. The difference images $\{f_k - f_k^{(0)}\}_{k=1}^K$ are used to improve the initial guess by "back projecting" each value in the difference images onto its receptive field in $F^{(0)}$, yielding an improved high-resolution face profile image $F^{(1)}$. This process is repeated iteratively to minimize the error function

$$e^{(n)} = \sqrt{\frac{1}{K}\sum_{k=1}^K \|f_k - f_k^{(n)}\|^2} \qquad (11.2)$$

The imaging process of f_k at the nth iteration is simulated by

$$f_k^{(n)} = (T_k(F^{(n)}) * h) \downarrow s \tag{11.3}$$

where $\downarrow s$ denotes a downsampling operator by a factor s, and * is the convolution operator. The iterative update scheme of the high-resolution image is expressed by

$$F^{(n+1)} = F^{(n)} + \frac{1}{K} \sum_{k=1}^{K} T_k^{-1}(((f_k - f_k^{(n)}) \uparrow s) * p) \tag{11.4}$$

where K is the number of low-resolution face profile images. $\uparrow s$ is an upsampling operator by a factor s, and p is a "back projection" kernel, determined by h. T_k is 2D motion parameters. The averaging process reduces additive noise.

In our system, we reconstruct a high-resolution face profile image from six adjacent video frames. We assume that six low-resolution face profile images have been localized and extracted from adjacent video frames. We then align these six low-resolution face profile images using affine transformation. Affine transformation works for in-plane, not out of plan rotations of the human face. The quality of the reconstructed image depends on how well the six profile images are registered. Finally, we apply the super-resolution algorithm given above to construct a high-resolution face profile image from the six aligned low-resolution face profile images. The resolution of the original low-resolution face profile images is 70×70 and the resolution of the reconstructed high-resolution face profile image is 140×140. Figure 11.2 shows the six low-resolution face profile images from six adjacent

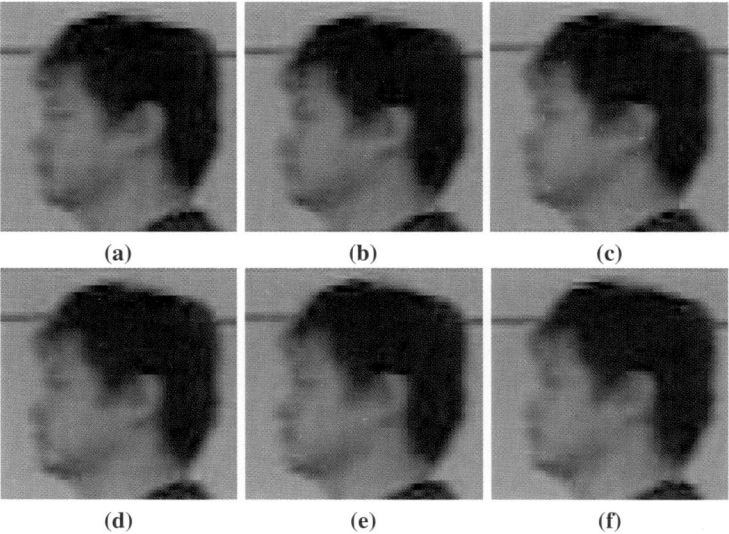

Fig. 11.2. Six low-resolution face profile images resized by using bilinear interpolation (**a**–**f**)

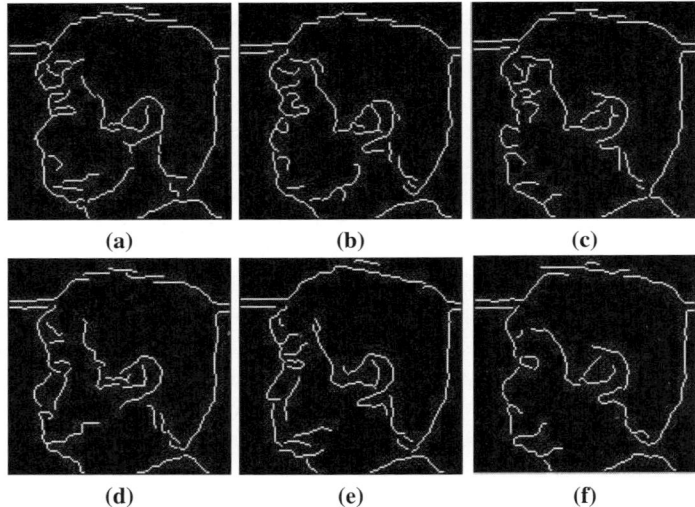

Fig. 11.3. The edge images of six low-resolution face profile images shown in Fig. 11.2

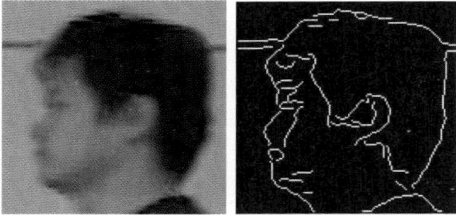

Fig. 11.4. The reconstructed high-resolution face profile and its edge image

video frames. For comparison, we resize the six low-resolution face profile images by using bilinear interpolation. Figure 11.3 shows the corresponding edge images of six low-resolution face profiles. Figure 11.4 shows the reconstructed high-resolution face profile image and its edge image. From these figures, we can see that the reconstructed high-resolution image is better than any of the six low-resolution images. It is clearly shown in the edge images that the edges of the high-resolution image are much smoother and more accurate than that of the low-resolution images. Using the reconstructed high-resolution image, we can extract better features for face profile matching.

11.2.2 Face Profile Recognition

Face profile is an important aspect for the recognition of faces, which provides a complementary structure of the face that is not seen in the frontal view. Though it inherently contains less discriminating power than frontal images, it is relatively easy to analyze and more foolproof. Within the last decade, several algorithms have been proposed for automatic person identification using face profile images. Most of these algorithms depend on the correct detection of all fiducial points and the determination of relationships among these fiducial points.

Harmon and Hunt [308] use manually entered profile traces from photographs of 256 male faces. They locate eight independent fiducials on the profiles and obtain the ninth fiducial by rotating a point from the chin about the pronasale until it intersects the profile above the pronasale. Later, Harmon et al. [309] increase the number of fiducials from nine to eleven, and achieve 96% recognition accuracy for 112 subjects, using a 17-dimensional feature vector. The most significant problem with tangency-based techniques is that there is not a line that is bitangent to the pronasale and chin for profiles with protruding lips [310]. Campos et al. [311] analyze the profile of the face using scale-space techniques to extract eight fiducials. This technique assumes that there will be nine zero-crossings on the profile, and this assumption could be invalidated by facial hair particularly moustaches and the hairline on the forehead. Dariush et al. [312] extract nine fiducials based on the observation that the curvature of the profile alternates between convex and concave, with the point of maximal absolute curvature in each segment corresponding to a fiducial. Akimoto et al. [313] use a template matching approach to find the position of the same five fiducials used by Galton [314]. The template consisting of approximately 50 line segments is used to represent a generic face profile.

In reality, some profiles are too difficult for all fiducials to be reliably extracted, so in these cases a feature vector approach based on the same fiducial points of different face profiles will fail. In this chapter, we use a curvature-based matching approach [300] for recognition, which does not focus on the extraction of all the fiducial points and the determination of relationship among these fiducial points [309, 311, 312]. We use the relationship of some fiducial points for their extraction, but not for an individual recognition. The Gaussian scale-space filter is first used to smooth the profile extracted from the high-resolution face profile image and then the curvature of the filtered profile is computed. Using the curvature value, the fiducial points, including the nasion and throat, can be reliably extracted using a fast and simple method after pronasale is determined. Finally, a DTW method is applied to compare the face profile portion from nasion to throat based on the curvature values.

Face Profile Representation

We apply a canny edge detector to the high-resolution face profile image. After edge linking and thinning, the profile of a face is extracted as the leftmost points different from background, which contain fiducial points like nasion, pronasale, and throat. The outline of a profile is treated as a 1D function, consisting of a set of points $T = (x, y)$, where x is a row index and y is a column index of a pixel. The Gaussian scale-space filter is applied to the face profile to eliminate the spatial quantization noise introduced during the digitization process, as well as other types of high frequency noise. The convolution between Gaussian kernel $g(x, \sigma)$ and signal

$f(x)$ depends both on x, the signal's independent variable, and on σ, the Gaussian's standard deviation. It is given by

$$F(x,\sigma) = f(x) \oplus g(x,\sigma) = \int_{-\infty}^{\infty} f(u) \frac{1}{\sigma\sqrt{2\pi}} e^{\frac{-(x-u)^2}{2\sigma^2}} du \quad (11.5)$$

where $\oplus$ denotes convolution with respect to x. The bigger the σ, the smoother the $F(x,\sigma)$. The curve T is parameterized as $T(u) = (x(u), y(u))$ by the arc length parameter u. An evolved version of T is $T_\sigma(u) = (X(u,\sigma), Y(u,\sigma))$, where $X(u,\sigma) = x(u) \oplus g(u,\sigma)$ and $Y(u,\sigma) = y(u) \oplus g(u,\sigma)$.

Curvature κ on T_σ is computed as

$$\kappa(u,\sigma) = \frac{X_u(u,\sigma)Y_{uu}(u,\sigma) - X_{uu}(u,\sigma)Y_u(u,\sigma)}{(X_u(u,\sigma)^2 + Y_u(u,\sigma)^2)^{1.5}} \quad (11.6)$$

where the first and second derivatives of X and Y can be computed as

$$X_u(u,\sigma) = x(u) \oplus g_u(u,\sigma) \quad X_{uu}(u,\sigma) = x(u) \oplus g_{uu}(u,\sigma)$$
$$Y_u(u,\sigma) = y(u) \oplus g_u(u,\sigma) \quad Y_{uu}(u,\sigma) = y(u) \oplus g_{uu}(u,\sigma)$$

where $g_u(u,\sigma)$ and $g_{uu}(u,\sigma)$ are the first derivative and the second derivative of Gaussian kernel.

Since the profiles include the hair and some other parts that are not reliable for matching, we extract a portion of profile starting from nasion to throat for effective matching. It is done by finding the fiducial points on the face profile. To localize the fiducial points, the curvature of a profile is first computed at an initial scale and the locations, where the local maxima of the absolute values occur, are chosen as corner candidates. These locations are tracked down and the fiducial points are identified at lower scales. The initial scale must be large enough to remove noise and small enough to retain the real corners. Our method has advantages in that it does not depend on too many parameters and does not require any thresholds. It is also fast and simple.

We define pronasale as the leftmost point above throat in the middle part of the profile and nasion as the first point that has local maximum of the absolute values above pronasale. The method of extracting the nasion and throat points is described as follows:

1. Compute the curvature of a profile at an initial scale, find local maxima of the absolute values as corner candidates, and track them down to lower scales.
2. Regard the rightmost point in the candidate set as the throat.
3. Regard the pronasale as one of the two leftmost candidate points in the middle part of the profile and then identify it using the curvature value around this point.
4. Assume that there are no candidate points between pronasale and nasion and identify the first candidate point above the pronasale as nasion.

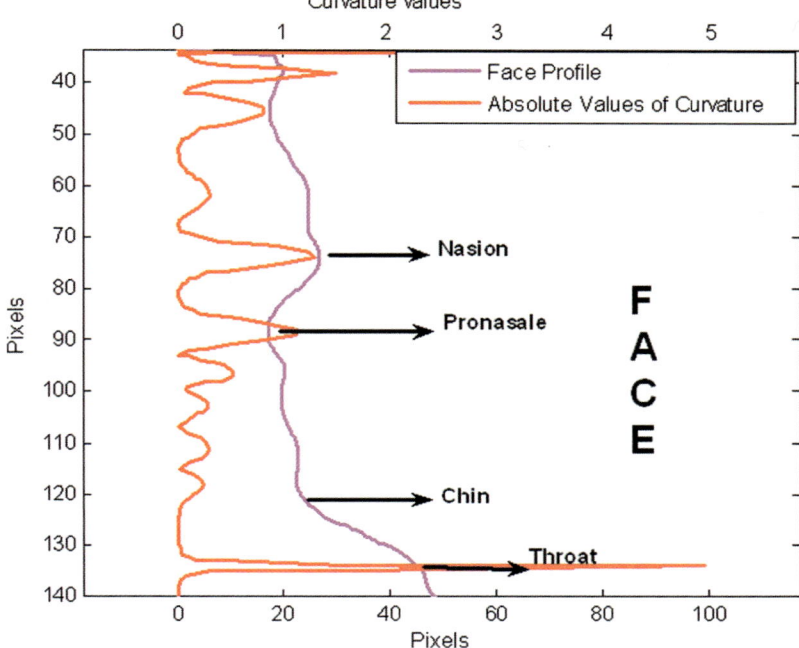

Fig. 11.5. The extracted face profile and the absolute values of curvature

Figure 11.5 shows the extracted face profile and the absolute values of curvature. It is clear that the locations of the fiducial points, including nasion, pronasale, and throat, have local maxima of the absolute values. Figure 11.6 shows the absolute values of curvature on face profiles belonging to four different people. We can see that different face profiles have different patterns of curvature. Therefore, we can use the absolute values of curvature as the feature to represent a face profile. Curvature features have some advantages in that they are invariant to rotation, translation, and uniform scaling.

Face Profile Matching Using Dynamic Time Warping

We use the Dynamic Time Warping (DTW) as the matching method to compute the similarity of two face profiles based on the absolute values of curvature, which are used to represent the shapes of face profiles. The DTW is an algorithm to calculate the optimal score and to find the optimal alignment between two strings. This method is a much more robust distance measure for time series than Euclidean distance, allowing similar shapes to match even if they are out of phase in the time axis [315]. We use the Needleman–Wunsch [316] global alignment algorithm to find the optimum alignment of two sequences when considering their entire length. For two strings $s[1\ldots n]$ and $t[1\ldots m]$, we compute $D(i,j)$ for entire sequences, where i ranges from 1 to m and j ranges from 1 to n. $D(i,j)$ is defined as

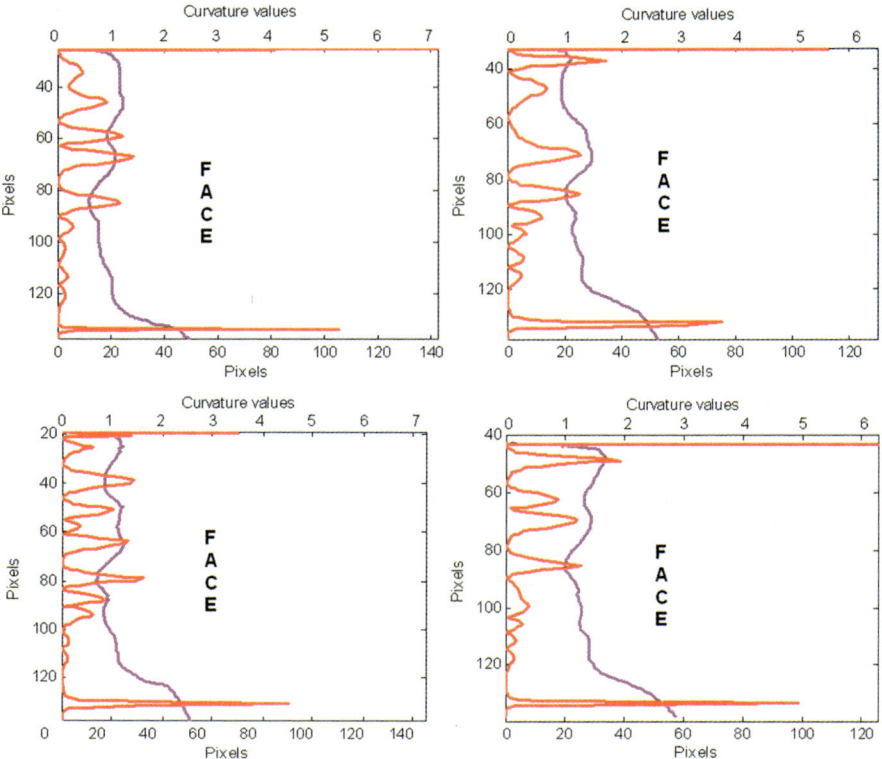

Fig. 11.6. Four examples of curvature features on face profiles

$$D(i,j) = \min\{D[i-1, j-1] + d(s[j], t[i]),$$
$$D[i-1, j] + gap,$$
$$D[i, j-1] + gap\} \qquad (11.7)$$

where $d(s[j], t[i])$ represents the similarity between two points on face profiles. Since the face profile is represented by the absolute values of curvature on the profile, $d(s[j], t[i])$ is calculated by Euclidean distance

$$d(s[j], t[i]) = ||s[j] - t[i]|| \qquad (11.8)$$

The penalty is defined for both horizontal and vertical gaps. It is small and yet exists just to control nondiagonal moves. Generally, the penalties should be set to less than 1/10th the maximum of the $d(s[j], t[i])$. In our method, we use the same constant penalty for both horizontal and vertical gaps. The maximum of $d(s[j], t[i])$ is ~ 5 and the gap penalties are set to 0.5 in our experiments. The final score $D(m, n)$ is the best score for the alignment.

A dynamic programming matrix is used to visualize the alignment. Figure 11.7 gives an example of DTW of two face profiles from the same person. From the

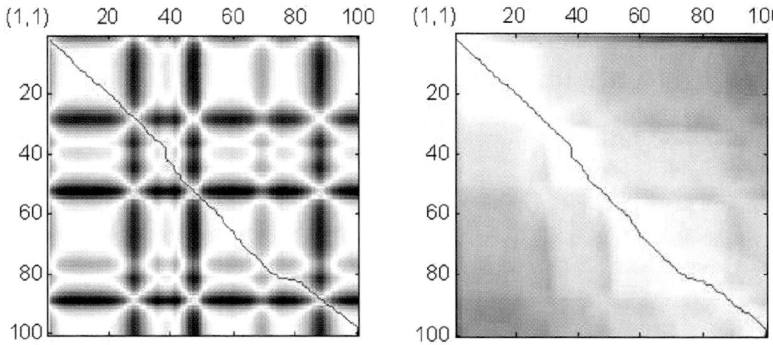

Fig. 11.7. The similarity matrix (*left*) and the dynamic programming matrix (*right*)

similarity matrix in Fig. 11.7, we can see a light stripe (high similarity values) approximately down the leading diagonal. From the dynamic programming matrix in Fig. 11.7, we can see the lowest-cost path between the opposite corners visibly follows the light stripe, which overlay the path on the similarity matrix. The least cost is the value in the bottom-right corner of the dynamic programming matrix. This is the value we would compare between different templates when we are doing classification. The unknown person is classified to the class which gets the least cost out of all the costs corresponding to all the classes.

11.2.3 Gait Recognition

In recent years, various techniques have been proposed for human recognition by gait. These techniques can be divided as model-based and model-free approaches. In this chapter, we focus on a model-free approach that does not recover a structural model of human motion. In the following, we provide related work on gait recognition.

Little and Boyd [317] describe the shape of the human motion with scale-independent features from moments of the dense optical flow, and recognize individuals by phase vectors estimated from the feature sequences. Sundaresan et al. [318] propose a hidden Markov models (HMMs)-based framework for individual recognition by gait. Huang et al. [319] extend the template matching method to gait recognition by combining transformation based on canonical analysis and eigenspace transformation for feature selection. Sarkar et al. [320] directly measure the similarity between the testing and training sequences by computing the correlation of corresponding time-normalized frame pairs. Collins et al. [321] first extract key frames from a sequence and then compute the similarity between two sequences using the normalized correlation.

While some gait recognition approaches [319] extract features from the correlation of all the frames in a walking sequence without considering their order, other approaches extract features from each frame and compose a feature sequence for the human walking sequence [317, 320, 321]. During the recognition procedure, these approaches either match the statistics collected from the feature sequence, or match the features between the corresponding pairs of frames in two sequences that are time-normalized with respect to their cycle lengths. The fundamental assumptions made here are (1) the order of poses in human walking cycles is the same, i.e., limbs move forward and backward in a similar way among normal people, and (2) differences exist in the phase of poses in a walking cycle, the extend of limbs, and the shape of the torso, etc. Under these assumptions, it is possible to represent the spatiotemporal information in a single 2D gait template, called *gait energy image* (discussed below), instead of an ordered image sequence.

Gait Frequency and Phase Estimation

Regular human walking can be considered as cyclic motion where human motion repeats at a stable frequency. Therefore, it is possible to divide the whole gait sequence into cycles and study them separately. We assume that silhouette extraction has been performed on original human walking sequences, and begin with the extracted binary silhouette image sequences. The silhouette preprocessing includes size normalization (proportionally resizing each silhouette image so that all silhouettes have the same height) and horizontal alignment (centering the upper half silhouette part with respect to its horizontal centroid). In a preprocessed silhouette sequence, the time series signal of lower half silhouette part size from each frame indicates the gait frequency and phase information. We estimate the gait frequency and phase by maximum entropy spectrum estimation [317] from the time series signal.

Gait Representation

Given the preprocessed binary gait silhouette image $B_t(x, y)$ at time t in a sequence, the gray-level gait energy image (GEI) is defined as follows [301]:

$$G(x, y) = \frac{1}{N} \sum_{t=1}^{N} B_t(x, y) \tag{11.9}$$

where N is the number of frames in the complete cycle(s) of a silhouette sequence, t is the frame number of the sequence (moment of time), and x and y are values in the 2D image coordinate. Figure 11.8 shows the sample silhouette images in a gait cycle from two people and the right most images are the corresponding GEIs. As expected, GEI reflects major shapes of silhouettes and their changes over the gait cycle. It accounts for human walking at different speeds. It is referred as the gait energy image because (a) each silhouette image is the space-normalized energy image of human walking at this moment, (b) GEI is the time-normalized accumulative energy image of human walking in the complete cycle(s), and (c) a pixel with higher intensity value in GEI means that human walking occurs more frequently at this position (i.e., with higher energy).

Fig. 11.8. Two examples of normalized and aligned silhouette images. The silhouette images shown in a row are in a gait cycle. The right most images are the corresponding gait energy images (GEIs)

GEI has several advantages over the gait representation of binary silhouette sequence. GEI is not sensitive to incidental silhouette errors in individual frames. Moreover, with such a 2D template, we do not need to consider the normalized time moment of each frame, and the incurred errors can be therefore avoided.

Direct GEI Matching

Individuals are recognized by measuring the similarity between the gallery (training) and probe (testing) templates. Given GEIs of two gait sequences, $G_g(x, y)$ and $G_p(x, y)$, their distance can be measured by calculating their normalized matching error

$$D(G_g, G_p) = \frac{\sum_{x,y} |G_g(x,y) - G_p(x,y)|}{\sqrt{\sum_{x,y} G_g(x,y) \sum_{x,y} G_p(x,y)}}, \quad (11.10)$$

where $\sum_{x,y} |G_g(x,y) - G_p(x,y)|$ is the matching error between two GEIs, $\sum_{x,y} G_g(x,y)$ and $\sum_{x,y} G_p(x,y)$ are total energy in two GEIs, respectively. The unknown person is classified to the class which gets the smallest distance (matching error) out of all the distances (matching errors) corresponding to all the classes.

11.2.4 Integrating Face Profile and Gait for Recognition at a Distance

Face profile cues and gait cues are integrated by three schemes. Commonly used classifier combination schemes [302] are obtained based on Bayesian theory, where the representations are assumed to be conditionally statistically independent. We employ *Sum* rule and *Product* rule in our fusion system, with which the similarity scores obtained individually from the face profile classifier and the gait classifier are combined. Before combination of the results of face profile classifier and the results of gait classifier, it is necessary to map distances obtained from different classifiers to the same range of values. We use exponential transformation here. Given that the distance for a probe X are $S_1, S_2, \ldots, S_c$, we obtain the normalized match scores as

$$\hat{S}_i = \frac{exp(-S_i)}{\sum_{i=1}^{c} exp(-S_i)} \quad i = 1, 2, \ldots, c \qquad (11.11)$$

After normalization, the match scores of face and gait from the same class are fused using different fusion methods. Let $\hat{S}_i^F$ and $\hat{S}_i^G$ be the normalized face match scores and the normalized gait match scores, respectively. The unknown person is classified to class k if

$$R\{\hat{S}_k^F, \hat{S}_k^G\} = \max R\{\hat{S}_i^F, \hat{S}_i^G\} \qquad (11.12)$$

where $R\{,\}$ means a fusion method. *Sum* and *Product* rules [302] are used in our experiments. Distances representing dissimilarity become match scores representing similarity by using (11.11), so the unknown person is classified to the class which gets the largest integrated match score out of all the integrated match scores corresponding to all the classes.

The last one is an indexing-verification scheme. In a biometric fusion system, a less accurate, but fast and simple classifier can pass on a smaller set of candidates to a more accurate, but time-consuming and complicated classifier. In our system, the face profile classifier works as a filter to pass on a smaller set of candidates to the next stage of gait classifier. Then, the gait classifier compares similarity among these candidates based on GEIs. The result of the gait classifier is the result of the fusion system.

11.3 Experimental Results

11.3.1 Data

The data are obtained by a Sony DCR-VX1000 digital video camera recorder. We collect 28 video sequences of 14 people walking in the outdoor conditions and exposing a side view to the camera. The camera operates at about 30 frames s^{-1}. The resolution of each frame is 720 × 480. The distance between people and the video camera is about 10 ft. Each of people has two sequences, one for training and the other one for testing. Each sequence includes one person. Figure 11.9 shows some video frames of four people.

11.3.2 Experiments

From each sequence, we construct one high-resolution face profile image from six low-resolution face profile images that are extracted from six adjacent video frames, and one GEI from a complete walking cycle that includes about 20 video frames. Since there are two sequences for each of 14 people, we have 14 high-resolution face profile images and 14 GEIs in the gallery, and another 14 high-resolution face profile images and 14 GEIs in the probe. The resolution of low-resolution face profile images is 70 × 70 and the resolution of reconstructed

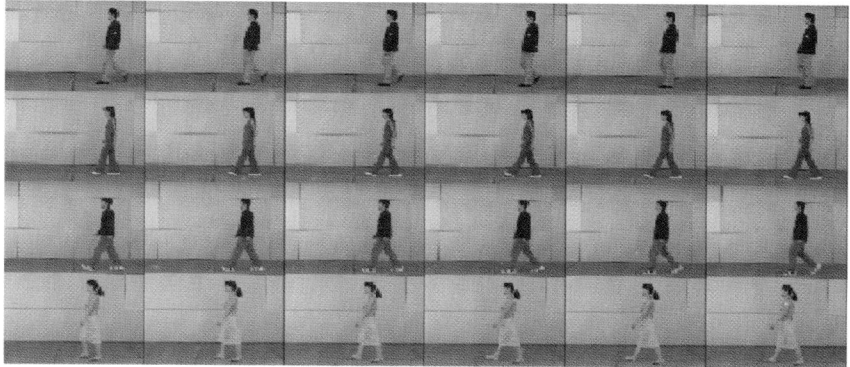

Fig. 11.9. Four examples of video sequences

Table 11.1. Experimental results

combination scheme	recognition rate		
	gait (%)	face profile (%)	integration (%)
no combination	85.7	64.3	
Sum rule			100
Product rule			92.9
indexing-verification			92.9

high-resolution face profile images is 140×140. The resolution of each GEI is 128×88.

Recognition metric is used to evaluate the performance of our method, the quality of extracted features, and their impact on identification. It is defined as the ratio of the number of the correctly recognized people to the number of all the people. The results for our database are shown in Table 11.1. We can see that 64.3% people are correctly recognized (5 errors out of 14 persons) by face profile and 85.7% people are correctly recognized by gait (2 errors out of 14 persons), respectively. For the fusion schemes, the best performance is achieved by the *Sum* rule at 100% accuracy. The *Product* rule and the indexing-verification scheme obtain the same recognition rate at 92.9%. When we use the indexing-verification scheme, we choose the first three matching results of the face profile classifier as candidates. Then, the gait classifier measures the similarity between the corresponding GEI of the testing people and the corresponding GEI of the training people in the candidate list. The unknown person is finally classified as the most similar class among the candidates.

From Table 11.1, we can see that there are two people who are not correctly recognized by gait, but when the face profile classifier is integrated, the recognition rate is improved. It is because gait recognition based on GEI is not only affected by the walking style of a person, but also by the shape of a human body. Environmental and clothing changes cause the difference in the shape of the training sequence and the testing sequence for the same person. However, the face profiles of these two people do not change so much in the training and the testing sequences. It shows that face profile is a useful cue for the fusion system. Figure 11.10 shows the corresponding GEIs of two people who are misclassified by the gait classifier. Figure 11.11 shows the corresponding face profiles of two people who are misclassified by the gait classifier. Note the difference in the training and testing GEIs in Fig. 11.10 and

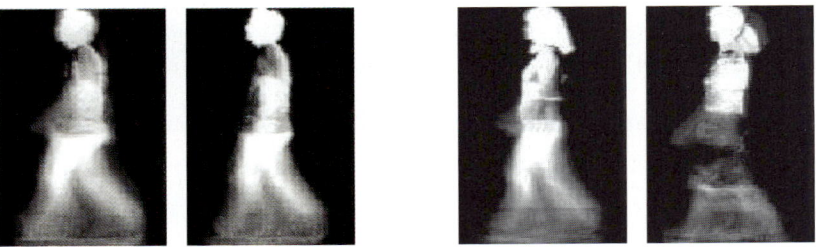

Fig. 11.10. GEIs of two people misclassified by the gait classifier. For each person, the training GEI and the testing GEI are shown for comparison

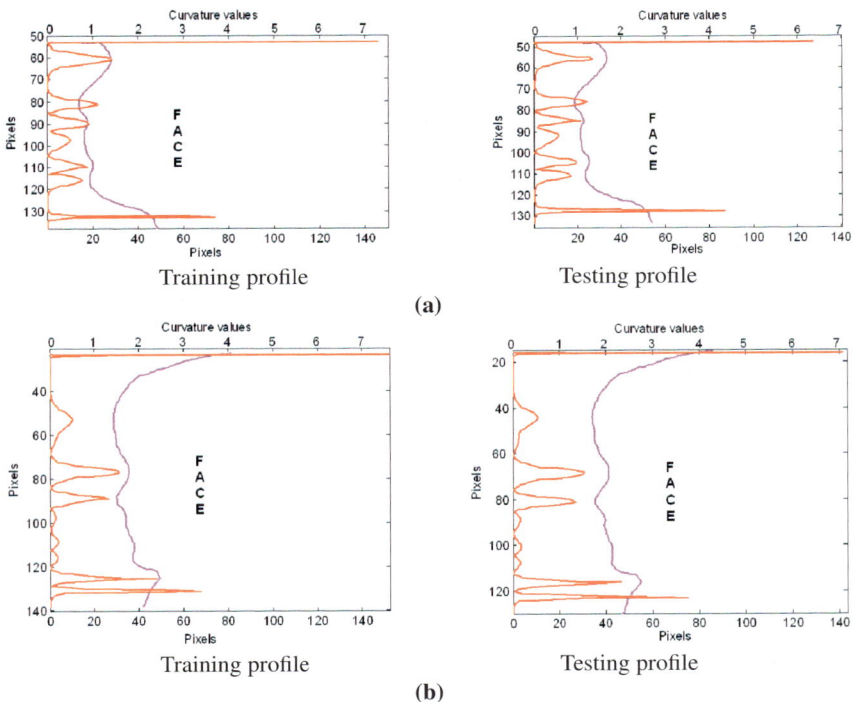

Fig. 11.11. Face profile of two people misclassified by the gait classifier. For each person, the training profile and the testing profile are shown for comparison

the similarity in the training and testing face profiles in Fig. 11.11. Since the face profile classifier is comparatively sensitive to the variation of facial expression and noise, the face profile classifier cannot get a good recognition rate by itself. When the gait classifier is combined with the face profile classifier, the better performance is achieved.

From the experiments, we can see that the fusion system using face profile and gait is promising. The fusion system has better performance than either of the individual classifier. It shows that our fusion system is relatively robust. Although the experiments are only done on a small database, our system has the potential since it integrates cues of face profile and cues of gait reasonably, which are independent biometrics.

11.4 Conclusions

This chapter introduces a video-based system combining face profile and gait for human recognition in a single camera scenario. For optimal face profile recognition, we extract face profile features from a high-resolution face profile image constructed from multiple video frames instead of a low-resolution face profile images directly obtained from a single video frame. For gait recognition, we use GEI, a spatiotemporal compact representation to characterize human walking properties. Serval schemes are considered for fusion of face profile and gait. The experimental results show that the integration of information from face profile and gait is effective for individual recognition in video. The performance improvement is archived when appropriate fusion rules are used. The idea of constructing the high-resolution face profile image from multiple video frames and generating the GEI is promising for human recognition in video.

Several issues that concern real-world applications require further research in the future. These include the extraction of accurate face profile from video frames in a crowded surveillance application, extraction of reliable silhouettes of moving people in the presence of environmental and clothing changes, and real-time operation of the fusion system. Moreover, for face profile recognition, the outer contour of the side face is sensitive to local distortion and noise. In our recent work [322, 323], we have used the side face, which includes entire side views of eye, nose, and mouth (discarding facial hair). Since it possesses both the shape and the intensity information, it is found to have more discriminating power for recognition than a face profile.

Part IV

Generic Approaches to Multibiometric Systems

12 Fusion Techniques in Multibiometric Systems

Arun Ross and Anil K. Jain

12.1 Introduction

Biometrics is the science of establishing the identity of an individual based on the inherent physical or behavioral traits associated with the person [324–326]. Biometric systems utilize fingerprints, iris, face, hand geometry, palmprint, finger vein structure, gait, voice, signature, keyboard typing pattern, etc. in order to recognize a person (Fig. 12.1). A typical biometric system operates by capturing the biometric trait of a person via an appropriately designed acquisition module and comparing the recorded trait with the biometric samples (or templates) in a database in order to determine the identity of the person (*identification*) or to validate a claimed identity (*verification*). For example, a face biometric system captures the face image of an individual, extracts a feature set from the segmented face, compares this feature set against the templates stored in the database and renders a decision regarding the identity of the individual. Thus, a generic biometric system may be viewed as a pattern recognition system in which the raw biometric data (or signal) constitutes the input pattern that is assigned a class label [327]. In an identification system, the class label pertains to the identity of the individual while in a verification system the class label is a match (genuine) or a nonmatch (impostor). In both modes of operation, a *reject* label is emitted when the system is unable to determine a valid class.

A generic biometric system has four important modules (a) the sensor module which captures the biometric trait[1] in the form of raw *data*; (b) the feature extraction module which processes the data to extract a *feature set* that is a compact representation of the trait; (c) the matching module which employs a matcher or a classifier to compare the extracted feature set with the *templates* residing in the database to generate match scores; and (d) the decision module which uses the matching scores to either determine an identity or validate a claimed identity.

The need for establishing identity in a reliable manner has spurred active research in the field of biometrics [328]. The deployment of biometric systems in border security programs (e.g., US-VISIT[2]), criminal investigations (e.g., IAFIS[3]),

---

[1] The terms biometric trait, biometric modality, and biometric indicator are used interchangeably in this chapter.
[2] United States Visitor and Immigrant Status Indicator Technology.
[3] Integrated Automated Fingerprint Identification System.

186 A. Ross and A.K. Jain

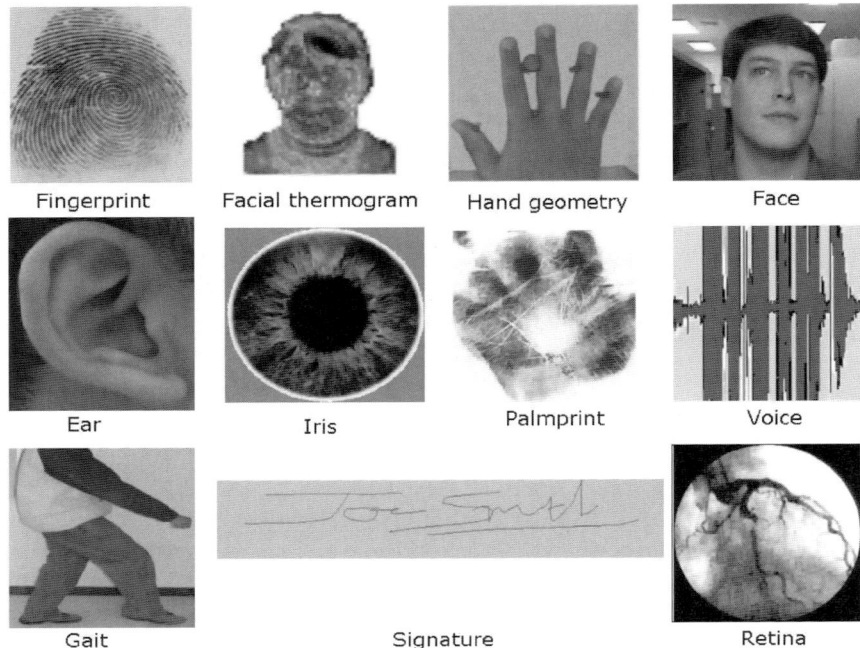

Fig. 12.1. Examples of biometric traits that can be used for authenticating an individual's identity

logical access points (e.g., computer login) and surveillance applications (e.g., face recognition in public spaces) further underscores the importance of designing and implementing large-scale authentication systems that can consistently render the correct decision under various operational scenarios. Furthermore, as the number of enrolled subjects increases over time, it is imperative that the matching accuracy of these systems is not compromised. Indeed, the problem of biometric recognition may be viewed as a *Grand Challenge*, given the expectations of high matching accuracy, ease of usability, and efficient scalability in a variety of applications accessed by different segments of the general population [329].

Most biometric systems that are presently in use, typically use a single biometric trait to establish identity (i.e., they are unibiometric systems). For example, the Schiphol Privium scheme at Amsterdam's Schipol airport employs iris scan smart cards to speed up the immigration process; the Ben Gurion International Airport at Tel Aviv employs automated hand geometry-based identification kiosks to enable Israeli citizens and frequent international travelers to rapidly negotiate the passport inspection process; some financial institutions in Japan have installed palm-vein authentication systems in their ATMs to validate the identity of a customer conducting a transaction; in Disney World, Orlando, the fingerprint information of season pass holders is used to ensure that multiple individuals do not fraudulently use a single pass; customers phoning in to schedule product shipments through Union Pacific's railcar system are authenticated by a speaker recognition software. With the proliferation of biometric-based solutions in civilian and law enforcement applications, it is important that the vulnerabilities and limitations of these systems are clearly understood. Some of the challenges commonly encountered by biometric systems are listed below.

1. *Noise in sensed data.* The biometric data being presented to the system may be contaminated by noise due to imperfect acquisition conditions or subtle variations in the biometric itself. For example, a scar can change a subject's fingerprint while the common cold can alter the voice characteristics of a speaker. Similarly, unfavorable illumination conditions may significantly affect the face and iris images acquired from an individual. Noisy data can result in an individual being incorrectly labeled as an impostor thereby increasing the false reject rate (FRR) of the system.
2. *Non-universality.* The biometric system may not be able to acquire meaningful biometric data from a subset of individuals resulting in a failure-to-enroll (FTE) error. For example, a fingerprint system may fail to image the friction ridge structure of some individuals due to the poor quality of their fingerprints. Similarly, an iris recognition system may be unable to obtain the iris information of a subject with long eyelashes, drooping eyelids or certain pathological conditions of the eye.[4] Exception processing will be necessary in order to accommodate such users into the authentication system.
3. *Upper bound on identification accuracy.* The matching performance of a unibiometric system cannot be continuously improved by tuning the feature extraction and matching modules. There is an implicit upper bound on the number of distinguishable patterns (i.e., the number of distinct biometric feature sets) that can be represented using a template. The capacity of a template is constrained by the variations observed in the feature set of each subject (i.e., intraclass variations) and the variations between feature sets of different subjects (i.e., interclass variations). Table 12.1 lists the error rates associated with four biometric modalities – fingerprints, face, voice, iris – as suggested by recent public tests. These statistics suggest that there is a tremendous scope for performance improvement especially in the context of large-scale authentication systems (also see [330]).
4. *Spoof attacks.* Behavioral traits such as voice [331] and signature [332] are vulnerable to spoof attacks by an impostor attempting to mimic the traits corresponding to legitimately enrolled subjects. Physical traits such as fingerprints can also be spoofed by inscribing ridge-like structures on synthetic material such as gelatine and play-doh [333, 334]. Targeted spoof attacks can undermine the security afforded by the biometric system and, consequently, mitigate its benefits [335].

Some of the limitations of a unibiometric system can be addressed by designing a system that consolidates *multiple* sources of biometric information. This can be

---

[4] http://news.bbc.co.uk/2/hi/uk_news/politics/3693375.stm.

Table 12.1. The false accept and false reject error rates (FAR and FRR) associated with the fingerprint, face, voice, and iris modalities

biometric trait	test	test conditions	false reject rate (%)	false accept rate (%)
fingerprint	FVC 2004 [339]	exaggerated skin distortion, rotation	2	2
fingerprint	FpVTE 2003 [340]	US Government operational data	0.1	1
face	FRVT 2002 [341]	varied lighting, outdoor/indoor, time	10	1
voice	NIST 2004 [342]	text independent, multilingual	5–10	2–5
iris	ITIRT 2005 [343]	indoor environment, multiple visits	0.99	0.94

The accuracy estimates of biometric systems depend on a number of test conditions including the sensor employed, acquisition protocol used, subject disposition, number of subjects, number of biometric samples per subject, demographic profile of test subjects, subject habituation, time lapse between data acquisition, etc.

accomplished by fusing, for example, multiple traits of an individual, or multiple feature extraction and matching algorithms operating on the same biometric. Such systems, known as multibiometric systems [336–338], can improve the matching accuracy of a biometric system while increasing population coverage and deterring spoof attacks. In this chapter, the various sources of biometric information that can be fused as well as the different levels of fusion that are possible are discussed.

12.2 Multibiometric Systems

Evidence accumulation and information fusion is an active area of research in several different fields including weather forecasting [344], UAV coordination [345], object tracking [346], robot navigation [347], and land-mine detection [348]. The notion of problem solving by combining the decisions rendered by multiple "experts" (or algorithms) in a cooperative framework has received substantial attention in the literature [349–358]. Indeed, information fusion has a long history and the theory of multiple classifier systems (MCS) has been rigorously studied [359–362].

In the realm of biometrics, the consolidation of evidence presented by multiple biometric sources is an effective way of enhancing the recognition accuracy of an authentication system. For example, the Integrated Automated Fingerprint Identification System (IAFIS) maintained by the FBI integrates the information presented by multiple fingers to determine a match in the master file. Some of the earliest *multimodal*[5] biometric systems reported in the literature combined the face

---

[5] See Sect. 12.3 for a description of the terminology.

(image/video) and voice (audio) traits of individuals [363, 364]. Besides enhancing matching accuracy, the other advantages of multibiometric systems over traditional unibiometric systems are enumerated below [338].

1. Multibiometric systems address the issue of nonuniversality (i.e., limited population coverage) encountered by unibiometric systems. If a subject's dry finger prevents her from successfully enrolling into a fingerprint system, then the availability of another biometric trait, say iris, can aid in the inclusion of the individual in the biometric system. A certain degree of flexibility is achieved when a user enrolls into the system using several different traits (e.g., face, voice, fingerprint, iris, hand) while only a subset of these traits (e.g., face and voice) is requested during authentication based on the nature of the application under consideration and the convenience of the user.
2. Multibiometric systems can facilitate the filtering or indexing of large-scale biometric databases. For example, in a bimodal system consisting of face and fingerprint, the face feature set may be used to compute an index value for extracting a candidate list of potential identities from a large database of subjects. The fingerprint modality can then determine the final identity from this limited candidate list.
3. It becomes increasingly difficult (if not impossible) for an impostor to spoof multiple biometric traits of a legitimately enrolled individual. If each subsystem indicates the probability that a particular trait is a "spoof," then appropriate fusion schemes can be employed to determine if the user, in fact, is an impostor. Furthermore, by asking the user to present a random subset of traits at the point of acquisition, a multibiometric system facilitates a challenge–response type of mechanism, thereby ensuring that the system is interacting with a *live* user. Note that a challenge–response mechanism can be initiated in unibiometric systems also (e.g., system prompts "Please say 1-2-5-7," "Blink twice and move your eyes to the right," "Change your facial expression by smiling," etc.).
4. Multibiometric systems also effectively address the problem of noisy data. When the biometric signal acquired from a single trait is corrupted with noise, the availability of other (less noisy) traits may aid in the reliable determination of identity. Some systems take into account the *quality* of the individual biometric signals during the fusion process. This is especially important when recognition has to take place in adverse conditions where certain biometric traits cannot be reliably extracted. For example, in the presence of ambient acoustic noise, when an individual's voice characteristics cannot be accurately measured, the facial characteristics may be used by the multibiometric system to perform authentication. Estimating the quality of the acquired data is in itself a challenging problem but, when appropriately done, can reap significant benefits in a multibiometric system.
5. These systems also help in the *continuous* monitoring or tracking of an individual in situations when a single trait is not sufficient. Consider a biometric

system that uses a 2D camera to procure the face and gait information of a person walking down a crowded aisle. Depending upon the distance and pose of the subject with respect to the camera, both these characteristics may or may not be simultaneously available. Therefore, either (or both) of these traits can be used depending upon the location of the individual with respect to the acquisition system thereby permitting the continuous monitoring of the individual.
6. A multibiometric system may also be viewed as a fault tolerant system which continues to operate even when certain biometric sources become unreliable due to sensor or software malfunction, or deliberate user manipulation. The notion of fault tolerance is especially useful in large-scale authentication systems involving a large number of subjects (such as a border control application).

The design of a multibiometric system is defined by several different factors including (a) the human–computer interface (HCI) used to acquire biometric information from an individual[6]; (b) the tradeoff between the additional cost incurred in introducing multiple biometric sources and the perceived improvement in matching accuracy; (c) the sources of biometric information used to provide evidence; (d) the level of fusion, i.e., the type of information to be fused; and (e) the fusion methodology adopted.

12.3 Taxonomy of Multibiometric Systems

A multibiometric system relies on the evidence presented by multiple sources of biometric information. Based on the nature of these sources, a multibiometric system can be classified into one of the following six categories [338]: multisensor, multialgorithm, multi-instance, multisample, multimodal, and hybrid.

1. *Multisensor systems.* Multisensor systems employ multiple sensors to capture a single biometric trait of an individual. For example, a face recognition system may deploy multiple 2D cameras to acquire the face image of a subject [365]; an infrared sensor may be used in conjunction with a visible-light sensor to acquire the subsurface information of a person's face [366–368]; a multispectral camera may be used to acquire images of the iris, face or finger [369, 370]; or an optical as well as a capacitive sensor may be used to image the fingerprint of a subject [371]. The use of multiple sensors, in some instances, can result in the acquisition of complementary information that can enhance the recognition ability of the system. For example, based on the nature of illumination due to ambient lighting, the infrared and visible-light images of a person's face can present different levels of information resulting in enhanced matching accuracy. Similarly, the performance of a 2D face matching system can be improved by utilizing the shape information presented by 3D range images.
2. *Multialgorithm systems.* In some cases, invoking multiple feature extraction and/or matching algorithms on the same biometric data can result in improved

---

[6] This is an important consideration for multimodal and multiunit systems (see Sect. 12.3).

matching performance. Multialgorithm systems consolidate the output of multiple feature extraction algorithms, or that of multiple matchers operating on the same feature set. These systems do not necessitate the deployment of new sensors and, hence, are cost-effective compared to other types of multibiometric systems. But on the other hand, the introduction of new feature extraction and matching modules can increase the computational complexity of these systems. Ross et al. [372] describe a fingerprint recognition system that utilizes minutiae as well as texture information to represent and match fingerprint images. The inclusion of the texture-based algorithm introduces additional processing time associated with the application of Gabor filters on the input fingerprint image. However, the performance of the hybrid matcher is shown to exceed that of the individual matchers. Lu et al. [373] discuss a face recognition system that combines three different feature extraction schemes (principal component analysis (PCA), independent component analysis (ICA), and linear discriminant analysis (LDA)). The authors postulate that the use of different feature sets makes the system robust to a variety of intraclass variations normally associated with the face biometric. Experimental results indicate that combining multiple face classifiers can enhance the identification rate of the biometric system.

3. *Multi-instance systems.* These systems use multiple instances of the same body trait and have also been referred to as multiunit systems in the literature. For example, the left and right index fingers, or the left and right irises of an individual, may be used to verify an individual's identity [374, 375]. The US-VISIT border security program presently uses the left- and right-index fingers of visitors to validate their travel documents at the port of entry. FBI's IAFIS combines the evidence of all ten fingers to determine a matching identity in the database. These systems can be cost-effective if a single sensor is used to acquire the multiunit data in a sequential fashion (e.g., US-VISIT). However, in some instances, it may be desirable to obtain the multiunit data simultaneously (e.g., IAFIS) thereby demanding the design of an effective (and possibly more expensive) acquisition device.

4. *Multisample systems.* A single sensor may be used to acquire multiple samples of the same biometric trait in order to account for the variations that can occur in the trait, or to obtain a more complete representation of the underlying trait. A face system, for example, may capture (and store) the frontal profile of a person's face along with the left and right profiles in order to account for variations in the facial pose. Similarly, a fingerprint system equipped with a small size sensor may acquire multiple dab prints of an individual's finger in order to obtain images of various regions of the fingerprint. A mosaicing scheme may then be used to stitch the multiple impressions and create a composite image. One of the key issues in a multisample system is determining the *number* of samples that have to be acquired from an individual. It is important that the procured samples represent the *variability* as well as the *typicality* of the individual's biometric data. To this end, the desired relationship between the samples has to be established before-hand in order to optimize the benefits of the integration

strategy. For example, a face recognition system utilizing both the frontal- and side-profile images of an individual may stipulate that the side-profile image should be a three-quarter view of the face [376, 377]. Alternately, given a set of biometric samples, the system should be able to automatically select the "optimal" subset that would best represent the individual's variability. Uludag et al. [378] discuss two such schemes in the context of fingerprint recognition. The first method, called DEND, employs a clustering strategy to choose a template set that best represents the intraclass variations, while the second method, called MDIST, selects templates that exhibit maximum similarity with the rest of the impressions.

5. *Multimodal systems.* Multimodal systems establish identity based on the evidence of multiple biometric traits. For example, some of the earliest multimodal biometric systems utilized face and voice features to establish the identity of an individual [364, 379, 380]. Physically uncorrelated traits (e.g., fingerprint and iris) are expected to result in better *improvement* in performance than correlated traits (e.g., voice and lip movement). The cost of deploying these systems is substantially more due to the requirement of new sensors and, consequently, the development of appropriate user interfaces. The identification accuracy can be significantly improved by utilizing an increasing number of traits although the *curse-of-dimensionality* phenomenon would impose a bound on this number. The curse-of-dimensionality limits the number of attributes (or features) used in a pattern classification system when only a small number of training samples is available [327]. The number of traits used in a specific application will also be restricted by practical considerations such as the cost of deployment, enrollment time, throughput time, expected error rate, user habituation issues, etc.

6. *Hybrid systems.* Chang et al. [381] use the term *hybrid* to describe systems that integrate a subset of the five scenarios discussed above. For example, Brunelli et al. [364] discuss an arrangement in which two speaker recognition algorithms are combined with three face recognition algorithms at the match score and rank levels via a HyperBF network. Thus, the system is multialgorithmic as well as multimodal in its design. Similarly, the NIST BSSR1 dataset [382] has match scores pertaining to two different face matchers operating on the frontal face image of an individual (multialgorithm), and a fingerprint matcher operating on the left- and right-index fingers of the same individual (multi-instance).

It is also possible to combine biometric information with nonbiometric entities such as tokens in order to enhance the matching performance. For example, [383] discuss a dual factor authenticator that combines a pseudorandom number (present in a token) with a facial feature set in order to produce a set of user-specific compact codes known as BioCode. The pseudorandom number and the facial feature sets are fixed in length and an iterated inner product is used to generate the BioCode. When an individual's biometric information is suspected to be compromised, then the token containing the random data is replaced, thereby revoking the previous authenticator. The use of biometric and nonbiometric authenticators in tandem is a powerful way of enhancing security. However, some of the inconveniences associated with traditional authenticators remain (such as "Where did I leave my token?").

Another category of multibiometric systems combine primary biometric identifiers (such as face and fingerprint) with soft biometric attributes (such as gender, height, weight, eye color, etc.). Soft biometric traits cannot be used to distinguish individuals reliably since the same attribute is likely to be shared by several different people in the target population. However, when used in conjunction with primary biometric traits, the performance of the authentication system can be significantly enhanced [384]. Soft biometric attributes also help in filtering (or indexing) large biometric databases by limiting the number of entries to be searched in the database. For example, if it is determined (automatically or manually) that the subject is an "Asian Male," then the system can constrain its search to only those identities in the database labeled with these attributes. Alternately, soft biometric traits can be used in surveillance applications to decide if at all primary biometric information has to be acquired from a certain individual. Automated techniques to estimate soft biometric characteristics is an ongoing area of research and is likely to benefit law enforcement and border control biometric applications.

12.4 Levels of Fusion

In a biometric system, the amount of available information gets compressed as one progresses along the various modules of the system. Based on the type of information available in a certain module, different levels of fusion can be defined. Sanderson and Paliwal [385] categorize the various levels of fusion into two broad categories: preclassification or fusion *before* matching and postclassification or fusion *after* matching (see Fig. 12.2). Such a categorization is necessary since the amount of information available for fusion reduces drastically once the matcher has been invoked. Preclassification fusion schemes typically require the development of new matching techniques (since the matchers used by the individual sources may no longer be relevant) thereby introducing additional challenges. Preclassification schemes include fusion at the sensor (or raw data) and the feature levels while postclassification schemes include fusion at the match score, rank, and decision levels. A brief description of each of these fusion levels is presented in this section.

12.4.1 Sensor-Level Fusion

The raw biometric data (e.g., a face image) acquired from an individual represents the richest source of information although it is expected to be contaminated by noise (e.g., nonuniform illumination, background clutter, etc.). Sensor-level fusion refers to the consolidation of (a) raw data obtained using multiple sensors or (b) multiple snapshots of a biometric using a single sensor. Mosaicing multiple impressions of the same finger is a good example of fusion at this level. Jain and Ross [386] discuss a mosaicing scheme that creates a composite fingerprint image from the evidence presented by multiple dab prints. The algorithm uses the minutiae points

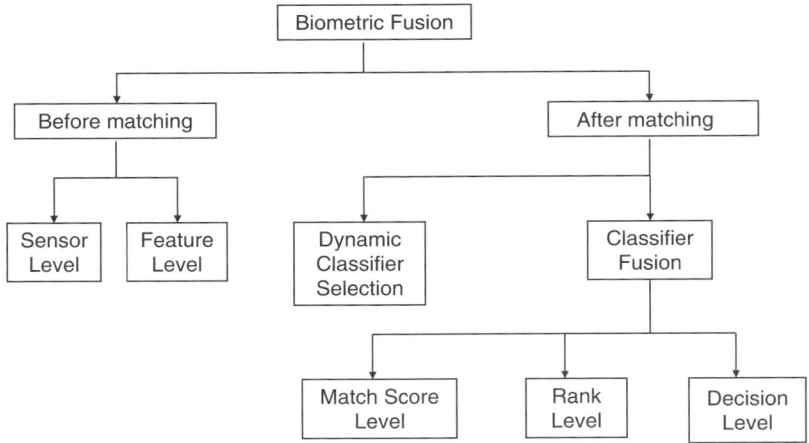

Fig. 12.2. Fusion can be accomplished at various levels in a biometric system. Most multibiometric systems fuse information at the match score level or the decision level. More recently researchers have begun to fuse information at the sensor and feature levels. In biometric systems operating in the identification mode, fusion can be done at the rank level

to first approximately register the two images using a simple affine transformation. The iterative closest point (ICP) algorithm is then used to register the ridge information corresponding to the two images after applying a low-pass filter to the individual images and normalizing their histograms. The normalization ensures that the pixel intensities of the individual dab prints are comparable. Blending is accomplished by merely concatenating the two registered images. The performance using the mosaiced image templates was shown to exceed that of the individual dab print templates.

Ratha et al. [387] describe a mosaicing scheme to integrate multiple snapshots of a fingerprint as the user *rolls* the finger on the surface of the sensor. Thus, a specific temporal order is imposed on the image frames when constructing the composite image. The authors investigate five different blending algorithms to construct a composite mosaiced image from the individual gray-scale images. They evaluate the efficacy of these five schemes by observing the size of the mosaiced print and its quality (in terms of the number of valid minutiae points detected).

Zhang et al. [388] employ a two-stage process to register the image slices obtained from a sweep-based fingerprint sensor. In the first stage, the minimum mean absolute error criterion is used to coarsely align a pair of image slices. In the next stage, a phase correlation scheme is used to refine the registration parameters. Choi et al. [389] propose the use of a novel enrollment scheme to capture a sequence of fingerprint images. Their scheme requires the finger to roll and slide horizontally on the sensing surface of the device. They use a block matching algorithm to account for local distortions between successive image pairs. A two-pass mesh-warping algorithm based on cubic splines is then used to fine align the images before stitching them into a single mosaiced entity. Ross et al. [390] use a thin-plate spline function to register two impressions of a subject's finger prior to blending them using a simple averaging scheme. Rather than operating on the gray-scale fingerprint image, the proposed mosaicing algorithm uses the enhanced fingerprint image to perform blending.

Mosaicing has also been attempted by researchers in face recognition where multiple 2D images representing different poses are stitched to generate a single image. Yang et al. [391] propose an algorithm to create panoramic face mosaics. Their acquisition system consists of five cameras that simultaneously obtain five different views of a subject's face. In order to determine the corresponding points in multiple face views, the authors place ten colored markers on the face. Based on these control points, their algorithm uses a sequence of fast linear transformations on component images to generate a face mosaic. Finally, a local smoothing process is carried out to smooth the mosaiced image. Two different schemes were used to represent the panoramic image: one in the spatial domain and the other in the frequency domain. The frequency domain representation resulted in an identification accuracy of 97.46% while the spatial domain representation provided 93.21% accuracy on a database of 12 individuals.

Liu and Chen [392] propose a face mosaicing technique that uses a statistical model to represent the mosaic. Given a sequence of face images captured under an orthographic camera model, each frame is unwrapped onto a certain portion of the surface of a sphere via a spherical projection. A minimization procedure using the Levenberg–Marquardt algorithm is employed to optimize the distance between an unwrapped image and the sphere. The representational (statistical) model comprises of a mean image and a number of eigen-images. The novelty of this technique is (a) the use of spherical projection, as opposed to cylindrical projection, which works better when there is head motion in both the horizontal and vertical directions and (b) the computation of a representational model using both the mean image and the eigen-images rather than a single template image. In [393], the authors propose another algorithm in which the human head is approximated with a 3D ellipsoidal model. The face, at a certain pose, is viewed as a 2D projection of this 3D ellipsoid. All 2D face images of a subject are projected onto this ellipsoid via geometrical mapping to form a texture map which is represented by an array of local patches. Matching is accomplished by adopting a probabilistic model to compute the distance of patches from an input face image. The authors report an identification accuracy of 90% on the CMU PIE database [394].

Singh et al. [395] present a face mosaicing technique that uses terrain transform to align multiple face images of a subject. Multiresolution splining is then used to blend the registered face images into a single entity. The authors state that the use of multiresolution splining during image mosaicing ensures that the biometric features of the face are not perturbed thereby protecting the salient biometric features in the mosaiced image. While 2D face mosaicing has been shown to enhance the matching accuracy of a face recognition system, more research is needed in this direction to establish the pros and cons of mosaicing over other fusion methods (such as score-level fusion, for example).

It is also possible to combine the 2D texture of a person's face with the corresponding 3D scan (i.e., the range image) in order to create a 3D texture. Two such 3D surfaces can be compared by first aligning them using landmark points, such as automatically detected high curvature points, and then comparing the texture associated with local patches. The local patches are usually defined using triangular meshes. Hsu [396] describes a face modeling algorithm that uses the 2D and 3D images of a person's face obtained during enrollment to modify a generic 3D face model and derive a user-specific 3D model. The generic 3D model is based on Waters' animation model [397] and contains 256 vertices and 441 triangular facets (for one-half of the face) that define various facial attributes. During the enrollment stage, the 2D and 3D images of a person's face are acquired using a Minolta Vivid 700 digitizer that generates a registered 200×200 range map and a 400×400 color image. A global alignment procedure is employed to approximately align the facial measurements of the user with the generic 3D model. A local alignment scheme is then invoked that perturbs features such as the eyes, nose, mouth, chin, and face boundary of the generic 3D model so that they fit the actual facial measurements of the individual. Next, a combination of displacement propagation and 2.5D active contours is used to smooth the face model and to refine the local features present in the model resulting in a user-specific 3D representation of the face. The availability of this model permits the generation of new (previously unseen) 2D images of a person's face (e.g., at different poses, illumination, head-tilt, etc.) without actually employing a scanner to capture such images. Hsu [396] uses this approach to compare 2D images of a person's face acquired during authentication with the user-specific 3D model residing in the template database.

12.4.2 Feature-Level Fusion

In feature-level fusion, the feature sets originating from multiple biometric algorithms are consolidated into a single feature set by the application of appropriate feature normalization, transformation, and reduction schemes. The primary benefit of feature-level fusion is the detection of correlated feature values generated by different biometric algorithms and, in the process, identifying a salient set of features that can improve recognition accuracy. Eliciting this feature set typically requires the use of dimensionality reduction methods [398, 399] and, therefore, feature-level fusion assumes the availability of a large number of training data. Also, the feature sets being fused are typically expected to reside in commensurate vector space in order to permit the application of a suitable matching technique upon consolidating the feature sets.

Feature-level fusion is challenging for the following reasons:

1. The relationship between the feature spaces of different biometric systems may not be known.
2. The feature sets of multiple modalities may be incompatible. For example, the minutiae set of fingerprints and the eigen-coefficients of face are irreconcilable. One is a variable length feature set (i.e., it varies across images) whose

individual values parameterize a minutia point; the other is a fixed length feature set (i.e., all images are represented by a fixed number of eigen-coefficients) whose individual values are scalar entities.

3. If the two feature sets are fixed length feature vectors, then one could consider concatenating them to generate a new feature set. However, concatenating two feature vectors might lead to the curse-of-dimensionality problem [400] where increasing the number of features might actually degrade the system performance especially in the presence of small number of training samples. Although the curse-of-dimensionality is a well-known problem in pattern recognition, it is particularly pronounced in biometric applications because of the time, effort and cost required to collect large amounts of biometric (training) data.

4. Most commercial biometric systems do not provide access to the feature sets used in their products. Hence, very few biometric researchers have focused on integration at the feature level and most of them generally prefer fusion schemes that use match scores or decision labels.

If the length of each of the two feature vectors to be consolidated is fixed across all users, then a feature concatenation scheme followed by a dimensionality reduction procedure may be adopted. Let $\mathbf{X} = \{x_1, x_2, \ldots, x_m\}$ and $\mathbf{Y} = \{y_1, y_2, \ldots, y_n\}$ denote two feature vectors ($\mathbf{X} \in R^m$ and $\mathbf{Y} \in R^n$) representing the information extracted from two different biometric sources. The objective is to fuse these two feature sets in order to yield a new feature vector, $\mathbf{Z}$, that would better represent an individual. The vector $\mathbf{Z}$ of dimensionality $k, k < (m + n)$, can be generated by first concatenating vectors $\mathbf{X}$ and $\mathbf{Y}$, and then performing feature selection or feature transformation on the resultant feature vector in order to reduce its dimensionality. The key stages of such an approach are described below (also see Fig. 12.3).

Feature Normalization

The individual feature values of vectors $\mathbf{X} = \{x_1, x_2, \ldots, x_m\}$ and $\mathbf{Y} = \{y_1, y_2, \ldots, y_n\}$ may exhibit significant differences in their range as well as form (i.e., distribution). Concatenating such diverse feature values will not be appropriate in many cases. For example, if the x_is are in the range $[0, 100]$ while the y_is are in the range $[0, 1]$, then the distance between two concatenated feature vectors will be more sensitive to the x_is than the y_is. The goal of feature normalization is to modify the location (mean) and scale (variance) of the features values via a transformation function in order to map them into a common domain. Adopting an appropriate normalization scheme also helps address the problem of outliers in feature values. While a variety of normalization schemes can be used, two simple schemes are discussed here: the min–max and median normalization schemes.

Let x and x' denote a feature value before and after normalization, respectively. The min–max technique computes x' as

$$x' = \frac{x - \min(F_x)}{\max(F_x) - \min(F_x)}, \qquad (12.1)$$

where F_x is the function which generates x, and $\min(F_x)$ and $\max(F_x)$ represent the minimum and maximum of all possible x values that will be observed, respectively. The min–max technique is effective when the minimum and the maximum values of the component feature values are known beforehand. In cases where such information is not available, an estimate of these parameters has to be obtained from the available set of training data. The estimate may be affected by the presence of outliers in the training data and this makes min–max normalization sensitive to outliers. The median normalization scheme, on the other hand, is relatively robust to the presence of noise in the training data. In this case, x' is computed as

$$x' = \frac{x - median(F_x)}{median(|\,(x - median(F_x))\,|)}. \tag{12.2}$$

The denominator is known as the median absolute deviation (MAD) and is an estimate of the scale parameter of the feature value. Although, this normalization scheme is relatively insensitive to outliers, it has a low efficiency compared to the mean and standard deviation estimators. Normalizing the feature values via any of these techniques results in modified feature vectors $\mathbf{X}' = \{x'_1, x'_2, \ldots x'_m\}$ and $\mathbf{Y}' = \{y'_1, y'_2, \ldots y'_n\}$. Feature normalization may not be necessary in cases where the feature values pertaining to multiple sources are already comparable.

Feature Selection or Transformation

Concatenating the two feature vectors, $\mathbf{X}'$ and $\mathbf{Y}'$, results in a new feature vector, $\mathbf{Z}' = \{x'_1, x'_2, \ldots x'_m, y'_1, y'_2, \ldots y'_n\}$, $\mathbf{Z}' \in R^{m+n}$. The curse-of-dimensionality dictates that the new vector of dimensionality $(m + n)$ need not necessarily result in an improved matching performance compared to that obtained by $\mathbf{X}'$ and $\mathbf{Y}'$ alone. The feature selection process is a dimensionality reduction scheme that entails choosing a minimal feature set of size $k, k < (m + n)$, such that a criterion (objective) function applied to the training set of feature vectors is optimized. There are several feature selection algorithms in the literature, and any one of these could be used to reduce the dimensionality of the feature set $\mathbf{Z}'$. Examples include sequential forward selection (SFS), sequential backward selection (SBS), sequential forward floating search (SFFS), sequential backward floating search (SBFS), "plus l take away r" and branch-and-bound search (see [399, 401] for details). Feature selection techniques rely on an appropriately formulated criterion function to elicit the optimal subset of features from a larger feature set. In the case of a biometric system, this criterion function could be the equal error rate (EER); the d-prime measure; the area of overlap between genuine and impostor training scores; the average GAR at predetermined FAR values in the ROC/DET curves corresponding to the training set; or the area under the ROC curve (AUC).

Dimensionality reduction may also be accomplished using feature *transformation* methods where the vector $\mathbf{Z}'$ is subjected to a linear or a nonlinear mapping that projects it to a lower dimensional subspace. Examples of such transformations include the use of PCA, ICA, multidimensional scaling (MDS), Kohonen Maps and neural networks [398]. The application of a feature selection or feature transformation procedure results in a new feature vector $\mathbf{Z} = \{z_1, z_2, \ldots z_k\}$ which can now be used to represent the identity of an individual.

Ross and Govindarajan [402] apply feature-level fusion to three different scenarios (a) multialgorithm, where two different face recognition algorithms based on PCA and LDA are combined; (b) multisensor, where three different color channels of a face image are independently subjected to LDA and then combined; and (c) multimodal, where the face and hand geometry feature vectors are combined. The general procedure adopted in [402] is summarized below.

1. Let $\{\mathbf{X}_i, \mathbf{Y}_i\}$ and $\{\mathbf{X}_j, \mathbf{Y}_j\}$ be the feature vectors obtained at two different time instances i and j. Here, $\mathbf{X}$ and $\mathbf{Y}$ represent the feature vectors derived from two different information sources. The corresponding fused feature vectors may be denoted as $\mathbf{Z}_i$ and $\mathbf{Z}_j$, respectively.
2. Let s_X and s_Y be the normalized match scores generated by comparing $\mathbf{X}_i$ with $\mathbf{X}_j$ and $\mathbf{Y}_i$ with $\mathbf{Y}_j$, respectively, and let $s_{match} = (s_X + s_Y)/2$ be the fused match score obtained using the simple sum rule.
3. A pair of fused feature vectors, $\mathbf{Z}_i$ and $\mathbf{Z}_j$, are then compared using two different distance measures: the Euclidean distance (s_{euc}) and the thresholded absolute distance (TAD) (s_{tad}). Thus,

$$s_{euc} = \sum_{r=1}^{k} (z_{i,r} - z_{j,r})^2 \qquad (12.3)$$

$$s_{tad} = \sum_{r=1}^{k} I(|z_{i,r} - z_{j,r}|, t). \qquad (12.4)$$

Here, $I(u, t) = 1$, if $u > t$ (and 0, otherwise), t is a prespecified threshold, and k is the dimensionality of the fused feature vector. The TAD measure determines the *number* of normalized feature values that differ by a magnitude greater than t. The s_{euc} and s_{tad} values are consolidated into one feature level score, s_{feat}, via the simple sum rule (Fig. 12.3). This retains information both at the match score level (s_{match}) as well as the feature level (s_{feat}).
4. Finally, the simple sum rule is used to combine s_{match} and s_{feat} in order to obtain the final score s_{tot}.

The authors compare the matching performances obtained using s_{match} and s_{tot} in all three scenarios. Results indicate that feature level fusion is advantageous in some cases. The feature selection scheme ensures that redundant or correlated feature values are detected and removed before invoking the matcher. This is probably one of the key benefits of performing fusion at the feature level [403]. Therefore, it is important that vendors of biometric systems grant access to feature level information to permit the development of effective fusion strategies.

Chibelushi et al. [379] discuss a scheme to combine the features associated with the voice (audio) and lip shape (video) of an individual in an identification system.

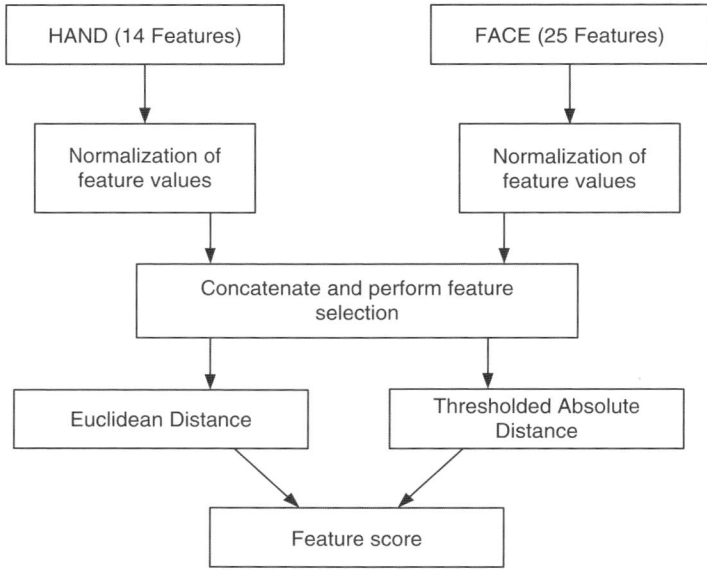

Fig. 12.3. The procedure adopted in [402] to perform feature level fusion

Fourteen mel-frequency cepstral coefficients (MFCC) and 12 geometric features are extracted from the audio and video streams to represent the voice and shape of the lips, respectively. The PCA and LDA transformations are used to reduce the dimensionality of the concatenated feature set. The authors demonstrate that the use of feature level fusion in their system is equivalent to increasing the signal-to-noise ratio (SNR) of the audio signal thereby justifying the use of lip shape in the fusion module. Other examples of feature level fusion can be found in [404] (face and iris) and [405] (hand geometry and palmprint).

12.4.3 Score-Level Fusion

A match score represents the result of comparing two feature sets extracted using the same feature extractor. A *similarity* score denotes how "similar" the two feature sets are, while a *distance* score denotes how "different" they are.[7]

In score-level fusion the match scores output by multiple biometric matchers are combined to generate a new match score (a scalar) that can be subsequently used by the verification or identification modules for rendering an identity decision. Fusion at this level is the most commonly discussed approach in the biometric literature primarily due to the ease of accessing and processing match scores (compared to the raw biometric data or the feature set extracted from the data). Fusion methods at this

---

[7] Consequently, a high similarity score between a pair of feature sets indicates a good match whereas a high distance score indicates a poor match.

level can be broadly classified into three categories [338]: density-based schemes, transformation-based schemes, and classifier-based schemes.

Density-Based Fusion Schemes

Let $\mathbf{s} = [s_1, s_2, \ldots, s_R]$ denote the scores emitted by multiple matchers, with s_j representing the match score of the jth matcher, $j = 1, \ldots, R$. Further, let the labels ω_0 and ω_1 denote the genuine and impostor classes, respectively. Then, by Bayes decision theory [327], the probability of error can be minimized by adopting the following decision rule.[8]

Assign $\mathbf{s} \to \omega_i$ if

$$P(\omega_i \,|\, \mathbf{s}) > P(\omega_j \,|\, \mathbf{s}), i \neq j, \quad and \quad i, j = 0, 1. \tag{12.5}$$

Here, the *a posteriori* probability $P(\omega_i \,|\, \mathbf{s}), i = 0, 1$, can be derived from the class-conditional density function $p(\mathbf{s} \,|\, \omega_i)$ using the Bayes formula, i.e.,

$$P(\omega_i \,|\, \mathbf{s}) = \frac{p(\mathbf{s} \,|\, \omega_i) P(\omega_i)}{p(\mathbf{s})}, \tag{12.6}$$

where $P(\omega_i)$ is the *a priori* probability of observing class ω_i and $p(\mathbf{s})$ denotes the probability of encountering $\mathbf{s}$. Thus, (12.5) can be re-written as

Assign $\mathbf{s} \to \omega_i$ if

$$\frac{p(\mathbf{s} \,|\, \omega_i)}{p(\mathbf{s} \,|\, \omega_j)} > \tau, i \neq j, \quad and \quad i, j = 0, 1 \tag{12.7}$$

where $\frac{p(\mathbf{s} \,|\, \omega_i)}{p(\mathbf{s} \,|\, \omega_j)}$ is known as the *likelihood ratio* and $\tau = \frac{P(\omega_j)}{P(\omega_i)}$ is a predetermined threshold. The density $p(\mathbf{s} \,|\, \omega_i)$ is typically estimated from a training set of match score vectors, using parametric or nonparametric techniques [406]. However, a large number of training samples is necessary to reliably estimate the joint-density function $p(\mathbf{s} \,|\, \omega_i)$ especially if the dimensionality of the feature vector $\mathbf{s}$ is large. In the absence of sufficient number of training samples (which is typically the case when the multibiometric system is first deployed or if its parameters are subsequently adjusted), it is commonly assumed that the scalar scores $s_i, s_2, \ldots s_R$ are generated by R independent random processes. This assumption permits the density function to be expressed as

$$p(\mathbf{s} \,|\, \omega_i) = \prod_{j=1}^{R} p(s_j \,|\, \omega_i), \tag{12.8}$$

where the joint-density function is now replaced by the product of its marginals. The marginal densities, $p(s_j \,|\, \omega_i)$, $j = 1, 2, \ldots R$, $i = 0, 1$, are estimated from

---

[8] This is known as the Bayes decision rule or the minimum-error-rate classification rule under the 0–1 loss function [327].

a training set of genuine and impostor scores corresponding to each of the R biometric matchers. Equation (12.8) results in the *product rule* which combines the scores generated by the R matchers as,

$$s_{prod} = \prod_{j=1}^{R} \frac{p(s_j \mid \omega_0)}{p(s_j \mid \omega_1)}. \tag{12.9}$$

Kittler et al. [360] modify the product rule by further assuming that the *a posteriori* probability $P(\omega_i \mid \mathbf{s})$ of class ω_i does not deviate much from its *a priori* probability $P(\omega_i)$ resulting in the *sum rule*:

$$s_{sum} = \frac{\sum_{j=1}^{R} p(s_j \mid \omega_0)}{\sum_{j=1}^{R} p(s_j \mid \omega_1)}. \tag{12.10}$$

Similar expressions can be derived for combining the match scores using the max, min and median rules [338, 360]. All the aforementioned rules implicitly assume that the match scores are *continuous* random variables. Dass et al. [407] relax this assumption and represent the univariate density functions (i.e., the marginals in (12.8)) as a mixture of discrete as well as continuous components. The resulting density functions are referred to as generalized densities. The authors demonstrate that the use of generalized density estimates (as opposed to continuous density estimates) significantly enhances the matching performance of the fusion algorithm. Furthermore, they use copula functions [408, 409] to model the correlation structure between the match scores $s_1, s_2, \ldots, s_R$ and, subsequently, define a novel fusion rule known as the *copula fusion rule*.

Transformation-Based Fusion Schemes

Density-based schemes, as stated earlier, require a large number of training samples (i.e., genuine and impostor match scores) in order to accurately estimate the density functions. This may not be possible in most multibiometric systems due to the time, effort, and cost involved in acquiring labeled multibiometric data in an operational environment. In such situations, it may be necessary to *directly* combine the match scores generated by multiple matchers using simple fusion operators (such as the simple sum of scores or order statistics) without first interpreting them in a probabilistic framework. However, such an approach is meaningful only when the scores output by the matchers are comparable. To facilitate this, a score normalization process is essential to transform the multiple match scores into a common domain. The process of score normalization entails changing the location and the scale parameters of the underlying match score distributions in order to ensure compatibility between multiple score variables. A few of the commonly discussed score normalization methods are described here.

The simplest normalization technique is the *min–max* normalization. Min–max normalization is best suited for the case where the bounds (maximum and minimum values) of the scores produced by a matcher are known. In this case, we can easily transform the minimum and maximum scores to 0 and 1, respectively. However, even if the match scores are not bounded, we can estimate the minimum and maximum values for the given set of training match scores and then apply the min–max normalization. Let s_j^i denote the ith match score output by the jth matcher, $i = 1, 2, \ldots, N$; $j = 1, 2, \ldots, R$ (R is the number of matchers and N is the number of match scores available in the training set). The min–max normalized score, ns_j^t, for the test score s_j^t is given by

$$ns_j^t = \frac{s_j^t - \min_{i=1}^{N} s_j^i}{\max_{i=1}^{N} s_j^i - \min_{i=1}^{N} s_j^i}. \tag{12.11}$$

When the minimum and maximum values are estimated from the given set of match scores, this method is not robust (i.e., the method is sensitive to outliers in the data used for estimation). Min–max normalization retains the original distribution of scores except for a scaling factor and transforms all the scores into a common range $[0, 1]$. Distance scores can be transformed into similarity scores by subtracting the normalized score from 1.

Decimal scaling can be applied when the scores of different matchers are on a logarithmic scale. For example, if one matcher has scores in the range $[0, 10]$ and the other has scores in the range $[0, 1000]$, the following normalization could be applied to transform the scores of both the matchers to the common $[0, 1]$ range.

$$ns_j^t = \frac{s_j^t}{10^{n_j}}, \tag{12.12}$$

where $n_j = \log_{10} \max_{i=1}^{N} s_j^i$. In the example with two matchers where the score ranges are $[0, 10]$ and $[0, 1000]$, the values of n would be 1 and 3, respectively. The problems with this approach are the lack of robustness and the implicit assumption that the scores of different matchers vary by a logarithmic factor.

The most commonly used score normalization technique is the *z-score* normalization that uses the arithmetic mean and standard deviation of the training data. This scheme can be expected to perform well if the average and the variance of the score distributions of the matchers are available. If we do not know the values of these two parameters, then we need to estimate them based on the given training set. The z-score normalized score is given by

$$ns_j^t = \frac{s_j^t - \mu_j}{\sigma_j}, \tag{12.13}$$

where μ_j is the arithmetic mean and σ_j is the standard deviation for the jth matcher. However, both mean and standard deviation are sensitive to outliers and hence, this method is not robust. z-score normalization does not guarantee a common numerical range for the normalized scores of the different matchers. If the distribution of the scores is not Gaussian, z-score normalization does not preserve the distribution of the given set of scores. This is due to the fact that mean and standard deviation are the optimal location and scale parameters only for a Gaussian distribution. While mean and standard deviation are reasonable estimates of location and scale, respectively, they are not optimal for an arbitrary match score distribution.

The *median* and *MAD* statistics are less sensitive to outliers as well as points in the extreme tails of the distribution. Hence, a normalization scheme using median and MAD would be relatively robust and is given by

$$ns_j^t = \frac{s_j^t - med_j}{MAD_j}, \qquad (12.14)$$

where $med_j = median_{i=1}^N s_j^i$ and $MAD_j = median_{i=1}^N |s_j^i - med_j|$. However, the median and the MAD estimators have a low efficiency compared to the mean and the standard deviation estimators, i.e., when the score distribution is not Gaussian, median and MAD are poor estimates of the location and scale parameters. Therefore, this normalization technique does not preserve the input score distribution and does not transform the scores into a common numerical range.

Cappelli et al. [410] use a *double sigmoid function* for score normalization in a multibiometric system that combines different fingerprint matchers. The normalized score is given by

$$ns_j^t = \begin{cases} \dfrac{1}{1 + \exp\left(-2\left(\frac{s_j^t - \tau}{\alpha_1}\right)\right)} & \text{if } s_j^t < \tau, \\[2ex] \dfrac{1}{1 + \exp\left(-2\left(\frac{s_j^t - \tau}{\alpha_2}\right)\right)} & \text{otherwise,} \end{cases} \qquad (12.15)$$

where τ is the reference operating point and α_1 and α_2 denote the left and right edges of the region in which the function is linear. The double sigmoid function exhibits linear characteristics in the interval $(\tau - \alpha_1, \tau - \alpha_2)$. While the double sigmoid normalization scheme transforms the scores into the $[0, 1]$ interval, it requires careful tuning of the parameters τ, α_1 and α_2 to obtain good efficiency. Generally, τ is chosen to be some value falling in the region of overlap between the genuine and impostor score distributions, and α_1 and α_2 are set so that they correspond to the extent of overlap between the two distributions toward the left and right of τ, respectively. This normalization scheme provides a linear transformation of the scores in the region of overlap, while the scores outside this region are transformed nonlinearly. The double sigmoid normalization is very similar to the min–max normalization followed by the application of a two-quadrics (QQ) or a logistic (LG) function as suggested by [411]. When the values of α_1 and α_2 are large, the double sigmoid normalization closely resembles the QQ-min–max normalization. On the other hand, we can make the double sigmoid normalization approach toward LG-min–max normalization by assigning small values to α_1 and α_2.

The *tanh-estimators* introduced by Hampel [412] are robust and highly efficient. The tanh normalization is given by

$$ns_j^t = \frac{1}{2}\left\{\tanh\left(0.01\left(\frac{s_j^t - \mu_{GH}}{\sigma_{GH}}\right)\right) + 1\right\}, \tag{12.16}$$

where μ_{GH} and σ_{GH} are the mean and standard deviation estimates, respectively, of the genuine score distribution as given by Hampel estimators. Hampel estimators are based on the following influence (ψ)-function:

$$\psi(u) = \begin{cases} u & 0 \le |u| < a, \\ a * sign(u) & a \le |u| < b, \\ a * sign(u) * \left(\frac{c - |u|}{c - b}\right) & b \le |u| < c, \\ 0 & |u| \ge c, \end{cases} \tag{12.17}$$

where

$$sign\{u\} = \begin{cases} +1, & \text{if } u \ge 0, \\ -1, & \text{otherwise.} \end{cases} \tag{12.18}$$

The Hampel influence function reduces the influence of the scores at the tails of the distribution (identified by a, b, and c) during the estimation of the location and scale parameters. Hence, this method is not sensitive to outliers. If many of the points that constitute the tail of the distributions are discarded, the estimate is robust but not efficient (optimal). On the other hand, if all the points that constitute the tail of the distributions are considered, the estimate is not robust but its efficiency increases. Therefore, the parameters a, b, and c must be carefully chosen depending on the amount of robustness required which in turn depends on the amount of noise in the available training data.

Mosteller and Tukey [413] introduce the biweight location and scale estimators that are robust and efficient. But, the *biweight estimators* are iterative in nature (initial estimates of the biweight location and scale parameters are chosen, and these estimates are updated based on the training scores), and are applicable only for Gaussian data. A summary of the characteristics of the different normalization techniques discussed here is shown in Table 12.2. The min–max, decimal scaling, and z-score normalization schemes are efficient, but are not robust to outliers. On the other hand, the median normalization scheme is robust but inefficient. Only the double sigmoid and tanh-estimators have both the desired characteristics, namely, robustness and efficiency.

Table 12.2. Summary of score normalization techniques

normalization technique	robustness	efficiency
min–max	no	high
decimal scaling	no	high
z-score	no	high
median and MAD	yes	moderate
double sigmoid	yes	high
tanh-estimators	yes	high

Once the match scores output by multiple matchers are transformed into a common domain they can be combined using simple fusion operators such as the sum of scores, product of scores, or order statistics (e.g., maximum/minimum of scores or median score).

Classifier-Based Fusion Schemes

In the verification mode of operation, the match scores generated by the multiple matchers may be input to a trained pattern classifier, such as a neural network, in order to determine the class label (genuine or impostor). In this approach, the goal is to directly estimate the class rather than to compute an intermediate scalar value. Classifier-based fusion schemes assume the availability of a large representative number of genuine and impostor scores during the training phase of the classifier when its parameters are computed. The component scores do not have to be transformed into a common domain prior to invoking the classifier.

In the biometric literature several classifiers have been used to consolidate the match scores of multiple matchers. Brunelli and Falavigna [364] use a HyperBF network to combine matchers based on voice and face features. Verlinde and Cholet [414] compare the relative performance of three different classifiers, namely, the k-Nearest Neighbor classifier using vector quantization, the decision tree classifier, and a classifier based on the logistic regression model when fusing the match scores originating from three biometric matchers. Experiments on the M2VTS database [415] show that the total error rate (sum of the false accept and false reject rates) of the multimodal system is an order of magnitude less than that of the individual matchers. Chatzis et al. [416] use classical k-means clustering, fuzzy clustering, and median radial basis function (MRBF) algorithms for fusion at the match score level. The proposed system combines the output of five different face and voice matchers. Each matcher provides a match score and a quality metric indicating the reliability of the match score. These values are concatenated to form a ten-dimensional vector that is input to the classifiers. Ben-Yacoub et al. [417] evaluate a number of classification schemes for fusion including support vector machine (SVM) with polynomial kernels, SVM with Gaussian kernels, C4.5 decision trees, multilayer perceptron, Fisher linear discriminant, and Bayesian classifier. Experimental evaluations on the XM2VTS database [418] consisting of 295 subjects suggest the benefit of score level fusion. Bigun et al. [380] propose a novel algorithm based on the Bayesian classifier that takes into account the estimated accuracy of the individual classifiers (i.e., matchers) during the fusion process. Sanderson and Paliwal [385] use a support vector machine (SVM) to combine the scores of face and speech experts. In order to address noisy input, they design structurally noise-resistant classifiers based on a piece-wise linear classifier and a modified Bayesian classifier. Wang et al. [419] view the match scores obtained from face and iris recognition modules as a two-dimensional feature vector and use Fisher's discriminant analysis and a neural network classifier to classify this match score vector. Ross and Jain [420] use decision tree and linear discriminant classifiers for classifying the match scores pertaining to the face, fingerprint, and hand geometry modalities.

12.4.4 Rank-Level Fusion

When a biometric system operates in the identification mode, the output of the system can be viewed as a ranking of the enrolled identities. In this case, the output indicates the set of possible matching identities sorted in decreasing order of confidence. The goal of rank level fusion schemes is to consolidate the ranks output by the individual biometric subsystems in order to derive a consensus rank for each identity. Ranks provide more insight into the decision-making process of the matcher compared to just the identity of the best match, but they reveal less information than match scores. However, unlike match scores, the rankings output by multiple biometric systems are comparable. As a result, no normalization is needed and this makes rank level fusion schemes simpler to implement compared to the score level fusion techniques.

Let us assume that there are M users enrolled in the database and let the number of matchers be R. Let $r_{j,k}$ be the rank assigned to user k by the jth matcher, $j = 1, \ldots, R$ and $k = 1, \ldots, M$. Let s_k be a statistic computed for user k such that the user with the lowest value of s is assigned the highest consensus (or reordered) rank. Ho et al. [421] describe the following three methods to compute the statistic s.

1. *Highest rank method.* In the highest rank method, each user is assigned the highest rank (minimum r value) as computed by different matchers, i.e., the statistic for user k is

$$s_k = \min_{j=1}^{R} r_{j,k}. \qquad (12.19)$$

   Ties are broken randomly to arrive at a strict ranking order. This method is useful only when the number of users is large compared to the number of matchers, which is typically the case in large-scale authentication systems. If this condition is not satisfied, the system will encounter several ties thereby rendering the final ranking uninformative. An advantage of the highest rank method is that it can utilize the strength of each matcher effectively. Even if only one matcher assigns a high rank to the correct identity, it is still very likely that this user will receive a high rank after reordering.

2. *Borda count method.* The Borda count method uses the sum of the ranks assigned by the individual matchers to calculate the value of s, i.e., the statistic for user k is

$$s_k = \sum_{j=1}^{R} r_{j,k}. \qquad (12.20)$$

   The magnitude of the Borda count for each user is a measure of the degree of agreement among the different matchers on whether the input belongs to that user. The Borda count method assumes that the ranks assigned to the users by the matchers are statistically independent and that all the matchers perform equally well.

3. *Logistic regression method.* The logistic regression method is a generalization of the Borda count method where a weighted sum of the individual ranks is calculated, i.e., the statistic for user k is

$$s_k = \sum_{j=1}^{R} w_j r_{j,k}. \tag{12.21}$$

The weight, w_j, to be assigned to the jth matcher, $j = 1, \ldots, R$, is determined by logistic regression [422]. The logistic regression method is useful when the different biometric matchers have significant differences in their accuracies. However, this method requires a training phase to determine the weights.

12.4.5 Decision-Level Fusion

Many commercial off-the-shelf (COTS) biometric matchers provide access only to the final recognition decision. When such COTS matchers are used to build a multibiometric system, only decision level fusion is feasible. Methods proposed in the literature for decision level fusion include "AND" and "OR" rules [423], majority voting [424], weighted majority voting [359], Bayesian decision fusion [361], the Dempster–Shafer theory of evidence [361] and behavior knowledge space [425].

Let M denote the number of possible decisions (also known as *class labels* or simply *classes* in the pattern recognition literature; these three terms are used interchangeably in the following discussion) in a biometric system. Also, let $\omega_1, \omega_2, \ldots \omega_M$ indicate the classes associated with each of these decisions.

1. *"AND" and "OR" rules.* In a multibiometric verification system, the simplest method of combining decisions output by the different matchers is to use the "AND" and "OR" rules. The output of the "AND" rule is a "match" only when all the biometric matchers agree that the input sample matches with the template. On the contrary, the "OR" rule outputs a "match" decision as long as at least one matcher decides that the input sample matches with the template. The limitation of these two rules is their tendency to result in extreme operating points. When the "AND" rule is applied, the false accept rate (FAR) of the multibiometric system is extremely low (lower than the FAR of the individual matchers) while the FRR is high (greater than the FRR of the individual matchers). Similarly, the "OR" rule leads to higher FAR and lower FRR than the individual matchers. When one biometric matcher has a substantially higher EER compared to the other matcher, the combination of the two matchers using "AND" and "OR" rules may actually degrade the overall performance [423]. Due to this phenomenon, the "AND" and "OR" rules are rarely used in practical multibiometric systems.
2. *Majority voting.* The most common approach for decision level fusion is majority voting where the input biometric sample is assigned to that identity on which a majority of the matchers agree. If there are R biometric matchers, the input sample is assigned an identity when at least k of the matchers agree on that identity, where

$$k = \begin{cases} \frac{R}{2} + 1 \text{ if } R \text{ is even,} \\ \frac{R+1}{2} \text{ otherwise.} \end{cases} \tag{12.22}$$

When none of the identities is supported by k matchers, a reject decision is output by the system. Majority voting assumes that all the matchers perform equally well. The advantages of majority voting are (a) no apriori knowledge about the matchers is needed and (b) no training is required to come up with the final decision. A theoretical analysis of the majority voting fusion scheme was done by [426] who established limits on the accuracy of the majority vote rule based on the number of matchers, the individual accuracy of each matcher and the pairwise dependence between the matchers.

3. *Weighted majority voting.* When the matchers used in a multibiometric system are not of similar recognition accuracy (i.e., imbalanced matchers/classifiers), it is reasonable to assign higher weights to the decisions made by the more accurate matchers. In order to facilitate this weighting, the labels output by the individual matchers are converted into degrees of support for the M classes as follows.

$$s_{j,k} = \begin{cases} 1, & \text{if output of the } j^{th} \text{ matcher is class } \omega_k, \\ 0, & \text{otherwise,} \end{cases} \quad (12.23)$$

where $j = 1, \ldots, R$ and $k = 1, \ldots, M$. The discriminant function[9] for class ω_k computed using weighted voting is

$$g_k = \sum_{j=1}^{R} w_j s_{j,k}, \quad (12.24)$$

where w_j is the weight assigned to the jth matcher. A test sample is assigned to the class with the highest score (value of discriminant function).

4. *Bayesian decision fusion.* The Bayesian decision fusion scheme relies on transforming the discrete decision labels output by the individual matchers into continuous probability values. The first step in the transformation is the generation of the confusion matrix for each matcher by applying the matcher to a training set **D**. Let CM^j be the $M \times M$ confusion matrix for the jth matcher. The (k,r)th element of the matrix CM^j (denoted as $cm^j_{k,r}$) is the number of instances in the training data set where a pattern whose true class label is ω_k is assigned to the class ω_r by the jth matcher. Let the total number of data instances in **D** be N and the number of elements that belong to class ω_k be N_k. Let c_j be the class label assigned to the test sample by the jth matcher. The value $cm^j_{k,c_j}/N_k$ can be considered as an estimate of the conditional probability $P(c_j \mid \omega_k)$ and N_k/N can be treated as an estimate of the prior probability of class ω_k. Given the vector of decisions made by R matchers $\mathbf{c} = [c_1, \ldots, c_R]$, we are interested

---

[9] The discriminant function is used to classify an input pattern. Typically, a discriminant function is defined for each pattern class and the input pattern is assigned to the class whose discriminant function gives the maximum response.

in calculating the posterior probability of class ω_k, i.e., $P(\omega_k \,|\, \mathbf{c})$. According to the Bayes rule,

$$P(\omega_k \,|\, \mathbf{c}) = \frac{P(\mathbf{c} \,|\, \omega_k)\, P(\omega_k)}{P(\mathbf{x})}, \qquad (12.25)$$

where $k = 1, \ldots, M$. The denominator in (12.25) is independent of the class ω_k and can be ignored for the decision making purpose. Therefore, the discriminant function for class ω_k is

$$g_k = P(\mathbf{c} \,|\, \omega_k)\, P(\omega_k). \qquad (12.26)$$

The Bayes decision fusion technique chooses that class which has the largest value of discriminant function calculated using (12.26). To simplify the computation of $P(\mathbf{c} \,|\, \omega_k)$, one can assume conditional independence between the different matchers. Under this assumption, the decision rule is known as naive Bayes rule and $P(\mathbf{c} \,|\, \omega_k)$ is computed as

$$P(\mathbf{c} \,|\, \omega_k) = P(c_1, \ldots, c_R \,|\, \omega_k) = \prod_{j=1}^{R} P(c_j \,|\, \omega_k). \qquad (12.27)$$

The accuracy of the naive Bayes decision fusion rule has been found to be fairly robust even when the matchers are not independent [427].

5. *Dempster–Shafer theory of evidence.* The Dempster–Shafer theory of evidence is based on the concept of assigning degrees of belief for uncertain events. Note that the degree of belief for an event is different from the probability of the event. This subtle difference is explained in the following example. Suppose we know that a biometric matcher has a reliability of 0.95, i.e., the output of the matcher is reliable 95% of the time and unreliable 5% of the time. Suppose that the matcher outputs a "match" decision. We can assign a 0.95 degree of belief to the "match" decision and a zero degree of belief to the "nonmatch" decision. The zero belief does not rule out the "nonmatch" decision completely, unlike a zero probability. Instead, the zero belief indicates that there is no reason to believe that the input does not match successfully against the template. Hence, we can view belief theory as a generalization of probability theory. Indeed, belief functions are more flexible than probabilities when our knowledge about the problem is incomplete.

Rogova [428] and Kuncheva et al. [429] propose the following methodology to compute the belief functions and to accumulate the belief functions according to the Dempster's rule. For a given input pattern, the decisions made by R classifiers for a M-class problem is represented using a $R \times M$ matrix known as a decision profile (DP) [429] which is given by,

$$DP = \begin{bmatrix} s_{1,1} & \cdots & s_{1,k} & \cdots & s_{1,M} \\ \cdots & & & & \\ s_{j,1} & \cdots & s_{j,k} & \cdots & s_{j,M} \\ \cdots & & & & \\ s_{R,1} & \cdots & s_{R,k} & \cdots & s_{R,M} \end{bmatrix},$$

where $s_{j,k}$ is the degree of support provided by the jth matcher to the kth class. At the decision level, the degree of support is expressed as

$$s_{j,k} = \begin{cases} 1, & \text{if output of the } j^{th} \text{ matcher is class } \omega_k, \\ 0, & \text{otherwise,} \end{cases} \quad (12.28)$$

where $j = 1, \ldots, R$ and $k = 1, \ldots, M$. The decision template (DT^k) of each class ω_k is the average DP for all the training instances that belong to the class ω_k. When the degrees of support defined in (12.28) are used, one can easily see that the elements of the decision template DT^k are related to the elements of the confusion matrices of the R matchers in the following manner.

$$DT^k_{j,r} = \frac{CM^j_{k,r}}{N_k}, \quad (12.29)$$

where N_k is the number of instances in the training set $\mathbf{D}$ that belong to class ω_k, $j = 1, \ldots, R$ and $k, r = 1, \ldots, M$. For a given test pattern X^t, the decision profile DP^t is computed after the decisions of the R matchers are obtained. The similarity between DP^t and the decision templates for the various classes is calculated as follows.

$$\Phi_{j,k} = \frac{\left(1 + (||DT^k_j - DP^t_j||)^2\right)^{-1}}{\sum_{r=1}^{M} \left(\left(1 + (||DT^r_j - DP^t_j||)^2\right)^{-1}\right)}, \quad (12.30)$$

where DT^k_j represents the jth row of DT^k belonging to class ω_k, DP^t_j represents the jth row of DP^t belonging to the test pattern X^t, and $||.||$ denotes the matrix norm. For every class $k = 1, \ldots, M$ and for every matcher $j = 1, \ldots, R$, we can compute the degree of belief as

$$b_{j,k} = \frac{\Phi_{j,k} \left[\prod_{r=1, r \neq k}^{M} (1 - \Phi_{j,r})\right]}{1 - \Phi_{j,k} \left[\prod_{r=1, r \neq k}^{M} (1 - \Phi_{j,r})\right]}. \quad (12.31)$$

The accumulated degree of belief for each class $k = 1, \ldots, M$ based on the outputs of R matchers is then obtained using the Dempster's rule as

$$g_k = \prod_{j=1}^{R} b_{j,k}. \quad (12.32)$$

The test pattern X^t is assigned to the class having the highest degree of belief g_k.

12.5 Summary

Multibiometric systems are expected to enhance the recognition accuracy of a personal authentication system by reconciling the evidence presented by multiple sources of information. In this chapter, the different sources of biometric information as well as the type of information that can be consolidated was presented. Different fusion strategies were also discussed. Typically, early integration strategies (e.g., feature-level) are expected to result in better performance than late integration (e.g., score-level) strategies. However, it is difficult to predict the performance gain due to each of these strategies prior to invoking the fusion methodology. While the *availability* of multiple sources of biometric information (pertaining either to a single trait or to multiple traits) may present a compelling case for fusion, the *correlation* between the sources has to be examined before determining their suitability for fusion. Combining uncorrelated or negatively correlated sources is expected to result in a better improvement in matching performance than combining positively correlated sources. This has been demonstrated by Kuncheva et al. [430] for fusion at the decision level using the majority vote scheme. Combining sources that make complementary errors is assumed to be beneficial. However, defining an appropriate diversity measure to predict fusion performance has been elusive thus far.

The development of robust HCIs is necessary to permit the efficient acquisition of multibiometric data from individuals (see [431] and the references therein). A HCI that is easy to use can result in rapid user habituation and promote the acquisition of high quality biometric data. With the increased use of biometrics in authentication solutions, it is only a matter of time before multibiometric systems are deployed in a variety of applications in government, military, and commercial systems.

13 Performance Prediction Methodology for Multibiometric Systems

Natalia A. Schmid and Joseph A. O'Sullivan

13.1 Introduction

Secure personal authentication and identification are challenging problems for modern society. While the rapid development of new information technologies makes life more organized and convenient, it also necessitates the development of new security systems to prevent unauthorized access and abuse. Biometrics-based authentication and identification systems are often designed for controlling access to facilities, personal computing and communication devices, financial accounts, and information databases.

For our purposes, study of biometrics is a study of physiological and behavioral characteristics of an individual [432]. To be qualified as a biometric, it must satisfy a number of requirements. The four most important are (1) *universality*: everyone must possess it; (2) *uniqueness*: no two individuals should have the same value of the characteristic; (3) *permanence*: it must not change with time; (4) *measurability*: it must be easy to measure. Several characteristics have been proposed, researched, and implemented for personal identification, including voice, infrared facial and hand vein thermograms, fingerprints, face, iris, ear, gait, keystroke dynamics, DNA, signature, acoustic emissions, odor, retinal scan, and hand and finger geometry. Most of these characteristics are believed to be unique. The uniqueness of physiological characteristics is a result of both genetic conditions and random physical processes occurring during the stage of embryonic development. Here, the physiological processes are modeled mathematically.

An engineering approach to the problem of identification is to state it as a recognition problem. Most information available to engineers about biometrics is in the form of images (iris scan, fingerprint, hand geometry) or signals (voice). Often systems (such as iris recognition based on IrisCode [433]) extract a robust set of features selected for good recognition performance. After features are extracted, the most common approach is to treat extracted features as deterministic parameters. For instance in a fingerprint identification problem, a set of minutia points is considered as one of the best deterministic representations in terms of performance, provided that the number of minutiae points is large enough. To make a fingerprint identification system based on minutiae points robust, bounding boxes used for counting the number of matching minutia points in two fingerprints are adjusted adaptively [434]. In face recognition, principal components analysis (PCA) and linear discriminant analysis (LDA) have been leading techniques [435, 436]. Face recognition techniques based on a three-dimensional representation are the current state-of-the-art techniques (see for example [437]). In hand geometry, features are Euclidean distances between different points on a hand [438]. An alternative to a deterministic approach is to model images, signals, or extracted features as realizations of stochastic processes. While for physical signatures, such as the distribution of magnetic particles on magnetic tape or ink, or the distribution of woodgrains on a wooden surface, estimation of stochastic models is more straightforward, for most biometric signatures it remains a challenging task. Common models of signatures as linear combinations of components may be considered to be stochastic models; the distributions of the coefficients are often closely approximated using standard parameterized distributions. The estimation of these models from training data is beyond the scope of this chapter. However, we note that the number of parameters, features, or components used in a representation must not be too large in order to avoid over-fitting the training data; which ones to use may be determined based on the information provided [439]. In this chapter, we proceed with an analysis assuming the stochastic models are known, providing the analytical tools to predict performance given these models.

A typical identification system is designed to operate in two modes: the off-line mode and the on-line mode. During the off-line mode, data are collected for each individual, and a template is extracted and stored in a database. The template may comprise a vector of features, coefficients in a components model, or parameters in a probability distribution for the data. During the on-line mode, input data are acquired, its template is extracted, and a match is performed against each template stored in the database. Performance of a recognition system is usually measured in terms of the probabilities of error as a function of design parameters (receiver operating characteristic (ROC) curves for a binary problem). Performance of practical biometrics-based identification systems is often evaluated by modeling the decision function as a Bernoulli distributed random variable with its mean estimated using training data and then evaluating the confidence intervals for the estimated probabilities of error [440–443].

Major challenges of biometric system design today include high-performance, efficient strategies for biometric fusion at the matching score level [444, 445]; adjustment of thresholds in decision functions with the goal of achieving optimal performance; fusion at the feature level [446]; defining level of individuality for a unimodal biometric [447]; collection of data under conditions of environmental and demographic variations for the purpose of system testing and performance prediction [432, 448, 449]. In this chapter, we address two major challenges. They are performance prediction and capacity analysis for multimodal biometrics-based identification systems. For our analysis, we model multimodal templates stored in the database of an identification system as noisy realizations of independent and identically distributed stationary ergodic random processes. The unknown input data presented for identification are matched against every template in the database. Here matching is equivalent to finding a noisy template that has a known joint distribution with the input data. The problem of identification is then a statistical hypothesis testing problem. The average probability of error is asymptotically approximated by an exponential function with explicit dependence on the number of parameters of the system. The minimum probability of error, as we show, is determined by the smallest exponent and is equal to the minimum Chernoff information among all distinct pairs of hypotheses. This result was earlier proved by Leang and Johnson [450] for data with independent and identically distributed components and by Schmid and O'Sullivan [451] for the case when signatures are modeled as realizations of stationary and ergodic processes. We extend the results to the case of multimodal signatures. The capacity of an identification system is obtained by examining a communication channel (with random coding) that is directly analogous to the recognition system.

Many of our results require a measure of the variability in the templates and information about the templates provided by the measured data to get large. One way to achieve this is for the dimension of the templates to increase (increasing number of features, components, or parameters), and the information per dimension to be lower bounded. Physically this may correspond to increasing resolution of an imaging sensor, the length of a biometric signature, or the number of samples while maintaining a lower bound on the noise per sample, to measuring and extracting new informative features, or making multiple measurements. As the information increases, the number of signatures that can be reliably distinguished increases.

The rest of this chapter is organized as follows. Section 13.2 introduces the model proposed to describe multimodal biometric data. In Sect. 13.3, the identification problem is stated as a multiple hypothesis testing problem. Several asymptotic results related to the probability of error are obtained by applying a large deviations approach in Sect. 13.3.1. In Sect. 13.4, the capacity of a biometrics-based identification system is determined. Examples with templates modeled as realizations of stationary Gaussian processes are given in Sect. 13.5. Section 13.6 presents the summary of the results.

13.2 Stochastic Model for Multimodal Biometric Signatures

The precision and efficiency of biometrics-based identification systems rely on the concept of individuality of biometric signatures. Since signatures of different individuals drawn from the same biometric modality contain a large number of common features, mostly due to the fact that the development of biometrics is guided by human genomics, we consider the following model for biometric signatures. Suppose that a biometric signature from a single modality is a function of two processes, a background process common for the considered signatures of all individuals and an individual signature process statistically independent of the background process and individual signature processes of the other users. Thus a biometric signature discretized to n samples of the mth user is given by

$$X_m^n = f(B^n, S_m^n), \qquad (13.1)$$

where B^n is an $n \times 1$ common background process, S_m^n is the $n \times 1$ individual signature of the mth user, and $f(\cdot)$ is an unknown random nonlinear function mixing background and individual processes. Since the general form of $f(\cdot)$ is hard to evaluate, we may appeal to its simplest additive form. In this form, the resulting biometric signature is given by

$$X_m^n = B^n + S_m^n. \tag{13.2}$$

In the case, when it is possible to estimate the common background signature and filter out its effect, biometric signatures are composed of independent individual signatures

$$X_m^n = S_m^n. \tag{13.3}$$

In this work we assume that the described preprocessing is possible and thus will consider signatures of type (13.3).

Biometric signatures acquired from two different modalities characterizing the same individual can often be treated as independent. Denote by $X_{A,m}^n$ the signature of individual m collected from modality A. Denote by $X_{B,m}^n$ the signature of the same individual collected from modality B. Then the combined signature X_m^{2n} of the individual can be obtained by concatenating $X_{A,m}^n$ and $X_{B,m}^n$. In the following we use X_m^{2n} to denote the combined signature vector.

Note that the model and results described in this work can be extended to include an arbitrary number of biometric modalities or representations of the same modality provided that the templates resulting from different representations or templates characterizing different modalities can be treated as independent.

13.3 Performance of a Multimodal Biometric Recognition System with M Templates

Suppose that M independent and identically distributed template random processes denoted by $\{X_{A,l}(k), l = 1, 2, \ldots\}$ (for biometric modality A), and $\{X_{B,l}(k), l = 1, 2, \ldots\}$ (for biometric modality B), $k = 1, \ldots, M$, and an input random process $\{Y_l, l = 1, 2, \ldots\}$, are available. M can be large but fixed. The random processes take values in polish spaces [452]; our examples are for real-valued processes. A noisy realization of the input process is acquired on-line and is tested for statistical dependence with each template in the data base. If the input process passes a test for statistical dependence with exactly one of the template processes, then the input process (and thus the corresponding individual) is positively identified. If the input process passes a test for statistical independence of all the template processes, the individual is announced to be unknown. Denote the joint distribution on the first n samples of each of the two multimodal processes and the input process by $P_{X^{2n}(k), Y^{2n}}$, and the corresponding marginal distributions by $P_{X^{2n}(k)}$, $k = 1, \ldots, M$, and $P_{Y^{2n}}$. The notation follows Gray [452]. In particular, $X^{2n}(k) = (X_{A,1}(k), X_{A,2}(k), \ldots, X_{A,n}(k), X_{B,1}(k), X_{B,2}(k), \ldots, X_{B,n}(k))$. Under hypothesis H_k, $k = 1, \ldots, M$, the input sequence Y^{2n} results from a joint probability distribution function with the signature sequence $X^{2n}(k)$ and is independent of all other signatures. Under H_0, the template processes and the input process are mutually independent. The optimal test statistic for the identification problem is an M-dimensional vector of log-likelihood ratios i_n^M. In this case, log-likelihood ratios have a special form and are called information densities:

$$i_n^M \equiv \frac{1}{n}\left[\log\frac{dP_{X^{2n}(1),Y^{2n}}}{dP_{X^{2n}(1)}\times dP_{Y^{2n}}},\ldots,\log\frac{dP_{X^{2n}(M),Y^{2n}}}{dP_{X^{2n}(M)}\times dP_{Y^{2n}}}\right]^T. \quad (13.4)$$

We assume, conditioned on the individual, that the data from modalities A and B are independent. The information density vector i_n^M can then be written as

$$i_n^M \equiv \frac{1}{n}\left[\log\frac{dP_{X_A^n(1),Y_A^n}}{dP_{X_A^n(1)}\times dP_{Y_A^n}}+\log\frac{dP_{X_B^n(1),Y_B^n}}{dP_{X_B^n(1)}\times dP_{Y_B^n}},\ldots,\right.$$
$$\left.\times\left[\log\frac{dP_{X_A^n(M),Y_A^n}}{dP_{X_A^n(M)}\times dP_{Y_A^n}}+\log\frac{dP_{X_B^n(M),Y_B^n}}{dP_{X_B^n(M)}\times dP_{Y_B^n}}\right]^T. \quad (13.5)$$

To state the consistency results for the test statistic in (13.4) we appeal to Assumptions 1 and 2 from [451]. The first assumption ensures that information densities are well defined. The second assumption provides a sufficient condition to guarantee convergence of components in (13.4) to their asymptotic expected values under the hypothesis of statistical independence.

Denote the asymptotic expected value of a component in (13.4) under the joint distribution by

$$\bar{I} = \limsup_{n\to\infty} E_{P_{X^{2n},Y^{2n}}} \frac{1}{n}\log\frac{dP_{X^{2n},Y^{2n}}}{dP_{X^{2n}}\times dP_{Y^{2n}}},$$

and under the product distribution by

$$\bar{D} = \limsup_{n\to\infty} E_{P_{X^{2n}}\times P_{Y^{2n}}} \frac{1}{n}\log\frac{dP_{X^{2n}}\times dP_{Y^{2n}}}{dP_{X^{2n},Y^{2n}}}.$$

The quantity $\bar{D}$ equals the relative entropy rate between the product distribution and the joint distribution and is thus a type of reverse mutual information rate (the mutual information rate $\bar{I}$ equaling the relative entropy rate between the joint distribution and the product distribution). We refer to $\bar{D}$ as the reverse information rate.

Because the biometric signatures characterizing modalities A and B of the same individual are statistically independent given that individual, the asymptotic expected values $\bar{I}$ and $\bar{D}$ can be written as

$$\bar{I} = \bar{I}_A + \bar{I}_B = \limsup_{n\to\infty} E_{P_{X_A^n,Y_A^n}} \frac{1}{n}\log\frac{dP_{X_A^n,Y_A^n}}{dP_{X_A^n}\times dP_{Y_A^n}}$$
$$+ \limsup_{n\to\infty} E_{P_{X_B^n,Y_B^n}} \frac{1}{n}\log\frac{dP_{X_B^n,Y_B^n}}{dP_{X_B^n}\times dP_{Y_B^n}}$$

and

$$\bar{D} = \bar{D}_A + \bar{D}_D = \limsup_{n \to \infty} E_{P_{X_A^n} \times P_{Y_A^n}} \frac{1}{n} \log \frac{dP_{X_A^n} \times dP_{Y_A^n}}{dP_{X_A^n, Y_A^n}}$$
$$+ \limsup_{n \to \infty} E_{P_{X_B^n} \times P_{Y_B^n}} \frac{1}{n} \log \frac{dP_{X_B^n} \times dP_{Y_B^n}}{dP_{X_B^n, Y_B^n}}.$$

If the assumptions above are satisfied, then by a theorem by Gray [452, pp 178–179] and Theorem 2 of [453], each component in (13.4) converges almost surely either to the mutual information rate, $\bar{I}$, or to the negative of the reverse information rate, $-\bar{D}$, depending on the true hypothesis. Thus the vector in (13.4) under $(M+1)$ distinct hypotheses converges to the vector of asymptotic values $V_0, V_1, \ldots, V_M$ given by

$$V_k \equiv \lim_{n \to \infty} i_n^M(X^{2n}(1), \ldots, X^{2n}(M), Y^{2n})$$

$$= [-\bar{D}, \ldots, \bar{I}, \ldots, -\bar{D}]^T, \text{ a.e. } P_{X_A(k), Y_A} \times P_{X_B(k), Y_B} \times \prod_{j \neq k}^{M} P_{X_A(j)} \times P_{X_B(j)},$$

with $\bar{I}$ in the kth position, $k = 1, \ldots, M$ and

$$V_0 \equiv \lim_{n \to \infty} i_n^M(X^{2n}(1), \ldots, X^{2n}(M), Y^{2n})$$

$$= [-\bar{D}, \ldots, -\bar{D}, \ldots, -\bar{D}]^T, \text{ a.e. } \prod_{k=1}^{M} P_{X_A(k)} \times P_{X_B(k)} \times P_{Y_A} \times P_{Y_B}.$$

13.3.1 Exponential Error Rate Analysis

Under stronger conditions, components in (13.4) converge to $\bar{I}$ and $-\bar{D}$ exponentially with the rate determined by large deviation rate functions. Large deviation results in this section are obtained following the approach described in [454, 455].

Define the vector of the log-moment generating functions as

$$\phi^{(n)}(\theta) = \frac{1}{n} \left[\log E_0\{e^{(n<\theta, i_n^M>)}\}, \ldots, \log E_M\{e^{(n<\theta, i_n^M>)}\} \right], \quad (13.6)$$

where θ is an M-dimensional vector of parameters, E_k denotes the expectation under H_k, and $< \cdot, \cdot >$ is the notation for the inner product between two vectors. Consider a set of assumptions from [455, p 43]:

1. The limiting vector of log-moment generating functions

$$\bar{\phi}(\theta) = \lim_{n \to \infty} \phi^{(n)}(\theta)$$

   is well defined.
2. $\theta = \mathbf{0}$ is in the interior of the domain of $\bar{\phi}(\cdot)$.
3. $z\bar{\phi}(\cdot)$ is steep on its domain.

13 Performance Prediction Methodology for Multibiometric Systems

Define the kth component of the $(M+1)$-dimensional vector of large deviation rate functions $I(\mathbf{t})$ as the Fenchel–Legendre transform of the function $\bar{\phi}_k(\theta)$ (the kth component of $\bar{\phi}(\theta)$)

$$I_k(\mathbf{t}) = \sup_{\theta \in \mathcal{R}^M}\left[<\theta,\mathbf{t}> - \bar{\phi}_k(\theta)\right], \; k=0,\ldots,M,$$

where $\mathbf{t} \in \mathcal{R}^M$.

Each pair of components in the vector $I(\mathbf{t})$ is linearly related. Consider the pair $(I_l(\mathbf{t}), I_k(\mathbf{t}))$, where $l \neq k \neq 0$. The relationship between the corresponding log-moment generating functions is given by

$$\phi_k^{(n)}([\theta_1,\ldots,\theta_M]^T) = \phi_l^{(n)}([\theta_1,\ldots,(\theta_l-1),\ldots,(\theta_k+1),\ldots,\theta_M]^T). \quad (13.7)$$

Since this is true for each given n, (13.7) also holds in the limit as $n \to \infty$, and hence, the corresponding components in the vector $I(\mathbf{t})$ are related as

$$I_k(\mathbf{t}) = t_l - t_k + I_l(\mathbf{t}),$$

where t_l and t_k are the entries in the vector $\mathbf{t}$.

The log-moment generating functions under H_0 and H_k, $k=1,\ldots,M$, are related as

$$\phi_k^{(n)}([\theta_1,\ldots,\theta_M]^T) = \phi_0^{(n)}([\theta_1,\ldots,(\theta_k+1),\ldots,\theta_M]^T),$$

and hence, the relationship between the rate functions $I_k(\mathbf{t})$ and $I_0(\mathbf{t})$ is given by

$$I_k(\mathbf{t}) = I_0(\mathbf{t}) - t_k.$$

The functions $I_k(\mathbf{t})$ have the property that they are zero at their corresponding mean vectors V_k. That is, $I_k(V_k) = 0, k=0,1,\ldots,M$. The relationships between rate functions yield

$$I_0(V_k) = \bar{I} \quad (13.8)$$
$$I_l(V_k) = \bar{I} + \bar{D}, l \neq k. \quad (13.9)$$

A typical measure of recognition performance applied in practice is the Bayes probability of error. Alternatively, to quantify the rapid decrease in error as a function of n, exponential rate functions are studied (see [456, 457]).

Denote by $\mathcal{D}_n$, where n is the dimension of vectors $X_A^n(k), Y_A^n, X_B^n(k)$, and Y_B^n in the test statistic, the optimally divided decision region such that the Bayes probability of error

$$P_e = \sum_{k=0}^{M} \pi_k P_{e\,|\,H_k}, \quad (13.10)$$

is minimized. Here π_k are known and sum to one and $P_{e\,|\,H_k}$ is the conditional probability of error under $H_k, k=0,\ldots,M$. Denote by $U_{k,n}$ the decision region under hypothesis H_k and by $U_{k,n}^c$ the region complementary to it. $U_{k,n}$ are such that $\cup_{k=0}^{M} U_{k,n} = \mathcal{D}_n$. Denote by $\mathcal{D}$, U_k, and U_k^c the asymptotic optimal decision region, the asymptotic decision region under H_k, and the complementary to it region, respectively. Let U_k^c be an open set and $\bar{U}_k^c$ be its closure.

Theorem 1 (Ellis). *If Assumptions 1 and 2 are true, then*

$$\limsup_{n\to\infty} \frac{1}{n} \log P\left(i_n^M(k) \in \bar{U}_k^c \mid H_k\right) \leq - \inf_{\mathbf{t}\in\bar{U}_k^c} I_k(\mathbf{t}).$$

If Assumptions 1 and 3 are true, then

$$\liminf_{n\to\infty} \frac{1}{n} \log P\left(i_n^M(k) \in U_k^c \mid H_k\right) \geq - \inf_{\mathbf{t}\in U_k^c} I_k(\mathbf{t}).$$

See [454, pp 21–23] for comments on the proof.

In the following theorem we use Ellis's results to derive the dependence between the probability of error and the length of the vectors $X^{2n}(k), k = 1, \ldots, M$, and Y^{2n}. Note that this result requires knowledge of the asymptotic exponential rate function, but can be used to evaluate the probability of identification error for an arbitrary, but large value n. Ultimately, we obtain a simple approximation for the probability of recognition error in the following form $P_e \approx F(M, n, \mathbf{t}) \exp(-n \min_{k\in\{0,1,\ldots,M\}} I_k(\mathbf{t}))$, where $F(M, n, \mathbf{t})$ is a slowly varying function of the number of templates in the database M, a threshold $\mathbf{t}$, and the number of samples n.

Theorem 2. *The exponential rate of the error probability for the Bayes rule in the $(M+1)$-ary case equals the minimum Chernoff information among all distinct pairs of the original $(M+1)$ hypotheses. That is,*

$$P_e \doteq e^{-n \min_{k\neq l} I_k^{(k,l)}(0)}, \quad k, l = 0, \ldots, M.$$

The proof relies on (1) upper and lower bounds on P_e due to binary hypothesis testing problems, (2) application of Ellis's theorem to find the asymptotic rates of conditional errors, and (3) application of Chernoff theorems [458, pp 312–313]. The point $I_k^{(k,l)}(0) = I_l^{(k,l)}(0)$ is called Chernoff information and is equal to the rate that determines the minimum probability of error in the binary recognition problem.

Comment 1: For the identification problem above, the asymptotic decision region $\mathcal{D}$ is a simplex with the asymptotic expected values V_k, of the vector of information densities under the hypotheses $0, 1, \ldots, M$ as its vertices in the M-dimensional space.

Comment 2: Under the condition that the origin of the coordinate system lies in the region $\mathcal{D}$, the exponential rate for the Bayes rule in the $(M+1)$-ary case is determined by the minimum Chernoff information among all distinct pairs of hypotheses

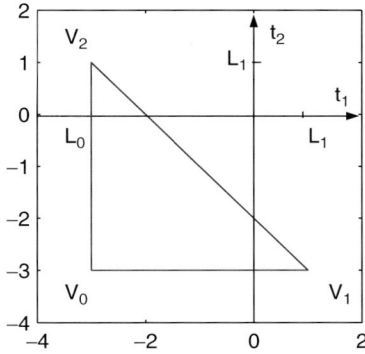

Fig. 13.1. Shown is the asymptotic decision region for $M = 2$ with $\bar{D} > \bar{I}$

including H_0. If the origin of the coordinate system does not lie in the region $\mathcal{D}$, the minimum probability of error is determined by the Chernoff information for the binary problem with the null-hypothesis tested against a non-null hypothesis.

This comment corresponds to the case $\bar{D} > \bar{I}$ (see Fig. 13.1).

Comment 3: As noted earlier, this extends the results of Leang and Johnson [450].

13.4 Recognition Capacity

In this section, we find the operational capacity for a multimodal biometrics-based identification system. Rather than fixing the number of hypotheses, the following analysis determines the exponential error rate under the condition that the number of templates grows exponentially with the length of the observed vector.

The results in this section are based on a straightforward analogy between a recognition system and a communication system that uses random coding. The templates in the recognition problem correspond to codewords in the communication problem. Suppose that $X^{2n}(1), X^{2n}(2), \ldots, X^{2n}(M)$, with $M = 2^{nR}$, templates (random codewords) are drawn i.i.d. from the distribution $P_{X^{2n}}$. Here R is the rate of the random code. Given that one of these random codewords is randomly selected as the truth, assume that an observation Y^{2n} is drawn from the corresponding conditional distribution determined by $P_{X^{2n},Y^{2n}}$, and that this distribution satisfies the assumptions from the previous sections. The observation Y^{2n} is directly analogous to the output of a communication system. The observation Y^{2n} must be matched against each of the 2^{nR} templates stored in the database. This matching process defines the recognition output and is analogous to channel decoding.

The probability of error is

$$P_e = \frac{1}{M} \sum_{m=1}^{M} P_{e \mid X^{2n}(m)}. \tag{13.11}$$

The minimum probability of error decision rule is the same as the maximum likelihood decision rule: select the most likely observation given $X^{2n}(m)$.

A rate R is achievable if a sequence of recognition systems can be defined with rates $R_n \leq R$ so that the probability of error goes to zero as n goes to infinity. Define the recognition capacity to be the supremum of achievable rates. The direct analogy to communication systems using random codes immediately yields the recognition capacity.

Theorem 3. *Under the assumptions above, the recognition capacity equals the sup-mutual information rate $\bar{I}(X;Y)$. That is, all rates less than $C = \bar{I}(X;Y)$ are achievable (the probability of error using the minimum probability of error decoding rule tends to zero as n tends to infinity); conversely, if rate R is achievable then $R \leq C$.*

13.5 Examples

The examples in this section illustrate theory developed above. The first example assumes a stationary Gaussian model for the data. While this model is a loose approximation for biometric data collected and processed according to currently available technologies, it is a valid model for description of physical patterns and structures consisting of a large number of tiny particles such as a wood structure with a large number of wood grains or a painted surface consisting of a large number of tiny particles of paint.

The second example can be applied to predict the performance of a multimodal biometric systems relying on PCA signatures of the face and iris biometrics.

13.5.1 M-ary Gaussian Example

Suppose that M independent and identically distributed noisy template Gaussian signatures $X^n(1), X^n(2), \ldots, X^n(M)$ and a noisy Gaussian candidate Y^n are available for identification. Each noisy template signature is modeled as an n-sample realization $Z^n(k)$, $k = 1, \ldots, M$ of a discrete-time stationary Gaussian random process with zero mean and Toeplitz covariance matrix $\mathbf{K}(S_Z)$ parameterized by the power spectral density S_Z and an n-sample realization $W^n(k)$ of a discrete-time additive stationary Gaussian noise with zero mean and covariance matrix $\mathbf{K}(S_W)$ parameterized by the power spectral density S_W

$$X^n(k) = Z^n(k) + W^n(k), \quad k = 1, \ldots, M.$$

A noisy Gaussian candidate is a realization of a Gaussian signature Z_Y^n and a vector of an additive Gaussian noise W_Y^n independent of and identically distributed with noise realizations $W^n(k)$, $k = 1, \ldots, M$, contained in the noisy signatures

$$Y^n = Z_Y^n + W_Y^n.$$

The hypothesis testing problem is stated as follows. Under the hypothesis H_k, $k = 1, \ldots, M$, $Z_Y^n = Z^n(k)$. Under the null hypothesis H_0, the candidate Z_Y^n is independent of all signatures $Z^n(k)$, $k = 1, \ldots, M$. The test statistic is given by (13.4) with the following entries

$$- [X^n(k), Y^n] \left(\mathbf{R}_1^{-1} - \mathbf{R}_0^{-1}\right) [X^n(k), Y^n]^T$$
$$- \log \det \left(\mathbf{R}_1 \mathbf{R}_0^{-1}\right), \quad k = 1, \ldots, M,$$

where $\mathbf{R}_1$ is the covariance matrix of the vector $[X^n(k), Y^n]$ when the noisy signature and the noisy candidate have the same signature part and $\mathbf{R}_0$ is the covariance matrix of the vector $[X^n(k), Y^n]$ when the signature part in the noisy signature and the noisy candidate are independent.

The asymptotic expected values V_k, $k = 0, 1, \ldots, M$ of the vector i_n^M under H_k are obtained by applying matrix analysis and the results from [459] related to the Toeplitz distribution theorem:

$$V_0 = \left[-\bar{D}, \ldots, -\bar{D}, \ldots, -\bar{D}\right]^T,$$
$$V_k = \left[-\bar{D}, \ldots, \bar{I}, \ldots, -\bar{D}\right]^T, \quad k = 1, \ldots, M, \tag{13.12}$$

where $\bar{I}$ is the kth entry in V_k and from [453]

$$\bar{D} = \frac{1}{4\pi} \int_{-\pi}^{\pi} \frac{2\hat{f}(\lambda)}{(1 - \hat{f}(\lambda))} d\lambda + \frac{1}{4\pi} \int_{-\pi}^{\pi} \log\left(1 - \hat{f}(\lambda)\right) d\lambda,$$
$$\bar{I} = -\frac{1}{4\pi} \int_{-\pi}^{\pi} \log\left(1 - \hat{f}(\lambda)\right) d\lambda.$$

The function $\hat{f}(\lambda)$ is analogous to the signal-to-noise ratio and is given by

$$\hat{f}(\lambda) = \frac{S_Z^2(\lambda)}{(S_Z(\lambda) + S_W(\lambda))^2}.$$

From Theorem 2, the minimum probability of error within the simplex defined by vertices $V_0, \ldots, V_M$ is achieved at the point of the pairwise Chernoff information for one of two binary hypothesis testing problems. The first binary problem is formed from the original null hypothesis and the original kth hypothesis. The large deviation rate function for this binary problem is the solution of the following equation:

$$I_1^{(0,1)}(0) = \sup_s \left[\frac{1}{4\pi} \int_{-\pi}^{\pi} \log\left(1 - s^2 \hat{f}(\lambda)\right) d\lambda \right. \tag{13.13}$$
$$\left. + \frac{s}{4\pi} \int_{-\pi}^{\pi} \log\left(1 - \hat{f}(\lambda)\right) d\lambda\right],$$

where the superscript in $I_1^{(0,1)}(0)$ contains the indices of two tested hypotheses. Details of the derivation for this case can be found in [453]. The second binary problem is formed from the original kth and lth hypotheses. Because of the symmetry of the problem, k can be set to 1 and l can be set to 2. The test statistic for this binary problem is the loglikelihood ratio:

$$\Lambda_n = -\frac{1}{2n} [X^n(1), Y^n, X^n(2)] \left(\tilde{\mathbf{R}}_1^{-1} - \tilde{\mathbf{R}}_2^{-1} \right)$$
$$\times [X^n(1), Y^n, X^n(2)]^T ,$$

where T denotes transpose and

$$\tilde{\mathbf{R}}_1 = \begin{bmatrix} \mathbf{K}_{ZW} & \mathbf{K}(S_Z) & 0 \\ \mathbf{K}(S_Z) & \mathbf{K}_{ZW} & 0 \\ 0 & 0 & \mathbf{K}_{ZW} \end{bmatrix}, \quad (13.14)$$

$$\tilde{\mathbf{R}}_2 = \begin{bmatrix} \mathbf{K}_{ZW} & 0 & 0 \\ 0 & \mathbf{K}_{ZW} & \mathbf{K}(S_Z) \\ 0 & \mathbf{K}(S_Z) & \mathbf{K}_{ZW} \end{bmatrix}, \quad (13.15)$$

with $\mathbf{K}_{ZW} = \mathbf{K}(S_Z) + \mathbf{K}(S_W)$.

The asymptotic log-moment generating function under the distribution of h_1 in this case is given by

$$\bar{\phi}_1 = -\frac{1}{4\pi} \int_{-\pi}^{\pi} \log \left[1 - (s^2 + (1+s)^2) \hat{f}(\lambda) \right] d\lambda$$
$$+ \frac{1}{4\pi} \int_{-\pi}^{\pi} \log \left[1 - \hat{f}(\lambda) \right] d\lambda.$$

The Chernoff information for this problem is the solution to the following equation:

$$I_1^{(1,2)}(0) = \sup_s [-\bar{\phi}_1],$$

where superscripts $(1, 2)$ in $I_1^{(1,2)}(0)$ are the indices of two tested hypotheses.

When $\hat{f}$ is a constant, the equations above can be readily solved. Suppose that the vectors in the test statistics have i.i.d. components and $\hat{f} = (\sigma^2/(\sigma^2 + \eta^2))^2$, where σ^2 is the variance of the signal part in tested Gaussian signatures and η^2 is the variance of the noise in tested Gaussian signatures. Then the optimal parameter that solves (13.13) is given by

$$s^* = -\frac{1}{\log(1-\hat{f})} - \sqrt{\frac{1}{\log^2(1-\hat{f})} + \frac{1}{\hat{f}}}. \quad (13.16)$$

For the second binary problem the optimal parameter is equal to $s^* = -0.5$. Then the exponent of the minimum probability of error $P(error)$ for the original M-ary problem is determined by the smallest of two Chernoff informations:

$$\min \left\{ I_1^{(0,1)}(0), I_1^{(1,2)}(0) \right\},$$

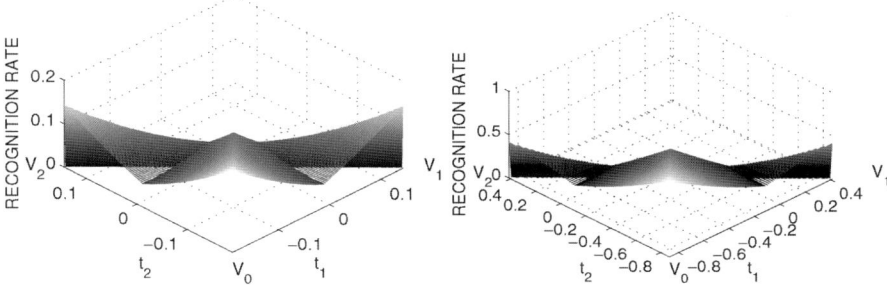

Fig. 13.2. The upper and the lower panels show the minimal component in the vector of the large deviation rate functions as a function of the thresholds t_1 and t_2 for $\sigma^2 = \eta^2 = 1$ and $\sigma^2 = 3$ and $\eta^2 = 1$, respectively

where

$$I_1^{(0,1)}(0) = \frac{1}{2}\log\left(1 - s^{*2}\hat{f}\right) + \frac{s^*}{2}\log(1 - \hat{f})$$

with s^* in (13.16) and

$$I_1^{(1,2)}(0) = \frac{1}{2}\log\left(1 - \frac{\hat{f}}{2}\right) - \frac{1}{2}\log\left(1 - \hat{f}\right).$$

Figure 13.2 shows the dependence of the recognition rate on the values of the thresholds t_1 and t_2 for two choices of σ^2 and η^2. The plot on the left panel is generated for $\sigma^2 = \eta^2 = 1$. The plot on the right panel is generated for $\sigma^2 = 3$ and $\eta^2 = 1$. Since for both cases $\bar{D} > \bar{I}$, by Comment 2 the minimal probability of error for both examples is achieved at the point $(0, -\bar{D})$ and is equal to the Chernoff information $I^{(0,1)}(0) = 0.0398$ and $I^{(0,1)} = 0.1349$, respectively.

13.5.2 Capacity of the Multimodal System Based on PCA Signatures of the Face and Iris

In this example we will show that by carefully modeling data we can predict performance of currently available in practice biometric recognition systems. The framework for approaching the modeling problem is demonstrated for PCA representation applied to combined Face and Iris data. Following this framework we will be able to evaluate the performance of various representation schemes for a number of biometric modalities.

Consider the case where both iris and face images are processed using global PCA method. Suppose that two sets of templates associated with the face modality, modality A, and iris modality, modality B, are available. The templates $X_A^n(m)$ and $X_B^n(m)$, two sets of weights in PCA representation of modalities A and B from user m, are modeled as realizations of two mutually independent Gaussian random vectors each of length n with mean zero and variances equal to the eigenvalues of the empirical covariance matrices formed using 108 sample vectors from each of biometric modalities, iris and face. Note that a modality of each individual is represented in training data by a single image. Vectors $X_c^n(m), m = 1, \ldots, M$, $c \in A, B$ are assumed to be mutually independent and identically distributed. Suppose that noisy template Y_A^n and Y_B^n from two independent modalities, face and iris, are presented for identification. The templates are realizations of Gaussian vectors with mean zero and variances equal to the eigenvalues of the empirical covariance matrices augmented with an estimated variance of noise in encoded face and iris data. Then the vector of information densities is given by $i_n^M = 1/n[i_n^M(1), \ldots, i_n^M(M)]^T$ with $i_n^M(m)$ being

$$i_n^M(m) = -\frac{1}{2n} \sum_{c \in \{A,B\}} \sum_{k=1}^{n} \left(\frac{X_{c,k}^2}{\sigma_c^2} - 2\frac{X_{c,k}Y_{c,k}}{\sigma_c^2} \right.$$
$$\left. + \frac{\lambda_{c,k}Y_{c,k}^2}{\sigma_c^2(\lambda_{c,k} + \sigma_c^2)} - \log\left(1 + \frac{\lambda_{c,k}}{\sigma_c^2}\right) \right), \quad (13.17)$$

where $\lambda_{c,k}$ is the kth eigenvalue of the modality $c \in \{A, B\}$ and σ_c^2 is the estimated noise present in the candidate observation Y_c^n. The PCA representation rate for iris and face (assuming that the biometric modalities are independent and resolution of images is high) is the asymptotic expected value of a single component (13.17) under the joint for this component distribution. After performing analysis of empirical dependencies of the eigenvalues on the parameters n and M, we have the expression for the representation rate

$$\bar{R}_{PCA} = \lim_{M \to \infty} \lim_{n \to \infty} \frac{1}{2n} \sum_{c \in \{A,B\}} \sum_{k=1}^{n} \log\left(1 + \frac{\lambda_{c,k}(M)}{\sigma_{c,k}^2(M)}\right),$$

where the dependence of eigenvalues and the noise value on the number of users in the database and the number of entries in representation vector are explicitly indicated. We have shown empirically (see Fig. 13.3) that the sequences

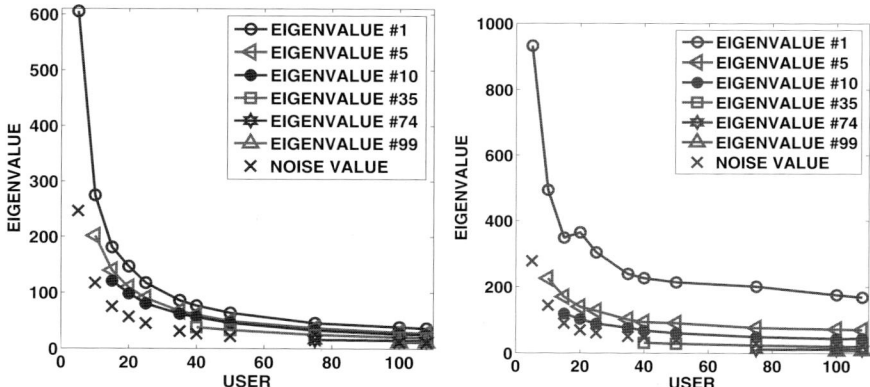

Fig. 13.3. Dependence of values of individual eigenvalues on the number of users used to estimate eigenvalues for iris (*left panel*) and face (*right panel*) biometrics

$\{\lambda_{c,k}(M)\}$ and $\{\sigma_c^2(M)\}$ have similar rates of decay as M and n increase. For this particular example, the rate of representation for the PCA encoded iris data is equal to 0.5. The rate of representation for the PCA encoded face data is equal to 0.189. Thus the capacity of the multimodal system that uses combined PCA representation for Face and Iris is 0.689.

13.6 Summary

Biometrics are physical signatures associated with individuals. Performance analysis of recognition systems based on physical signatures in general and biometrics in particular is often done experimentally, with results not easily generalized to alternative implementations. This chapter describes a framework for determining the performance of physical signature authentication based on likelihood models. The hypothesis testing approach yields the information density as the test statistic for deciding if two realizations of a random process are independent. Under the hypothesis that the realizations are dependent, the information density converges to the sup-mutual information rate between the observations. Under the independence hypothesis, the information density converges to the negative sup-relative entropy rate between the product and the joint distributions. Conditions are given under which the convergence rates are exponential, and rate functions are derived.

Performance of the identification system is analyzed applying the theory of large deviations. It is shown that the minimum probability of error is determined by the smallest component in the vector of large deviations rate functions. Capacity of a biometrics-based recognition system is defined.

The results rely on models for the data. For many anticipated applications, the derivation of these models may be the most challenging aspect. For example, in biometric applications, stochastic models for the biometric features being measured are required. In magnetic medium authentication, explicit stochastic micromagnetic models for the medium must be used. Typically, present implementations use suboptimal, nonparametric test statistics such as the correlation coefficient between the candidate signature presented for authentication and the previous signature measurement [460, 461]. Much work remains in quantifying the performance loss in using a suboptimal approach.

Acknowledgment

The authors would like to thank Francesco Nicolo for his assistance in preparing this manuscript. The authors are grateful to the editorial team for inviting them to contribute a chapter.

Part V

Acknowledgments, Biographies, References and Index items

14 Acknowledgments

Chapter 2: The authors like to acknowledge Keyur Patel and Manas Desai for designing a video-based face recognition system that was very helpful for experimenting with the proposed algorithms.

Chapter 4: This work was supported by the DOE University Research program in Robotics under grant #DOE-DEFG02-86NE37968 and NSF-CITeR grant #01-598B-UT. Special appreciations are to Trey Bohon, Tim Grundman, Kevin Lynn, and Doug Kiesler for their work in collecting the indoor face video database as part of their Senior Design project.

Chapter 5: The authors like to thank the United States Technical Support Working Group (TSWG), especially Prof. Jim Wayman and Dr. David Herrington for supporting this research effort.

Chapter 9: Portions of this work were supported by the following sponsors: the MIT Oxygen Alliance, ITRI, Intel Corporation, and the Queen's University Belfast Exchange Scheme.

Chapter 10: We are grateful for the support provided by the University of Houston; Department of Computer Science, University of Houston; Texas Learning and Computation Center, Univ. of Houston; Central Intelligence Agency; and Southwest Public Safety Technology Center, University of Houston.

Chapter 13: The authors like to thank Francesco Nicolo for his assistance in preparing this manuscript. The authors are grateful to the editorial team for inviting them to contribute a chapter.

15 Biographies

The following is the list of contributors and their biographies:

Besma R. Abidi received two M.S. degrees in image processing and remote sensing with honors from the National Engineering School of Tunis, Tunisia, in 1985 and 1986, respectively, and the Ph.D. degree from The University of Tennessee, Knoxville (UTK), in 1995. She is a Research Assistant Professor with the Department of Electrical and Computer Engineering, UTK, which she joined in 1998. She occupied a postdoctorate position with the Oak Ridge Institute of Science and Energy, Oak Ridge, TN, and was a Research Scientist at the Oak Ridge National Laboratory from 1998 to 2001. From 1985 to 1988, she was an Assistant Professor at the National Engineering School of Tunis. She has published over 70 papers and several book chapters. Her general areas of research are in image enhancement and restoration, sensor positioning and geometry, sensor fusion, two- and three-dimensional video tracking, and biometrics. She is a senior member of IEEE, member of the Computer Society, The Society of Women Engineers, SPIE, Tau Beta Pi, Eta Kappa Nu, Phi Kappa Phi, and The Order of the Engineer.

Mongi A. Abidi received the M.S. and Ph.D. degrees in Electrical Engineering from the University of Tennessee, Knoxville, in 1985 and 1987, respectively. He is W. Fulton Professor with the Department of Electrical and Computer Engineering at the University of Tennessee, Knoxville, which he joined in 1986. His interests include image processing, multisensor processing, three-dimensional imaging, and robotics. He has published over 230 papers in these areas and coedited a book entitled Data Fusion in Robotics and Machine Intelligence, (Academic 1992). He is the recipient of the 1994–1995 Chancellor's Award for Excellence in Research and Creative Achievement, and the 2001 Brooks Distinguished Professor Award. He is a member of the Computer Society, Pattern Recognition Society, SPIE, Tau Beta Pi, Phi Kappa Phi, and Eta Kappa Nu and The Order of the Engineer. He also received the First Presidential Principal Engineer Award.

Ognjen Arandjelovic is a Ph.D. student in the Machine Intelligence Laboratory at the University of Cambridge. He graduated top of his class from the Department of Engineering Science at the University of Oxford in 2003. His research interests include computer vision and machine learning, and their application in other scientific disciplines. He is a research scholar of Trinity College, Cambridge and a Fellow of the Cambridge Overseas Trust.

Vijayan Asari is an Associate Professor in Electrical and Computer Engineering at Old Dominion University, Norfolk, Virginia. He is the Director of the Computational Intelligence and Machine Vision Laboratory (ODU Vision Lab) in the ECE department. He received the Bachelor's degree in Electronics and Communication Engineering from the University of Kerala (College of Engineering, Trivandrum), India, the M.Tech. and Ph.D. degrees in Electrical Engineering from the Indian Institute of Technology, Madras. His research activities are in the areas of image processing, computer vision, pattern recognition, neural networks, and digital system architectures for application specific integrated circuits. He is a senior member of the IEEE.

Bir Bhanu received the S.M. and E.E. degrees in Electrical Engineering and Computer Science from the Massachusetts Institute of Technology, Cambridge, the Ph.D. degree in Electrical Engineering from the Image Processing Institute, University of Southern California, Los Angeles, and the M.B.A. degree from the University of California, Irvine. He has been the founding Professor of Electrical Engineering and served its first Chair at the University of California at Riverside (UCR). He has been the Cooperative Professor of Computer Science and Engineering and Director of Visualization and Intelligent Systems Laboratory (VISLab) since 1991. Currently, he also serves as the founding Director of an interdisciplinary Center for Research in Intelligent Systems (CRIS) at UCR. Previously, he was a Senior Honeywell Fellow at Honeywell Inc., Minneapolis, MN. He has been on the faculty of the Department of Computer Science at the University of Utah, Salt Lake City, and has worked at Ford Aerospace and Communications Corporation, CA, INRIA-France, and IBM San Jose Research Laboratory, CA. He has been the principal investigator of various programs for DARPA, NASA, NSF, AFOSR, ARO, and other agencies and industries in the areas of learning and vision, image understanding, pattern recognition, target recognition, biometrics, navigation, image databases, and machine vision applications. He is the coauthor of Evolutionary Synthesis of Pattern Recognition Systems (New York: Springer-Verlag, 2005), Computational Algorithms for Fingerprint Recognition (Norwell, MA: Kluwer, 2004), Genetic Learning for Adaptive Image Segmentation (Norwell, MA: Kluwer, 1994), and Qualitative Motion Understanding (Norwell, MA; Kluwer, 1992), and the coeditor of Computer Vision Beyond the Visible Spectrum, (New York: Springer-Verlag, 2004). He holds 11 US and international patents and over 250 reviewed technical publications in the areas of his interest. He has received two outstanding paper awards from the Pattern Recognition Society and has received industrial and university awards for research excellence, outstanding contributions, and team efforts. He has been on the editorial board of various journals and has edited special issues of several IEEE transactions and other journals. He has been General Chair for the IEEE Conference on Computer Vision and Pattern Recognition, IEEE Workshops on Applications of Computer Vision, IEEE Workshops on Learning in Computer Vision and Pattern Recognition; Chair for the DARPA Image Understanding Workshop, and Program Chair for the IEEE Workshops on Computer Vision Beyond the Visible Spectrum. He is a Fellow of the American Association for the Advancement of Science (AAAS), Institute of Electrical and Electronics Engineers (IEEE), International Association of Pattern Recognition (IAPR), and the International Society for Optical Engineering (SPIE).

Pradeep Buddharaju received his M.S. degree in the Department of Computer Science of the University of Houston and is currently enrolled as a doctoral student in the same department. His research interests include biometric authentication, security, surveillance, pattern recognition, and computer vision. He worked as a summer research intern at Honeywell Laboratories during 2004–2006. He is the recipient of the Motorola Outstanding Paper Award in the third International Summer School on Biometrics, Alghero, Italy, 2006. He also received the Best Teaching Assistant Award from the Department of Computer Science of the University of Houston for the year 2004. Mr. Buddharaju developed a novel method for face recognition that capitalizes on the unique and time-invariant physiological information of the face.

Rama Chellappa received the B.E. (Hons.) degree from the University of Madras, India, in 1975 and the M.E. (Distinction) degree from the Indian Institute of Science, Bangalore, in 1977. He received the M.S.E.E. and Ph.D. Degrees in Electrical Engineering from Purdue University, West Lafayette, IN, in 1978 and 1981 respectively. Since 1991, he has been a Professor of Electrical and Computer Engineering and an affiliate Professor of Computer Science at University of Maryland, College Park. He is also affiliated with the Center for Automation Research (Director) and the Institute for Advanced Computer Studies (Permanent member). Recently, he was named a Minta Martin Professor of Engineering. Prior to joining the University of Maryland, he was an Assistant (1981–1986) and Associate Professor (1986–1991) and Director of the Signal and Image Processing Institute (1988–1990) at University of Southern California, Los Angeles. Over the last 25 years, he has published numerous book chapters, peer-reviewed journal and conference papers in image and video processing, analysis and recognition. He has also coedited/coauthored six books on neural networks, Markov random fields, face/gait-based human identification and activity modeling. His current research interests are face and gait analysis, three-dimensional modeling from video, automatic target recognition from stationary and moving platforms, surveillance and monitoring, hyper spectral processing, image understanding, and commercial applications of image processing and understanding. He has served as an associate editor of the IEEE Transactions on Signal Processing, Pattern Analysis and Machine Intelligence, Image Processing, and Neural Networks. He was a Coeditor-in-Chief of Graphical Models and Image Processing. He also served as the Editor-in-Chief of IEEE Transactions on Pattern Analysis and Machine Intelligence during 2001–2004. He served as a member of the IEEE Signal Processing Society Board of Governors during 1996–1999 and as its Vice President of Awards and Membership during 2002–2004. He has received several awards, including NSF Presidential Young Investigator Award, an IBM Faculty Development Award, the 1990 Excellence in Teaching Award from the School of Engineering at USC, the 1992

Best Industry Related Paper Award from the International Association of Pattern Recognition (with Q. Zheng), and the 2000 Technical Achievement Award from IEEE Signal Processing Society. At University of Maryland, he was elected as a Distinguished Faculty Research Fellow (1996–1998) and as a Distinguished Scholar-Teacher (2003). He is a Fellow of IEEE and the International Association for Pattern Recognition. He has served as a General the Technical Program Chair for several IEEE International and National Conferences and Workshops. He is a Golden Core Member of IEEE Computer Society.

Amit K. Roy-Chowdhury has been an Assistant Professor of Electrical Engineering at the University of California, Riverside since January 2004. He completed his Ph.D. in 2002 from the University of Maryland, College Park, where he also worked as a Research Associate in 2003. Previously, he received his Master of Engineering in Systems Science and Automation from the Indian Institute of Science, Bangalore, India in 1997. His research interests are in the broad areas of image processing and analysis, computer vision, video communications and machine learning. Currently, he is working on problems of pose and illumination invariant video-based object recognition, event analysis in large video networks, and multiterminal video compression. He has over 50 papers in peer-reviewed journals, conferences and edited books. He is an author of the book titled "Recognition of Humans and their Activities Using Video". He is on the program committee of many major conferences in computer vision and image/signal processing and is a regular reviewer for the main journals in these areas.

Roberto Cipolla received the B.A. degree (Engineering) from the University of Cambridge in 1984 and the M.S.E. degree (Electrical Engineering) from the University of Pennsylvania in 1985. In 1991, he was awarded the D.Phil. degree (Computer Vision) from the University of Oxford. His research interests are in computer vision and robotics and include the recovery of motion and three-dimensional shape of visible surfaces from image sequences, visual tracking and navigation, robot hand-eye coordination, algebraic and geometric invariants for object recognition and perceptual grouping, novel man-machine interfaces using visual gestures, and visual inspection. He has authored three books, edited six volumes, and coauthored more than 200 papers.

Satyanadh Gundimada received the Bachelor's degree in Electronics and Communication Engineering from Osmania University, India, in 2001, the Master's degree in Electrical Engineering from Old Dominion University, in 2003. He is currently pursuing his Ph.D. degree in Electrical and Computer Engineering at Old Dominion University. His current research interests include pattern recognition, multisensor image fusion, computer vision and image processing.

Dr. Riad I. Hammoud is a research scientist, author, accomplished entrepreneur, futurist and advisor. He is currently a *senior research scientist* at the World Headquarters of Electronics and Safety Division of Delphi Corporation. Since February 2001, he holds a Ph.D. degree in "Computer Vision and Robotics" from

INRIA Rhone-Alpes, and a M.S. degree in "Control of Systems" from Université de Technologie de Compiègne, France. After his Ph.D., he launched a startup on the campus of Grenoble on "Interactive Video Tech". Around mid 2001, he moved to US and joined Rensselaer Polytechnic Institute (Troy, NY) and Indiana University (Bloomington, IN) as visiting and postdoctoral fellow. His research interests include automatic target classification in and beyond the visible spectrum, object tracking, biometrics, and real-time optimization techniques for safety, security and entertainment applications. His research is performed under confidential terms and has been funded by Alcatel Alsthom Research, INRIA, Honda, US Air Force, Indiana University, and Delphi Electronics & Safety. He published over 30 referred publications on object and image classification, video analysis, eye and pedestrian tracking, stereo vision, statistical modeling, face biometrics, surveillance and driver monitoring systems. He holds over 20 patents pending, defense publications and trade secrets. He authored several Springer books: "Interactive Video: algorithms and technologies", "Multi-Sensory and Multi-Modal face biometrics for personal identification", "Computer-Aided Eye Monitoring", and "Object Tracking and Classification Beyond the Visible Spectrum". He has been organizing and chairing several IEEE, SPIE and ACM International workshops and conference sessions (OTCBVS, ETRA, ATR, IVAN, IVRCIA). He is been serving on the reviewing committee of several journals in computer vision (PAMI, IJCV, CVIU), as well as national and international conferences (IEEE Intelligent Vehicles 2006, SPIE Defense and Security Symposiums, ACM ETRA Symposiums, IEEE Advanced Video and Signal Surveillance). He was appointed in 2004 as *guest editor* of a special issue of Springer it International Journal of Computer Vision (IJCV), and in 2005 as *guest editor* of a special issue of Elseiver *Computer Vision and Image Understanding Journal* (CVIU). He participated to the *Society Automotive Engineers* (SAE) conference, as a *panelist*, on 1st November 2005. Recently, he gave an invited talk at the Houston University, Texas, on "Video technologies on the move" (February 2006), and he participated to the defense jury of a Ph.D. student at the University of Sherbrooke, Canada (March 2006). He was nominated by US government as an *outstanding researcher/professor*, in May 2005. He received numerous awards from Delphi Corporation including the "best technical publicity/paper award" in 2006. He is the architect of the core algorithms of two vision-based safety products of Delphi Electronics & Safety: Driver Fatigue Monitoring and Driver Distraction Detection.

Ju Han received the Ph.D. degree in Electrical Engineering from University of California, Riverside in 2005. Currently, he is a postdoctoral fellow in the Life Science Division at Lawrence Berkeley National Laboratory. His research interests include biometrics, biological image understanding and computational biology.

Timothy J. Hazen received the S.B., S.M., and Ph.D. degrees in Electrical Engineering and Computer Science from the Massachusetts Institute of Technology (MIT) in 1991, 1993, and 1998, respectively. Since 1998 he has been a Research Scientist at the MIT Computer Science and Artificial Intelligence Laboratory where he works in the areas of automatic speech recognition, automatic person identification, multimodal speech processing, and conversational speech systems. He is also an Associate Editor of the IEEE Transactions on Audio, Speech and Language Processing.

Bernd Heisele received his Ph.D. from the University of Stuttgart in 1997. He is currently a Senior Scientist at the Honda Research Institute and a Visiting Researcher at the MIT Center for Biological and Computational Learning. His research focuses on the problems of detecting, tracking, and recognizing faces and other nonrigid moving objects.

Anil K. Jain is a University Distinguished Professor in the Department of Computer Science and Engineering at Michigan State University. He received his B.Tech. degree from Indian Institute of Technology, Kanpur and M.S. and Ph.D. degrees from Ohio State University. He received a distinguished alumni award from Ohio State University. His research interests include statistical pattern recognition, data clustering and biometric authentication. He received the 1996 IEEE Transactions on Neural Networks Outstanding Paper Award and best paper awards from the Pattern Recognition Society in 1987 and 1991. He was the Editor-in- Chief of the IEEE Transactions on Pattern Analysis and Machine Intelligence. He is a fellow of the IEEE, ACM, IAPR, SPIE, and AAAS. He has received a Fulbright Research Award, a Guggenheim fellowship, the Alexander von Humboldt Research Award and the 2003 IEEE Computer Society Technical Achievement Award. Holder of six patents in the area of fingerprint matching, he is the author of a number of books, including Handbook of Multibiometrics, Springer 2006, Biometric Systems, Technology, Design and Performance Evaluation, Springer 2005, Handbook of Face Recognition, Springer 2005, Handbook of Fingerprint Recognition, Springer 2003 (received the PSP award from the Association of American Publishers), BIOMETRICS: Personal Identification in Networked Society, Kluwer 1999, and Algorithms For Clustering Data, Prentice-Hall 1988. ISI has designated him as a highly cited researcher. He is an Associate editor of the IEEE Transactions on Information Forensics and Security and ACM Transactions on Knowledge Discovery in Data. He is a member of the National Academies panels on Whither Biometrics and Improvised Explosive Device.

Ioannis A. Kakadiaris received the Ptychion (B.Sc.) in Physics from the University of Athens, Greece, in 1989, the M.Sc. in Computer Science from Northeastern University, Boston, MA, in 1991, and the Ph.D. in Computer Science from University of Pennsylvania, Philadelphia, PA, in 1997. He joined the University of Houston (UH) in August 1997 after completing a postdoctoral fellowship at the University of Pennsylvania. He is the founder and Director of UH's Computational Biomedicine Laboratory (formerly the Visual Computing Lab) and Director of the Division of Bio-Imaging and Bio-Computation at the UH Institute for Digital Informatics and Analysis. In addition, he is the founder of Vision and Graphics Computing, Inc. His research interests include biomedical image analysis, biometrics, computer vision, and pattern recognition. He is the recipient of the year 2000 NSF Early Career Development Award, UH Computer Science Research

Excellence Award, UH Enron Teaching Excellence Award, James Muller VP Young Investigator Prize, and the Schlumberger Technical Foundation Award.

Pradeep Khosla is currently the Dean (2004) of the Carnegie Institute of Technology (the College of Engineering at Carnegie Mellon), the Philip and Marsha Dowd Professor in the College of Engineering and School of Computer Science (1998), and Founding co-Director of CyLab (2003). His previous positions include Assistant Professor (1986–1990), Associate Professor (1990–1994), and Professor (1994), Founding Director (1997–1999) of Institute for Complex Engineered Systems (ICES), Department Head of Electrical and Computer Engineering (1999–2004), and Director of Information Networking Institute (2000–2004). Prior to joining Carnegie-Mellon, he worked with Tata Consulting Engineers, and Siemens in the area of real-time control. He received B. Tech. (Hons.) from IIT (Kharagpur, India) in 1980, and both M.S. (1984) and Ph.D. (1986) degrees from Carnegie Mellon University. From January 1994 to August 1996 he was on leave from Carnegie Mellon and served as a DARPA Program Manager in the Software and Intelligent Systems Technology Office (SISTO), Defense Sciences Office (DSO), and Tactical Technology Office (TTO), where he managed advanced research and development programs, with a total budget exceeding $50M in FY96, in the areas of Information based Design and Manufacturing, Web based Information Technology Infrastructure, Real-Time Planning, and Distributed AI and Intelligent Systems, Real-Time Embedded Software, Sensor-based Control, and Collaborative Robotics. During his tenure as Founding Director (1997–1999), ICES grew to a total budget of more than $8M per year through strategic positioning to pursue interdisciplinary projects that involved faculty from six different colleges at Carnegie Mellon in the areas of Embedded Systems, Tissue Engineering, Design and Manufacturing, Design and Human Factors, and Networking. During his tenure as department head of Electrical and Computer Engineering, the department grew more than 80% in research volume, added 23 new faculty (tenure track and research), defined several strategic multidisciplinary initiatives, and the Computer Engineering graduate program was ranked number one for the first time, and the undergraduate program ranked third by US News and World Report in their 2002 rankings. In 2003 he founded Carnegie Mellon CyLab a university-wide research center with the goal of integrating technology (security, privacy, and next generation IT), policy, and economics of IT, to address multidisciplinary issues that require collaboration of experts across various disciplines. CyLab is a broad-based research, development, and community outreach oriented multidisciplinary center that supports and involves more than 40 faculty and 100 graduate students from five different colleges within Carnegie Mellon, and has an annual budget of more than $10M per year. As Director of the Information Networking Institute (INI), he increased its enrollment several fold, created the Master of Science in Information Security Technology and Management degree program, and defined international graduate degree, and research programs with the Athens Information Technology (AIT) Institute in Athens, Greece (CyLab, Athens), Cylab Korea, and CyLab Japan. He is involved in education both at the graduate and the undergraduate level.

He was a member of the committee that formulated a curriculum for the multidisciplinary Ph.D. program in Robotics at Carnegie Mellon. He was also a member of the Wipe, the Slate Clean Committee that created a new four year undergraduate ECE degree curriculum at CMU and proposed, amongst several other new ideas, the notion of teaching Engineering to freshman an idea that has been adopted widely by US and international universities. In support of the new curriculum he developed the Introductory Freshman level course "Introduction to Electrical and Computer Engineering" that emphasizes the notion of Teaching in Context. He is the coauthor of a text book and a laboratory manual for this freshman course. As Dean, he initiated the development of a professional M.S. program in Innovation Management and is providing leadership to redefine undergraduate education in engineering. He is a recipient of the Inlaks Foundation Fellowship in 1982, the Carnegie Institute of Technology Ladd award for excellence in research in 1989, two NASA Tech Brief awards (1992, 1993), the ASEE 1999 George Westinghouse Award for Education, the Siliconindia Leadership award for Excellence in Academics and Technology in 2000, and the W. Wallace McDowell award from IEEE Computer Society in 2001. He was elected Fellow of IEEE in January 1995, Fellow of AAAI in 2003, Fellow of AAAS in 2004, and member of the NAE in 2006. He served as Distinguished lecturer for the IEEE Robotics and Automation Society (1998–2003). In December 2002, he was appointed a member of the IT transition team of Pennsylvania Governor-elect Ed Rendell and in February 2003 he was appointed to the National Research Council Board on Manufacturing and Engineering Design for a three year term. His research has resulted in three books and more than 300 articles in journals, conferences, and book contributions. He has been a keynote and plenary speaker at several international conferences and workshops. He has served as member of the AdCom of the IEEE Robotics and Automation Society and the IEEE Systems, Man and Cybernetics Society, Chairman of the Education Committee of the IEEE Robotics and Automation Society, Professional Activities (PACE) Chair of the Robotics and Automation Society, member of Robotics and Intelligent Machines Coordinating Council (RIMCC), member of the Long Range Planning Committee of the Robotics and Automation Society, member of the Board of Directors of The Robotics Industries Association (RIA) from 1998 to 2002, and member of the Board of Directors of Pittsburgh Tissue Engineering Initiative (PTEI) during 2000–2001, and 2004 present. He served as Technical Editor of the IEEE Transactions on Robotics and Automation and Associate Editor for ASME Journal of Computers and Information Science in Engineering (JCISE), and IEEE Security and Privacy. He currently serves on editorial boards of IEEE Spectrum, and Oxford University Press series in Electrical and Computer Engineering. He is a consultant to several companies and Venture Capitalists and has served on the technology advisory boards of many start-ups and currently serves on several advisory boards including iNetworks, ITU Ventures, iPolicy, and Alcoa CIO's Advisory Board. He is a member of the Board of Directors of Quantapoint Inc., the Children's Institute, IIT Foundation, and MPC corporation. He also serves on the advisory boards of Institute for Systems Research (University of Maryland), College of Engineering (University of Waterloo), and is a member of the IT advisory committee, CSIRO, Australia. He has served as member of the Strategy Review Board for Ministry of Science and Technology, Taiwan; Council of Deans of the Aeronautics Advisory Committtee, NASA; and Senior Advisory Group, DARPA Program on Joint Unmanned Combat Air Systems. He is a cofounder of Quantapoint Inc. – a high-tech company based in Pittsburgh. Quantapoint specializes in high precision laser scanners that are used for creating high fidelity three-dimensional models.

B.V.K. Vijaya Kumar is a Professor in the Electrical and Computer Engineering Department at Carnegie Mellon University. His research interests include Automatic Target Recognition Algorithms, Biometric Recognition Methods and Coding and Signal Processing for Data Storage Systems. His publications include the book entitled Correlation Pattern Recognition (coauthored with Dr. Abhijit Mahalanobis and Dr. Richard Juday, Cambridge University Press, November 2005), eight book chapters and about 400 technical papers. He served as the Pattern Recognition Topical Editor for the Information Processing division of Applied Optics and is currently serving as an Associate Editor for IEEE Transactions on Information Forensics and Security. He has served on many conference program committees and was a cogeneral chair of the 2004 Optical Data Storage conference and a cogeneral chair of the 2005 IEEE AutoID Workshop. He is a senior member of IEEE, a Fellow of SPIE – The International Society of Optical Engineering, a Fellow of Optical Society of America (OSA) and a Fellow of the International Association of Pattern Recognition (IAPR). In 2003, Prof. Kumar received the Eta Kappa Nu award for Excellence in Teaching in the ECE Department at CMU and the Carnegie Institute of Technology's Dowd Fellowship for educational contributions.

Ji Ming is a Professor in the School of Electronics, Electrical Engineering and Computer Science at the Queen's University Belfast. He received a B.S. degree from Sichuan University, China, in 1982, an M.Phil. degree from Changsha Institute of Technology, China, in 1985, and a Ph.D. degree from Beijing Institute of Technology, China, in 1988, all in Electronic Engineering. He was Associate Professor with the Department of Electronic Engineering, Changsha Institute of Technology, from 1990 to 1993. From August 2005 to January 2006, he was a visiting scientist at the MIT Computer Science and Artificial Intelligence Laboratory. His research interests include speech and language processing, image processing, signal processing and pattern recognition.

Mohammed N. Murtuza, M.Sc. Candidate. Najam received his B.Sc. degree in Computer Science from Texas A&M University, College Station, TX, in 2003. Currently, he is a research assistant at the Computational Biomedicine Lab. His research interests include face recognition, ear recognition, three-dimensional biometric systems, genetic algorithms, and image processing.

Joseph A. O'Sullivan was born in St. Louis, MO, on 7th January 1960. He received the B.S., M.S., and Ph.D. all in Electrical Engineering from the University of Notre Dame in 1982, 1984, and 1986, respectively. In 1986, he joined the faculty in the Department of Electrical Engineering at Washington University, and is now Professor of Electrical and Systems Engineering. He has joint appointments in the Departments of Radiology and of Biomedical Engineering. He is Director of the Electronic Systems and Signals Research Laboratory and Associate Director of the Center for Security Technologies at Washington University. His is Chair of the Faculty Senate, Chair of the Faculty Senate Council, and faculty representative to the Board of Trustees at Washington University. He was Secretary of the Faculty Senate and of the Senate Council from 1995 to 1998. He was the publications editor for the IEEE Transactions on Information Theory from 1992 to 1995, was Associate Editor for Detection and Estimation, and was a Guest Associate Editor for the 2000 Special Issue on Information Theoretic Imaging. He was cochair of the 1999 Information Theory Workshop on Detection, Estimation, Classification, and Imaging. He is local arrangements chair for the IEEE 2003 Statistical Signal Processing Workshop. He will be cochair of the IEEE 2006 International Symposium on Information Theory. His research interests include information theory, information-theoretic imaging, automatic target recognition, CT imaging in the presence of known high density attenuators, information hiding, and hyperspectral imaging. He was chair of the St. Louis Section of the IEEE in 1994. He is a Fellow of the IEEE, a member of Eta Kappa Nu, SPIE, SIAM, and AAAS. He was awarded an IEEE third Millennium Medal.

Alex Park received S.B., M.Eng., and Ph.D. degrees in Electrical Engineering and Computer Science from the Massachusetts Institute of Technology in 2001, 2002, and 2006, respectively. The topic of his doctoral dissertation was the unsupervised acquisition of words from unlabeled speech. In addition to investigating the problem of unsupervised speech processing, he has also worked on speaker recognition, auditory modeling, and noise robust speech recognition. He is currently a postdoctoral researcher at the MIT Computer Science and Artificial Intelligence laboratory.

Georgios Passalis, Ph.D. candidate, received his B.Sc. from the Department of Informatics and Telecommunications, University of Athens. He subsequently received his M.Sc. from the Department of Computer Science, University of Houston. Currently, he is a Ph.D. Candidate at the University of Athens and Research Associate at the Computational Biomedicine Lab, University of Houston. His thesis is focused on the domains of Computer Graphics and Computer Vision. His research interests include object retrieval, face recognition, hardware accelerated voxelization, and object reconstruction.

Dr. Pavlidis holds Ph.D. and M.S. degrees in Computer Science from the University of Minnesota, an M.S. degree in Robotics from the Imperial College of the University of London, and a B.S. degree in Electrical Engineering from Democritus University in Greece. He joined the Computer Science Department at the University of Houston in September 2002, where he is currently a Professor and Director of the Computational Physiology Lab. His current research interests are in the area of computational medicine, where he is charting new territory. He has developed a series of methods to compute vital signs of subjects in an automated, contact-free, and passive manner. This new technology has found widespread applications in computational psychology and is expected to find additional applications in preventive medicine. The quantification of stress in particular, through the computation of periorbital blood perfusion, is his most well-known piece of research. It is this research that established him as one of the founders of modern lie detection technology. He is a Fulbright Fellow, a Senior Member of IEEE, and a member of the ACM. He also serves as Associate Editor for the Journal Pattern Analysis and Applications (Springer) and he has chaired numerous major IEEE conferences.

Narayanan Ramanathan received the B.E.(Hons.) degree in Electrical and Electronics Engineering from Birla Institute of Technology and Science, Pilani in 2002. He received the Masters degree in Electrical and Computer Engineering from University of Maryland College Park in 2004. He is currently pursuing Ph.D. in Electrical and Computer Engineering from University of Maryland College Park. His research interests span computer vision, pattern recognition and image processing. He was awarded the university gold medal for the Batch of 2002 from Birla Institute of Technology and Science, Pilani.

Arun Ross received the B.E. (Hons.) degree in Computer Science from the Birla Institute of Technology and Science, Pilani, India, in 1996, and the M.S. and Ph.D. degrees in Computer Science and Engineering from Michigan State University, East Lansing, in 1999 and 2003, respectively. Between 1996 and 1997, he was with the Design and Development Group of Tata Elxsi (India) Ltd., Bangalore, India. He also spent three summers (2000–2002) with the Imaging and Visualization Group at Siemens Corporate Research, Inc., Princeton, NJ, working on fingerprint recognition algorithms. He is currently an Assistant Professor in the Lane Department of Computer Science and Electrical Engineering, West Virginia University, Morgantown. His research interests include pattern recognition, classifier combination, machine learning, computer vision and biometrics. He is actively involved in the development of Pattern Recognition and Biometrics curricula at West Virginia University. He was an invited speaker at the Eighteenth Annual Kavli Frontiers of Science Symposium organized by the National Academy of Sciences in 2006. He also served as the US expert in multibiometrics at the ISO/IEC JTC1 SC37 meeting for biometric standards in 2004. Ross is the coauthor of the book "Handbook of Multibiometrics" published by Springer in 2006.

Prof. Marios Savvides is currently a Research Assistant Professor at the Electrical and Computer Engineering (ECE) Department at Carnegie Mellon University with a joint appointment at Carnegie Mellon CyLab. He obtained his B.Eng. (Hons.) degree in Microelectronics Systems Engineering from University of Manchester Institute of Science and Technology (UMIST) in England in 1997, his M.Sc. degree in Robotics in 2000 from the Robotics Institute in Carnegie Mellon University (CMU), Pittsburgh and his Ph.D. in May 2004 from the Electrical and Computer Engineering Department also at CMU where he focused on Biometrics Identification technology for his thesis "Reduced Complexity Face Recognition using Advanced Correlation

Filters and Fourier Subspace Methods for Biometric Applications". Prior to his current position he was a Systems Scientist in ECE and CyLab at CMU from May 2004 until January 2006. He was also the technical lead in CMU's face recognition and iris efforts in $FRGC + FRVT2006$ and $ICE1.0 + ICE2006$. His research interests are in Biometric recognition of Face, Iris, Fingerprint and Palmprint modalities. He is a member of IEEE and SPIE and serves on the Program Committee of several Biometric conferences including IEEE AutoID, SPIE Defense & Security Biometric Identification Technologies and International Conference on Image Analysis and Recognition (ICIAR) and the Biometrics Symposium of the annual Biometrics Consortium (BC 2006). He also served on the Biometrics Panel in SPIE's Defense & Security Symposium 2006 and is also listed in 2005 Edition of Marquis'x Who's Who in America. He has authored and coauthored over 75 conference and journal articles in the Biometrics area, has filed two patent applications and authored three book chapters in this field.

Dr. Natalia A. Schmid is an Assistant Professor in the Lane Department of Computer Science and Electrical Engineering at West Virginia University. She received D.Sc. degree in Electrical Engineering from Washington University in St. Louis. Her primary research interests include modern estimation and detection, statistical signal processing, and information theory with applications to biometrics, automatic target recognition, and medical imaging. She is currently involved in a number of projects related to biometrics and funded by the Center for Identification Technologies and Research, a joint NSF industry sponsored program. She is also working on designing and evaluating Automatic Target Recognition algorithms using the data collected by mini-UAVs exhibiting swarmed behavior. During her doctoral years at Washington University, she worked in the fields of Automatic Target Recognition (ATR) and general authentication. During her postdoctoral years at the University of Illinois in Urbana Champaign she was involved in work on Passive Radar Imaging. She publishes results of her work in recognized journals and conference proceedings. She was an invited speaker at two recent Workshops on Iris Recognition and Biometrics Quality organized and sponsored by the NIST.

Theoharis Theoharis, Associate Professor. He received his D.Phil. in computer graphics and parallel processing from the University of Oxford in 1988. He subsequently served as a research fellow (postdoctoral) at the University of Cambridge and as a consultant with Andersen Consulting. He is currently an Associate Professor with University of Athens and Adjunct Faculty with the Computational Biomedicine Lab, University of Houston. His main research interests lie in the fields of Computer Graphics, Visualization, Biometrics, and Archaeological Reconstruction.

George Toderici, Ph.D. candidate. George received his B.Sc. in Computer Science and Mathematics from the University of Houston. Currently, he is a Ph.D. Candidate at the University of Houston. He is a member of the Computational Biomedicine Lab focusing on face recognition research. George's research interests include machine learning, pattern recognition, object retrieval, and their possible applications on the GPU.

Eugene Weinstein is currently a Ph.D. student in the Computer Science Department at New York University, and an intern at Google's New York research lab. His current research interest is machine learning and its application to problems in speech, music, image, and text processing. Before coming to NYU, he was a research scientist at the Computer Science and Artificial Intelligence Laboratory at the Massachusetts Institute of Technology, where he worked on integration and core research efforts within Project Oxygen, a lab-wide pervasive computing project. He received his M.Eng. and B.S. degree in Computer Science and Engineering from MIT in 2001 and 2000, respectively. His Master's thesis work was on facilitating the development and deployment of robust natural language dialogue systems.

Lawrence B. Wolff is President and CEO of Equinox Corporation one of the leading companies providing electro-optical image fusion technology to the US military. He is also a Research Faculty member in the Department of Computer Science at Johns Hopkins University. From 1990 to 2002 he was a full-time faculty member there and has published over 100 articles in books, journals, conferences, workshops, and encyclopedias on aspects of physics-based computer vision and applied optics. In academia he pioneered a number of techniques applying polarization and multispectral measurement to computer vision. He received a Ph.D. and M.S. in Computer Science from Columbia University and a B.S. in Mathematics and Physics from Yale University. He has received a number of awards including the NSF Presidential Young Investigator award and was an Associate Editor for IEEE PAMI.

Yilei Xu (S'05) received his B.S. degree in Electrical Engineering from Peking University, Beijing, China, in 2004 and M.S. degree in Electrical Engineering from University of California at Riverside in 2006, where he is now pursuing the Ph.D. degree in Electrical Engineering Department. His main research interests include the computer vision, video processing and analysis, pattern recognition, machine learning. Now he is working on illumination modeling and motion estimation from video sequences.

Yi Yao received her B.S. and M.S. both in Electrical Engineering from Nanjing University of Aeronautics and Astronautics, China in 1996 and 2000, respectively. Currently she is a Ph.D. candidate in the Department of Electrical and Computer Engineering at the University of Tennessee, Knoxville. Her research interests include long range surveillance systems, object tracking, and multicamera surveillance systems.

Xiaoli Zhou received the B.S. and M.S. degrees from Beijing University of Posts and Telecommunications, China, in 1998 and 2001, respectively. Currently, she is a Ph.D. candidate in the Department of Electrical Engineering at the University of California, Riverside (UCR). She is working on her research at the Center for Research in Intelligent Systems at UCR. Her research interests are in computer vision, pattern recognition and image processing. Her recent research has been concerned with fusion of biometrics for human recognition at a distance in video.

References

1. W. Zhao, R. Chellappa, P.J. Phillips, and A. Rosenfeld. Face recognition: a literature survey. *ACM Transactions*, 35(4):399–458, 2003.
2. P.J. Phillips, P.J. Grother, R.J. Micheals, D.M. Blackburn, E. Tabassi, and J.M. Bone. Face recognition vendor test 2002: evaluation report. Technical Report NISTIR 6965 (http://www.frvt.org), 2003.
3. P.J. Phillips et al. Overview of the face recognition grand challenge. In *Computer Vision and Pattern Recognition*, 2005.
4. R. Basri and D.W. Jacobs. Lambertian reflectance and linear subspaces. *IEEE Transactions on Pattern Analysis and Machine Intelligence*, 25(2):218–233, 2003.
5. R. Ramamoorthi and P. Hanrahan. On the relationship between radiance and irradiance: determining the illumination from images of a convex Lambertian object. *Journal of the Optical Society of America A*, 18(10):2448–2459, 2001.
6. Y. Xu and A. Roy-Chowdhury. Integrating the effects of motion, illumination and structure in video sequences. In *Proceedings of IEEE International Conference on Computer Vision*, 2005.
7. Y. Xu and A. Roy-Chowdhury. Integrating motion, illumination and structure in video sequences, with applications in illumination-invariant tracking. *IEEE Transactions on Pattern Analysis and Machine Intelligence*, 2006 (in press).
8. K.C. Lee, J. Ho, M.H. Yang, and D.J. Kriegman. Video-based face recognition using probabilistic appearance manifolds. In *Computer Vision and Pattern Recognition*, volume 1, 2003, pp 313–320.
9. A. O'Toole et al. A video database of moving faces and people. *IEEE Transactions on Pattern Analysis and Machine Intelligence*, 27(5):812–816, 2005.
10. T. Sim, S. Baker, and M. Bsat. The CMU pose, illumination and expression database. *IEEE Transactions on Pattern Analysis and Machine Intelligence*, 25:1615–1618, 2003.
11. W. Zhao and R. Chellappa (Eds.). *Face Processing: Advanced Modeling and Methods*. New York: Academic, 2005.
12. K. Lee, J. Ho, and D.J. Kriegman. Acquiring linear subspaces for face recognition under variable lighting. *IEEE Transactions on Pattern Analysis and Machine Intelligence*, 27(5):684–698, 2005.
13. L. Zhang, S. Wang, and D. Samaras. Face synthesis and recognition from a single image under arbitrary unknown lighting using a spherical harmonic basis morphable model. In *Computer Vision and Pattern Recognition*, 2005.
14. V. Blanz, P. Grother, P. Phillips, and T. Vetter. Face recognition based on frontal views generated from non-frontal images. In *Computer Vision and Pattern Recognition*, 2005.
15. X. He, S. Yan, Y. Hu, P. Niyogi, and H.J. Zhang. Face recognition using laplacianfaces. *IEEE Transactions on Pattern Analysis and Machine Intelligence*, 27(5):328–340, 2005.

16. R. Gross, I. Matthews, and S. Baker. Eigen light-fields and face recognition across pose. In *Proceedings of the IEEE International Conference on Automatic Face and Gesture Recognition*, 2002.
17. R. Gross, I. Matthews, and S. Baker. Fisher light-fields for face recognition across pose and illumination. In *Proceedings of the German Symposium on Pattern Recognition*, 2002.
18. S. Zhou, R. Chellappa, and D. Jacobs. Characterization of human faces under illumination variations using rank, integrability, and symmetry constraints. In *European Conference on Computer Vision*, 2004.
19. M. Savvides, B.V.K. Vijaya Kumar, and P.K. Khosla. Corefaces – robust shift invariant PCA based correlation filter for illumination tolerant face recognition. In *Computer Vision and Pattern Recognition*, 2004.
20. M.A.O. Vasilescu and D. Terzopoulos. Multilinear independent components analysis. In *Computer Vision and Pattern Recognition*, 2005.
21. C. Xie, B.V.K. Vijaya Kumar, S. Palanivel, and B. Yegnanarayana. A still-to-video face verification system using advanced correlation filters. In *Proceedings of 1st International Conference on Biometric Authentication*, 2004.
22. K.W. Bowyer and K. Chang. A survey of 3D and Multimodal 3D+2D Face Recognition. In *Face Processing: Advanced Modeling and Methods*. New York: Academic, 2005.
23. B.K.P. Horn and B.G. Schunck. Determining optical flow. *Artificial Intelligence*, 17:185–203, 1981.
24. A. Pentland. Photometric motion. *IEEE Transactions on Pattern Analysis and Machine Intelligence*, 13(9):879–890, 1991.
25. G.D. Hager and P.N. Belhumeur. Efficient region tracking with parametric models of geometry and illumination. *IEEE Transactions on Pattern Analysis and Machine Intelligence*, 20(10):1025–1039, 1998.
26. S. Negahdaripour. Revised definition of optical flow: integration of radiometric and geometric cues for dynamic scene analysis. *IEEE Transactions on Pattern Analysis and Machine Intelligence*, 20(9):961–979, 1998.
27. D. Simakov, D. Frolova, and R. Basri. Dense shape reconstruction of a moving object under arbitrary, unknown lighting. In *IEEE International Conference on Computer Vision*, 2003.
28. L. Zhang, B. Curless, A. Hertzmann, and S.M. Seitz. Shape and motion under varying illumination: unifying structure from motion, photometric stereo, and multi-view stereo. In *Proceedings of IEEE International Conference on Computer Vision*, 2003.
29. D. Freedman and M. Turek. Illumination-invariant tracking via graph cuts. In *Proceedings of IEEE Conference on Computer Vision and Pattern Recognition*, 2005.
30. C. Tomasi and J. Shi. Good features to track. In *Proceedings of IEEE Conference on Computer Vision and Pattern Recognition*, 1994, pp 593–600.
31. H. Jin, P. Favaro, and S. Soatto. Real-time feature tracking and outlier rejection with changes in illumination. In *IEEE International Conference on Computer Vision*, 2001.
32. R.I. Hartley and A. Zisserman. *Multiple View Geometry in Computer Vision*. Cambridge: Cambridge University Press, 2000.
33. V. Lepetit and P. Fua. *Monocular Model-Based 3D Tracking of Rigid Objects*. Hanover, MA: Now Publishers, 2005.
34. M. Black and A. Jepson. EigenTracking: robust matching and tracking of articulated objects using a view-based representation. In *Proceedings of ECCV*, 1996, pp 329–342.

35. J. Lim, J. Ho, M.H. Yang, and D.J. Kriegman. Passive photometric stereo from motion. *Proceedings of IEEE International Conference on Computer Vision*, volume 2, 2005, pp 1635–1642.
36. T.F. Cootes, G.J. Edwards, and C.J. Taylor. Active appearance models. *IEEE Transactions on Pattern Analysis and Machine Intelligence*, 23(6):681–685, 2001.
37. S. Koterba, S. Baker, I. Matthews, C. Hu, H. Xiao, J. Cohn, and T. Kanade. Multi-view AAM fitting and camera calibration. In *IEEE International Conference on Computer Vision*, 2005.
38. P. Belhumeur and D. Kriegman. What is the set of images of an object under all possible lighting conditions? In *Computer Vision and Pattern Recognition*, 1996.
39. H.F. Chen, P.N. Belhumeur, and D.W. Jacobs. In search of illumination invariants. In *Proceedings of IEEE Conference on Computer Vision and Pattern Recognition*, volume 1, 2000, pp 254–261.
40. D.G. Lowe. Object recognition from local scale-invariant features. In *Proceedings of IEEE International Conference on Computer Vision*, 1999, pp 1150–1157.
41. R. Alferez and Y.F. Wang. Geometric and illumination invariants for object recognition. *IEEE Transactions on Pattern Analysis and Machine Intelligence*, 21(6):505–536, 1999.
42. D.A. Slater and G. Healey. The illumination-invariant matching of deterministic local structure in color images. *IEEE Transactions on Pattern Analysis and Machine Intelligence*, 19(10):1146–1151, 1997.
43. Y. Adini, Y. Moses, and S. Ullman. Face recognition: the problem of compensating for changes in illumination direction. *IEEE Transactions on Pattern Analysis and Machine Intelligence*, 19(7):721–732, 1997.
44. J. Ho and D Kriegman. On the effect of illumination and face recognition. In *Face Processing: Advanced Modeling and Methods*. New York: Academic, 2005.
45. R. Ramamoorthi. Modeling illumination variation with spherical harmonics. In *Face Processing: Advanced Modeling and Methods*. New York: Academic, 2005.
46. L.D. Lathauwer, B.D. Moor, and J. Vandewalle. A multilinear singular value decomposition. *SIAM Journal on Matrix Analysis and Applications*, 21(4):1253–1278, 2000.
47. V. Blanz and T. Vetter. Face recognition based on fitting a 3D morphable model. *IEEE Transactions on Pattern Analysis and Machine Intelligence*, 25(9):1063–1074, 2003.
48. A. Roy-Chowdhury, R. Chellappa, and R. Gupta. 3D face modeling from monocular video sequences. In *Face Processing: Advanced Modeling and Methods*. New York: Academic, 2005.
49. A. Roy-Chowdhury and R. Chellappa. Face reconstruction from monocular video using uncertainty analysis and a generic model. *Computer Vision and Image Understanding*, 91(1–2):188–213, 2003.
50. L.S. Mark, R.E. Shaw, and J.B. Pittenger. Natural constraints, scales of analysis, and information for the perception of growing faces. In *Social and Applied Aspects of Perceiving Faces*. Hillsdale, NJ: Lawrence Erlbaum Associates, 1998.
51. D. Thompson. *On Growth and Form*. New York: Dover, 1992 (original publication – 1917).
52. J.B. Pittenger and R.E. Shaw. Aging faces as viscal-elastic events: implications for a theory of nonrigid shape perception. *Journal of Experimental Psychology: Human Perception and Performance*, 1(4):374–382, 1975.
53. J.B. Pittenger, R.E. Shaw, and L.S. Mark. Perceptual information for the age level of faces as a higher order invariant to growth. *Journal of Experimental Psychology: Human Perception and Performance*, 5:478–493, 1979.

54. L.S. Mark, J.T. Todd, and R.E. Shaw. Perception of growth: a geometric analysis of how different styles of change are distinguished. *Journal of Experimental Psychology: Human Perception and Performance*, 7:855–868, 1981.
55. L.S. Mark and J.T. Todd. Describing geometric information about human growth in terms of geometric invariants. *Journal of Perception and Psychophysics*, 37:249–256, 1985.
56. J.T. Todd, L.S. Mark, R.E. Shaw, and J.B. Pittenger. The perception of human growth. *Scientific American*, 242(2):132–144, 1980.
57. Y.H. Kwon and N. da Vitoria Lobo. Age classification from facial images. *Computer Vision and Image Understanding*, 74:1–21, 1999.
58. Y.H. Kwon and N. da Vitoria Lobo. Age classification from facial images. In *IEEE Conference on Computer Vision and Pattern Recognition*, 1994, pp 762–767.
59. B. Tiddeman, D. Michael Burt, and D. Perret. Prototyping and transforming facial texture for perception research. *Computer Graphics and Applications, IEEE*, 21(5): 42–50, 2001.
60. M. Burt and D.I. Perrett. Perception of age in adult Caucasian male faces: computer graphic manipulation of shape and colour information. *Journal of Royal Society*, 259:137–143, 1995.
61. A. Lanitis, C.J. Taylor, and T.F. Cootes. Toward automatic simulation of aging effects on face images. *IEEE Transactions on Pattern Analysis and Machine Intelligence*, 24(4):442–455, 2002.
62. A. Lanitis, C. Draganova, and C. Christodoulou. Comparing different classifiers for automatic age estimation. *IEEE Transactions on Systems, Man and Cybernetics: Part B*, 34(1):621–628, 2004.
63. M. Gandhi. *A Method for Automatic Synthesis of Aged Human Facial Images*. Master's Thesis, McGill University, September 2004.
64. N. Ramanathan and R. Chellappa. Face verification across age progression. In *IEEE Conference on Computer Vision and Pattern Recognition*, volume 2, San Diego, CA, 2005, pp 462–469.
65. N. Ramanathan and R. Chellappa. Face verification across age progression. *IEEE Transactions on Image Processing*, 15(11):3349–3361, 2006.
66. N. Ramanathan and R. Chellappa. Modeling age progression in young faces. In *IEEE Conference on Computer Vision and Pattern Recognition*, volume 1, New York, NY, 2006, pp 387–394.
67. B. Moghaddam and A. Pentland. Probabilistic visual learning for object representation. *IEEE Transactions on Pattern Analysis and Machine Intelligence*, 19(7):696–710, 1997.
68. M. Loeve. *Probability Theory*. Princeton, NJ: Von Nostrand, 1955.
69. D.G. Stork, R.O. Duda, and P.E. Hart. *Pattern Classification*, 2nd edition. New York: Wiley, 2001.
70. G. McLachlan and T. Krishnan. *The EM Algorithm and Extensions*. New York: Wiley, 1996.
71. R. Alley. *Social and Applied Aspects of Perceiving Faces*. Hillsdale, NJ: Lawrence Erlbaum Associates, 1998.
72. L.G. Farkas. *Anthropometry of the Head and Face*. New York: Raven, 1994.
73. L.G. Farkas and I.R. Munro. *Anthropometric Facial Proportions in Medicine*. Springfield, IL: Charles C. Thomas, 1987.
74. D.M. Bates and D.G. Watts. *Nonlinear Regression and its Applications*. New York: Wiley, 1988.

75. F.L. Bookstein. Principal warps: thin-plate splines and the decomposition of deformations. *IEEE Transactions on Pattern Analysis and Machine Intelligence*, 11(6):567–585, 1989.
76. Face and Gesture Recognition Network: FG-NET aging database (online) available at http://sting.cycollege.ac.cy/~alanitis/fgnetaging/.
77. K. Messer, J. Matas, J. Kittler, J. Luettin, and G. Maitre. XM2VTSDB: the extended M2VTS database. In *Conference on Audio and Video-Based Biometric Person Authentication*, Washington, DC, March 1999.
78. E. Bailly-Bailliere, S. Bengio, F. Bimbot, M. Hamouz, J. Kittler, J. Mariethoz, J. Matas, K. Messer, V. Popovici, F. Poree, B. Ruiz, and J.-P. Thiran. The BANCA database and evaluation protocol. In *International Conference on Audio- and Video-Based Biometric Person Authentication*, Guilfore, UK, June 2003, pp 625–638.
79. P.J. Philips, H. Moon, P.J. Rauss, and S. Rizvi. The FERET evaluation methodology for face recognition algorithms. *IEEE Transactions on Pattern Analysis and Machine Intelligence*, 22(10):1090–1104, 2000.
80. W. Gao, B. Cao, S. Shan, D. Zhou, X. Zhang, and D. Zhao. The CASPEAL large-scale Chinese face database and evaluation protocols. Technical Report, Joint Research and Development Laboratory, CAS, 2004.
81. T. Sim, S. Baker, and M. Bsat. The CMU pose, illumination, and expression database. *IEEE Transactions on Pattern Analysis and Machine Intelligence*, 25(12):1615–1618, 2003.
82. A.J. O'Toole, J. Harms, S.L. Snow, D.R. Hurst, M.R. Pappas, J.H. Avyad, and H. Abdi. A video database of moving faces and people. *IEEE Transactions on Pattern Analysis and Machine Intelligence*, 27(5):812–816, 2005.
83. Y. Yao, B. Abidi, and M. Abidi. Digital imaging with extreme zoom: system design and image restoration. In *IEEE Conference on Vision Systems*, New York, January 2006.
84. P. Griffin. Understanding the face image format standards. In *ANSI/NIST Workshop*, Gaithersburg, MD, 2005.
85. N. Kalka, J. Zuo, N.A. Schmid, and B. Cukic. Image quality assessment for iris biometric. In *SPIE Symposium on Defense and Security, Conference on Human Identification Technology III*, Orlando, FL, April 2006.
86. Q. Xiong and C. Jaynes. Mugshot database acquisition in video surveillance networks using incremental auto-clustering quality measures. In *IEEE Conference on Advanced Video and Signal Based Surveillance*, Miami, FL, July 2003, pp 191–198.
87. G. Ramponi and A. Polesel. Rational unsharp masking technique. *Journal of Electronic Imaging*, 7(2):333–338, 1998.
88. G. Ramponi. A cubic unsharp masking technique for contrast enhancement. *Signal Processing*, 67(2):211–222, 1998.
89. T.G. Chan and C.K. Wong. Total variation blind deconvolution. *IEEE Transactions on Image Processing*, 7(3):370–375, 1998.
90. X. Fan, Q. Zhang, D. Liang, and L. Zhao. Face image restoration based on statistical prior and image blur measure. In *International Conference on Multimedia and Expo*, volume 3, Baltimore, MD, July 2003, pp 297–300.
91. I. Stainvas and N. Intrator. Blurred face recognition via a hybrid network architecture. In *International Conference on Pattern Recognition*, volume 2, Barcelona, Spain, September 2000, pp 805–808.
92. L. Liao and X. Lin. Blind image restoration with eigen-face subspace. *IEEE Transactions on Image Processing*, 14(11):1766–1772, 2005.

References

93. A. Santos, C. Ortiz de Solorzano, J.J. Vaquero, J.M. Pena, N. Malpica, and F. del Pozo. Evaluation of autofocus functions in molecular cytogenetic analysis. *Journal of Microscopy*, 188:264–272, 1997.
94. C.F. Batten. *Autofocusing and Astigmatism Correction in the Scanning Electron Microscope*. Master's Thesis, University of Cambridge, 2000.
95. E.P. Krotkov. *Active computer vision by cooperative focus and stereo*. Berlin Heidelberg New York: Springer, 1989.
96. N.K. Chern, P.A. Neow, and M.H. Ang. Practical issues in pixel-based autofocusing for machine vision. In *International Conference on Robotics and Automation*, Seoul, Korea, 2001, pp 2791–2796.
97. N.F. Zhang, M.T. Postek, R.D. Larrabee, A.E. Vladar, W.J. Keery, and S.N. Jones. Image sharpness measurement in scanning electron microscope part III. *Scanning*, 21(4):246–252, 1999.
98. M. Kristan, J. Pers, M. Perse, and S. Kovacic. A Bayes-spectral-entropy-based measure of camera focus using a discrete cosine transform. *Pattern Recognition Letters*, 27:1431–1439, 2006.
99. X. Li. Blind image quality assessment. In *IEEE ICIP*, Rochester, NY, September 2002, pp 449–452.
100. J. Dijk, M. van Ginkel, R.J. van Asselt, L.J. van Vliet, and P.W. Werbeek. A new sharpness measure based on Gaussian lines and edges. In *Proceedings of 8th Annual Conference on the Advanced School for Computing and Imaging*, June 2002, pp 39–43.
101. J. Caviedes and S. Gurbuz. No-reference sharpness metric based on local edge kurtosis. In *IEEE ICIP*, Rochester, NY, September 2002, pp 53–56.
102. J. Lin, Z. Zhang, and Q. Shi. Estimating the amount of defocus through a wavelet transform approach. *Pattern Recognition Letters*, 25(4):407–411, 2004.
103. S. Wang, R. Nathuji, R. Bettati, and W. Zhao. Providing statistical delay guarantees in wireless networks. In *ICDCS 2004*, March 2004.
104. Y. Yao, B. Abidi, and M. Abidi. Evaluation of sharpness measures and search algorithms for the auto-focusing of high magnification images. In *SPIE*, Orlando, FL, 2006.
105. K.S. Choi, J.S. Lee, and S.J. Ko. New autofocusing technique using the frequency selective weighted median filter for video camera. *IEEE Transactions on Consumer Electronics*, 45(3):820–827, 1999.
106. P.J. Phillips, P. Grother, R.J. Micheals, D.M. Blackburn, E. Tabassi, and M. Bone. Face recognition vendor test 2002: evaluation report. Technical Report.
107. Y. Yao, B. Abidi, N.D. Kalka, N. Schmid, and M. Abidi. High magnification and long distance face recognition: database acquisition, evaluation, and enhancement. In *Biometrics Symposium, Biometric Consortium Conference*, Baltimore, MD, September 2006.
108. V. Agarwal, A.V. Gribok, and M.A. Abidi. Image restoration using $\mathcal{L}_1$ norm penalty function. Technical Report, The University of Tennessee, Knoxville, 2005.
109. C.W. Groetsch. *The Theory of Tikhonov Regularization for Fredholm Integral Equations of the First Kind*. Boston: Pitman, 1984.
110. A.N. Tikhonov and V.Y. Arsenin. *Solutions of Ill-Posed Problems*. New York: Wiley, 1977.
111. T.F. Chan, G.H. Golub, and P. Mulet. A nonlinear primal–dual method for total variation-based image restoration. *SIAM Journal on Scientific Computing*, 20(6):1964–1977, 1999.
112. J.L. Wayman. Digital signal processing in biometric identification: a review. In *International Conference on Image Processing*, volume 1, 2002, pp I-37–I-40.

113. A.K. Jain, A. Ross, and S. Prabhakar. An introduction to biometric recognition. *IEEE Transactions on Circuits and Systems for Video Technology*, 14(1):4–20, 2004.
114. M.A. Turk and A.P. Pentland. Face recognition using eigenfaces. In *IEEE Computer Society Conference on Computer Vision and Pattern Recognition*, 1991, pp 586–591.
115. P.N. Belhumeur, J.P. Hespanha, and D.J. Kriegman. Eigenfaces vs. fisherfaces: recognition using class specific linear projection. *IEEE Transactions on Pattern Analysis and Machine Intelligence*, 19(7):711–720, 1997.
116. P.N. Belhumeur and D.J. Kriegman. What is the set of images of an object under all possible lighting conditions? In *IEEE Computer Society Conference on Computer Vision and Pattern Recognition*, 1996, pp 270–277.
117. P. Viola and M. Jones. Robust real-time face detection. In *Proceedings of 8th IEEE International Conference on Computer Vision (ICCV)*, volume 2, 2001, p 747.
118. H. Schneiderman. *A Statistical Approach to 3D Object Detection Applied to Faces and Cars*. Doctoral Dissertation, The Robotics Institute, Carnegie Mellon University, Pittsburgh, PA, 2000.
119. M. Savvides, B.V.K. Vijaya Kumar, and P.K. Khosla. Face verification using correlation filters. In *Proceedings of IEEE Automatic Advanced Identification Technologies*, Tarrytown, NY, June 2002, pp 19–21.
120. M. Savvides, K. Venkataramani, and B.V.K. Vijaya Kumar. Incremental updating of advanced correlation filters for biometric authentication systems. In *Proceedings of International Conference on Multimedia and Expo, 2003 (ICME '03)*, volume 3, 2003, pp 229–232.
121. M. Savvides, B.V.K. Vijaya Kumar, and P.K. Khosla. Cancellable biometric filters for face recognition. In *International Conference on Pattern Recognition*, volume 3, 2004, pp 922–925.
122. A.V. Lugt. Signal detection by complex spatial filtering. *IEEE Transactions on Information Theory*, 10:139–145, 1964.
123. C.F. Hester and D. Casasent. Multivariant technique for multiclass pattern recognition. *Applied Optics*, 19:1758–1761, 1980.
124. A. Mahalanobis, B.V.K. Vijaya Kumar, and D. Casasent. Minimum average correlation energy filters. *Applied Optics*, 26:3630–3633, 1987.
125. B.V.K. Vijaya Kumar. Minimum variance synthetic discriminant functions. *Journal of the Optical Society of America A*, 3:1579–1584, 1986.
126. P. Refregier. Optimal trade-off filters for noise robustness, sharpness of the correlation peak, and Horner efficiency. *Optics Letters*, 16:829–831, 1991.
127. B.V.K. Vijaya Kumar. Tutorial survey of composite filter designs for optical correlators. *Applied Optics*, 31:4773–4801, 1992.
128. M. Savvides. *Reduced Complexity Face Recognition Using Advanced Correlation Filters and Fourier Subspace Methods for Biometrics Applications*. Ph.D. Thesis, Carnegie Mellon University, Pittsburgh, PA, 2004.
129. B.V.K. Vijaya Kumar, A. Mahalanobis, and R.D. Juday. *Correlation Pattern Recognition*. Cambridge: Cambridge University Press, 2004.
130. A. Mahalanobis, B.V.K. Vijaya Kumar, S.R.F. Sims, and J.F. Epperson. Unconstrained correlation filters. *Applied Optics*, 33:3751–3759, 1994.
131. M. Hayes, J. Lim, and A. Oppenheim. Signal reconstruction from phase or magnitude. *IEEE Transactions on Acoustics, Speech, and Signal Processing*, 28(6):672–680, 1980.
132. M. Hayes, J. Lim, and A. Oppenheim. Phase-only signal reconstruction. In *IEEE International Conference on Acoustics, Speech, and Signal Processing (ICASSP)*, volume 5, 1980, pp 437–440.

133. S. Curtis, S.J. Lim, and A. Oppenheim. Signal reconstruction from one bit of Fourier transform phase. In *IEEE International Conference on Acoustics, Speech, and Signal Processing (ICASSP)*, volume 9, 1984, pp 487–490.
134. T. Quatieri Jr. and A. Oppenheim. Iterative techniques for minimum phase signal reconstruction from phase or magnitude. *IEEE Transactions on Acoustics, Speech, and Signal Processing*, 29(6):1187–1193, 1981.
135. M. Savvides and B.V.K. Vijaya Kumar. Quad phase minimum average correlation energy filters for reduced memory illumination tolerant face authentication. In *Lecture Notes in Computer Science*, volume 2688/2003. Berlin Heidelberg New York: Springer, 2003, pp 19–26.
136. T. Sim, S. Baker, and M. Bsat. The CMU pose, illumination, and expression (PIE) database of human faces. Technical Report, The Robotics Institute, Carnegie Mellon University, Pittsburgh, PA, 2001.
137. O. Arandjelović and R. Cipolla. Face recognition from video using the generic shape-illumination manifold. In *Proceedings of IEEE European Conference on Computer Vision (ECCV)*, May 2006.
138. T-J. Chin and D. Suter. A new distance criterion for face recognition using image sets. In *Proceedings of Asian Conference on Computer Vision (ACCV)*, 2006, pp 549–558.
139. J. Sivic, M. Everingham, and A. Zisserman. Person spotting: video shot retrieval for face sets. In *Proceedings of IEEE International Conference on Image and Video Retrieval (CIVR)*, 2005, pp 226–236.
140. P.J. Phillips, P. Grother, R.J. Micheals, D.M. Blackburn, E. Tabassi, and M. Bone. Face recognition vendor test 2002. Technical Report, National Institute of Standards and Technology, 2003. Evaluation Report, 1-56.
141. P.S. Penev. Dimensionality reduction by sparsification in a local-features representation of human faces. Technical Report, The Rockefeller University, 1999.
142. T. Kanade. Picture processing by computer complex and recognition of human faces. Technical Report, 1973.
143. I.J. Cox, J. Ghosn, and P.N. Yianilos. Feature-based face recognition using mixture-distance. In *Proceedings of IEEE Conference on Computer Vision and Pattern Recognition (CVPR)*, 1996, pp 209–216.
144. L.B. Wolff, D.A. Socolinsky, and C.K. Eveland. Quantitative measurement of illumination invariance for face recognition using thermal infrared imagery. In *IEEE Workshop on Computer Vision Beyond the Visible Spectrum*, 2001.
145. D. Socolinsky, A. Selinger, and J. Neuheisel. Face recognition with visible and thermal infrared imagery. *Computer Vision and Image Understanding*, 91(1–2):72–114, 2003.
146. F. Prokoski. History, current status, and future of infrared identification. In *IEEE Workshop on Computer Vision Beyond the Visible Spectrum*, Hilton Head, SC, 2000.
147. D. Socolinsky and A. Selinger. Thermal face recognition in an operational scenario. In *Proceedings of IEEE Conference on Computer Vision and Pattern Recognition (CVPR)*, volume 2, 2004, pp 1012–1019.
148. S. Kong, J. Heo, B. Abidi, J. Paik, and M. Abidi. Recent advances in visual and infrared face recognition – a review. *Computer Vision and Image Understanding*, 97:103–135, 2004.
149. D. Socolinsky and A. Selinger. Comparative study of face recognition performance with visible and thermal infrared imagery. In *Proceedings of IEEE International Conference on Pattern Recognition (ICPR)*, 2002, pp 217–222.
150. A. Selinger and D. Socolinsky. Appearance-based facial recognition using visible and thermal imagery: a comparative study. Technical Report 02-01, Equinox Corporation, 2002.

151. A. Srivastava and X. Liu. Statistical hypothesis pruning for recognizing faces from infrared images. *Image and Vision Computing*, 21(7):651–661, 2003.
152. P. Buddharaju, I. Pavlidis, and I. Kakadiaris. Face recognition in the thermal infrared spectrum. In *IEEE CVPR Workshop on Object Tracking and Classification Beyond the Visible Spectrum*, 2004, p 133.
153. G. Friedrich and Y. Yeshurun. Seeing people in the dark: face recognition in infrared images. In *Proceedings of IAPR British Machine Vision Conference (BMVC)*, 2003, pp 348–359.
154. R. Brunelli and D. Falavigna. Personal identification using multiple cues. *IEEE Transactions Pattern Analysis and Machine Intelligence*, 17(10):955–966, 1995.
155. A. Ross and A. Jain. Information fusion in biometrics. *Pattern Recognition Letters*, 24(13):2115–2125, 2003.
156. X. Chen, P. Flynn, and K. Bowyer. Visible-light and infrared face recognition. In *Workshop on Multimodal User Authentication*, 2003, pp 48–55.
157. J. Heo, B. Abidi, S.G. Kong, and M. Abidi. Performance comparison of visual and thermal signatures for face recognition. In *Biometric Consortium Conference*, Arlington, VA, September 2003.
158. S. Ben-Yacoub, Y. Abdeljaoued, and E. Mayoraz. Fusion of face and speech data for person identity verification. *IEEE Transactions on Neural Networks*, 10(5):1065–1074, 1999.
159. L. Hong and A. Jain. Integrating faces and fingerprints for personal identification. *IEEE Transactions Pattern Analysis and Machine Intelligence*, 20(12):1295–1307, 1998.
160. K. Chang, K.W. Bowyer, S. Sarka, and B. Victor. Comparison and combination of ear and face image in appearance-based biometrics. *IEEE Transactions Pattern Analysis and Machine Intelligence*, 25(9):1160–1165, 2003.
161. J. Heo, S. Kong, B. Abidi, and M. Abidi. Fusion of visual and thermal signatures with eyeglass removal for robust face recognition. In *IEEE CVPR Workshop on Object Tracking and Classification Beyond the Visible Spectrum*, 2004, p 122.
162. Identix Ltd, FaceIt (http://www.FaceIt.com/).
163. H. Hotelling. Relations between two sets of variates. *Biometrika*, 28:321–372, 1936.
164. T. Kim, O. Arandjelović, and R. Cipolla. Learning over sets using boosted manifold principal angles (BoMPA). In *Proceedings of IAPR British Machine Vision Conference (BMVC)*, volume 2, 2005, pp 779–788.
165. Å. Björck and G.H. Golub. Numerical methods for computing angles between linear subspaces. *Mathematics of Computation*, 27(123):579–594, 1973.
166. E. Oja. *Subspace Methods of Pattern Recognition*. Letchworth/New York: Research Studies/Wiley, 1983.
167. O. Arandjelović and A. Zisserman. On film character retrieval in feature-length films. In *Interactive Video: Algorithms and Technologies*. Berlin Heidelberg New York: Springer, 2006.
168. T.L. Berg, A.C. Berg, J. Edwards, M. Maire, R. White, Y.W. Teh, E. Learned-Miller, and D.A. Forsyth. Names and faces in the news. In *Proceedings of IEEE Conference on Computer Vision and Pattern Recognition (CVPR)*, volume 2, 2004, pp 848–854.
169. D. Cristinacce, T.F. Cootes, and I. Scott. A multistage approach to facial feature detection. In *Proceedings of IAPR British Machine Vision Conference (BMVC)*, volume 1, 2004, pp 277–286.
170. P.F. Felzenszwalb and D. Huttenlocher. Pictorial structures for object recognition. *International Journal of Computer Vision (IJCV)*, 61(1):55–79, 2005.

171. L. Trujillo, G. Olague, R. Hammud, and B. Hernandez. Automatic feature localization in thermal images for facial expression recognition. In *Proceedings of IEEE International Workshop on Object Tracking and Classification Beyond the Visible Spectrum*, 2005, 3 pp.
172. A. Fitzgibbon and A. Zisserman. On affine invariant clustering and automatic cast listing in movies. In *Proceedings of IEEE European Conference on Computer Vision (ECCV)*, 2002, pp 304–320.
173. Y. Adini, Y. Moses, and S. Ullman. Face recognition: the problem of compensating for changes in illumination direction. *IEEE Transactions on Pattern Analysis and Machine Intelligence*, 19(7):721–732, 1997.
174. H. Wang, S.Z. Li, and Y. Wang. Face recognition under varying lighting conditions using self quotient image. In *Proceedings of IEEE International Conference on Automatic Face and Gesture Recognition (FGR)*, 2004, pp 819–824.
175. O. Arandjelović and R. Cipolla. A new look at filtering techniques for illumination invariance in automatic face recognition. In *Proceedings of IEEE International Conference on Automatic Face and Gesture Recognition (FGR)*, 2006, pp 449–454.
176. T.C. Walker and R.K. Miller. *Health Care Business Market Research Handbook*, 5th edition. Norcross, GA: Richard K. Miller & Associates, 2001.
177. A.M. Martinez. Recognizing imprecisely localized, partially occluded and expression variant faces from a single sample per class. *IEEE Transactions on Pattern Analysis and Machine Intelligence (PAMI)*, 24(6):748–763, 2002.
178. K.K. Sung and T. Poggio. Example-based learning for view-based human face detection. *IEEE Transactions on Pattern Analysis and Machine Intelligence (PAMI)*, 20(1):39–51, 1998.
179. O. Arandjelović and R. Cipolla. Face set classification using maximally probable mutual modes. In *Proceedings of IEEE International Conference on Pattern Recognition (ICPR)*, August 2006, pp 511–514.
180. K. Fukui and O. Yamaguchi. Face recognition using multi-viewpoint patterns for robot vision. In *International Symposium of Robotics Research*, 2003.
181. A. Jain, R. Bolle, and S. Pankanti. *Biometrics: Personal Identification in Networked Society*, 1st edition. Dordecht: Kluwer, 1999.
182. W. Zhao, R. Chellappa, P.J. Phillips, and A. Rosenfeld. Face recognition: a literature survey. *ACM Computing Surveys (CSUR)*, 35(4):399–458, 2003.
183. I. Pavlidis and P. Symosek. The imaging issue in an automatic face/disguise detection system. In *Proceedings of IEEE Workshop on Computer Vision Beyond the Visible Spectrum: Methods and Applications*, Hilton Head Island, SC, June 2000, pp 15–24.
184. F. Prokoski. History, current status, and future of infrared identification. In *Proceedings of IEEE Workshop on Computer Vision Beyond the Visible Spectrum: Methods and Applications*, Hilton Head Island, SC, June 2000, pp 5–14.
185. D.A. Socolinsky and A. Selinger. A comparative analysis of face recognition performance with visible and thermal infrared imagery. In *Proceedings of 16th International Conference on Pattern Recognition*, volume 4, Quebec, Canada, 2002, pp 217–222.
186. J. Wilder, P.J. Phillips, C. Jiang, and S. Wiener. Comparison of visible and infrared imagery for face recognition. In *Proceedings of the 2nd International Conference on Automatic Face and Gesture Recognition*, Killington, VT, October 1996, pp 182–187.
187. D.A. Socolinsky, L.B. Wolff, J.D. Neuheisel, and C.K. Eveland. Illumination invariant face recognition using thermal infrared imagery. In *Proceedings of the IEEE Computer Society Conference on Computer Vision and Pattern Recognition (CVPR 2001)*, volume 1, Kauai, HI, 2001, pp 527–534.

188. A. Selinger and D.A. Socolinsky. Face recognition in the dark. In *Proceedings of the Joint IEEE Workshop on Object Tracking and Classification Beyond the Visible Spectrum*, Washington, DC, June 2004.
189. R. Cutler. Face recognition using infrared images and eigenfaces (http://www.cs.umd.edu/rgc/face/face.htm), 1996.
190. X. Chen, P.J. Flynn, and K.W. Bowyer. PCA-based face recognition in infrared imagery: baseline and comparative studies. In *Proceedings of the IEEE International Workshop on Analysis and Modeling of Faces and Gestures*, Nice, France, 17 October 2003, pp 127–134.
191. A. Srivastava and X. Liu Statistical hypothesis pruning for recognizing faces from infrared images. *Journal of Image and Vision Computing*, 21(7):651–661, 2003.
192. P. Buddharaju, I. Pavlidis, and I. Kakadiaris. Face recognition in the thermal infrared spectrum. In *Proceedings of the Joint IEEE Workshop on Object Tracking and Classification Beyond the Visible Spectrum*, Washington, DC, June 2004.
193. J. Heo, S.G. Kong, B.R. Abidi, and M.A. Abidi. Fusion of visual and thermal signatures with eyeglass removal for robust face recognition. In *Proceedings of the Joint IEEE Workshop on Object Tracking and Classification Beyond the Visible Spectrum*, Washington, DC, June 2004.
194. A. Gyaourova, G. Bebis, and I. Pavlidis. Fusion of infrared and visible images for face recognition. In *Proceedings of the 8th European Conference on Computer Vision*, Prague, Czech Republic, May 2004.
195. D.A. Socolinsky and A. Selinger. Thermal face recognition in an operational scenario. In *Proceedings of the IEEE Computer Society Conference on Computer Vision and Pattern Recognition*, volume 2, Washington, DC, June 2004, pp 1012–1019.
196. J.G. Wang, E. Sung, and R. Venkateswarlu. Registration of infrared and visible-spectrum imagery for face recognition. In *Proceedings of the 6th IEEE International Conference on Automatic Face and Gesture Recognition*, Seoul, Korea, May 2004, pp 638–644.
197. X. Chen, P. Flynn, and K. Bowyer. IR and visible light face recognition. *Computer Vision and Image Understanding*, 99(3):332–358, 2005.
198. S.G. Kong, J. Heo, B.R. Abidi, J. Paik, and M.A. Abidi. Recent advances in visual and infrared face recognition – a review. *Computer Vision and Image Understanding*, 97(1):103–135, 2005.
199. Lin C.L. and Fan K.C. Biometric verification using thermal images of palm-dorsa vein patterns. *IEEE Transactions on Circuits and Systems for Video Technology*, 14(2):199–213, 2004.
200. S.K. Im, H.S. Choi, and S.W. Kim. A direction-based vascular pattern extraction algorithm for hand vascular pattern verification. *ETRI Journal*, 25(2):101–108, 2003.
201. T. Shimooka and K. Shimizu. Artificial immune system for personal identification with finger vein pattern. In *Proceedings of the 8th International Conference on Knowledge-Based Intelligent Information and Engineering Systems. Lecture Notes in Computer Science*, volume 3214, September 2004, pp 511–518.
202. N. Miura, A. Nagasaka, and T. Miyatake. Feature extraction of finger vein patterns based on iterative line tracking and its application to personal identification. *Systems and Computers in Japan*, 35(7):61–71, 2004.
203. F.J. Prokoski and R. Riedel. Infrared identification of faces and body parts. In *Biometrics: Personal Identification in Networked Society*. Dordecht: Kluwer, 1998 (Chapter 9).
204. P. Buddharaju, I.T. Pavlidis, and P. Tsiamyrtzis. Physiology-based face recognition. In *Proceedings of the IEEE Advanced Video and Signal Based Surveillance*, Como, Italy, September 2005.

205. C. Manohar. *Extraction of Superficial Vasculature in Thermal Imaging.* Master's Thesis, University of Houston, Houston, TX, December 2004.
206. B.J. Moxham, C. Kirsh, B. Berkovitz, G. Alusi, and T. Cheeseman. *Interactive Head and Neck (CD-ROM).* London: Primal Pictures, 2002.
207. I. Pavlidis, P. Tsiamyrtzis, C. Manohar, and P. Buddharaju. Biometrics: face recognition in thermal infrared. In *Biomedical Engineering Handbook.* Boca Raton: CRC, 2006.
208. L. Di Stefano and A. Bulgarelli. Simple and efficient connected components labeling algorithm. In *Proceedings of the International Conference on Image Analysis and Processing,* Venice, Italy, September 1999, pp 322–327.
209. D. Maltoni, D. Maio, A.K. Jain, and S. Prabhakar. *Handbook of Fingerprint Recognition.* Berlin Heidelberg New York: Springer, 2003.
210. M.A. Oliveira and N.J. Leite. Reconnection of fingerprint ridges based on morphological operators and multiscale directional information. In *Proceedings of 17th Brazilian Symposium on Computer Graphics and Image Processing,* Curitiba, PR, Brazil, October 2004, pp 122–129.
211. B.K. Jang and R.T. Chin. One-pass parallel thinning: analysis, properties, and quantitative evaluation. *IEEE Transactions on Pattern Analysis and Machine Intelligence,* 14(11):1129–1140, 1992.
212. S. Yang and I.M. Verbauwhede. A secure fingerprint matching technique. In *Proceedings of the 2003 ACM SIGMM Workshop on Biometrics Methods and Applications,* Berkley, CA, 2003, pp 89–94.
213. X. Jiang and W.Y. Yau. Fingerprint minutiae matching based on the local and global structures. In *Proceedings of the 15th International Conference on Pattern Recognition,* volume 2, Barcelona, Catalonia, Spain, September 2000, pp 1038–1041.
214. M. Turk and A.P. Pentland. Eigenfaces for recognition. *Journal of Cognitive Neuroscience,* 3(1):71–86, 1991.
215. W. Zhao, R. Chellappa, and A. Rosenfeld. Face recognition: a literature survey. *ACM Computing Surveys,* 35:399–458, 2003.
216. P.N. Belhumeur, J.P. Hespanha, and D.J. Kriegman. Eigenfaces vs. fisherfaces: recognition using class specific linear projection. *IEEE Transactions on Pattern Analysis and Machine Intelligence ,* 19:711–720, 1997.
217. M. Turk and A. Pentland. Eigenfaces for recognition. *Journal of Cognitive Neuroscience,* 3(1):71–86, 1991.
218. A. Pentland, B. Moghaddam, and T. Starner. View-based and modular eigenspaces for face recognition. In *Proceedings of IEEE Conference on Computer Vision and Pattern Recognition (CVPR),* 1994, pp 84–91.
219. W. Zhao, A. Krishnaswamy, R. Chellappa, D. Swets, and L. Weng. Discriminant analysis of principal components for face recognition. In *Proceedings of IEEE International Conference on Automatic Face and Gesture Recognition,* 1998, pp 336–341.
220. J. Huang, P. Yuen, C. Chen, W. Sheng, and J.H. Lai. Kernel subspace LDA with optimized kernel parameters on face recognition. In *Proceedings of IEEE International Conference on Automatic Face and Gesture Recognition,* 2004, pp 327–332.
221. M.-H. Yang, N. Ahuja, and D. Kriegman. Face recognition using kernel eigenfaces. *Advances in NIPS 14,* 2002.
222. M.-H. Yang, N. Ahuja, and D. Kriegman. Kernel eigenfaces vs. kernel fisherfaces: face recognition using kernel methods. In *Proceedings of IEEE International Conference on Automatic Face and Gesture Recognition (FGR),* 2002, pp 215–220.
223. J. Yang, Z. Jin. J.-Y. Yang, D. Zhang, and A.F. Frangi. Essence of kernel fisher discriminant: KPCA plus LDA. *Pattern Recognition,* 37(10):2097–2100, 2004.

224. D.A. Socolinsky and A. Selinger. Thermal face recognition in an operational scenario. In *Proceedings of IEEE Conference on Computer Vision and Pattern Recognition (CVPR)*, 2004, pp 1012–1019.
225. X. Chen, P. Flynn, and K. Bowyer. IR and visible light face recognition. *Computer Vision and Image Understanding*, 99(3):332–358, 2005.
226. S. Singh, A. Gyaourova, G. Bebis, and I. Pavlidis. Infrared and visible image fusion for face recognition. In *Proceedings of the International Society for Optical Engineering, Biometric Technology for Human Identification*, 2004, pp 585–596.
227. A.V. Oppenheim and J.S. Lim. The importance of phase in signals. In *Proceedings of IEEE*, volume 69, 1981, pp 529–541.
228. P. Kovesi. Edges are not just steps. In *Proceedings of Asian Conference on Computer Vision (ACCV)*, 2002, pp 23–25.
229. 3dMD (http://www.3dmd.com/), 2006.
230. R. Gottumukkal and V. Asari. An improved face recognition technique based on modular PCA approach. *Pattern Recognition Letters*, 25(4):429–436, 2004.
231. H. Li, B.S. Manjunath, and S.K. Mitra. Multi-sensor image fusion using the wavelet transform. In *Proceedings of IEEE International Conference on Image Processing (ICIP)*, volume 1, 1994, pp 51–55.
232. S. Davis and P. Mermelstein. Comparison of parametric representations for monosyllabic word recognition. *IEEE Transactions on Acoustics, Speech, and Signal Processing*, 28(4):357–366, 1980.
233. D. Reynolds, T. Quatieri, and R. Dunn. Speaker verification using adapted Gaussian mixture models. *Digital Signal Processing*, 10(1–3):19–41, 2000.
234. A. Park and T. Hazen. ASR dependent techniques for speaker identification. In *Proceedings of International Conference on Spoken Language Processing*, Denver, CO, September 2002, pp 1337–1340.
235. A. Park and T. Hazen. A comparison of normalization and training approaches for ASR-dependent speaker identification. In *Proceedings of International Conference on Spoken Language Processing*, Jeju Island, Korea, October 2004.
236. W. Zhao, R. Chellappa, P. Phillips, and A. Rosenfeld. Face recognition: a literature survey. *ACM Computing Surveys*, 35(4):399–458, 2004.
237. B. Heisele, P. Ho, and T. Poggio. Face recognition with support vector machines: global versus component-based approach. In *Proceedings of International Conference on Computer Vision*, volume 2, Vancouver, Canada, July 2001, pp 688–694.
238. P. Viola and M. Jones. Rapid object detection using a boosted cascade of simple features. In *Proceedings of IEEE Computer Society Conference on Computer Vision and Pattern Recognition*, Kauai, HI, December 2001, pp 511–518.
239. B. Heisele, T. Serre, S. Prentice, and T. Poggio. Hierarchical classification and feature reduction for fast face detection with support vector machines. *Pattern Recognition*, 36:2007–2017, 2003.
240. V. Vapnik. *The Nature of Statistical Learning Theory*. Berlin Heidelberg New York: Springer, 1995.
241. T. Hazen, E. Weinstein, and A. Park. Towards robust person recognition on handheld devices using face and speaker identification technologies. In *Proceedings of International Conference on Multimodal Interfaces*, Vancouver, Canada, November 2003.
242. T. Hazen, E. Weinstein, R. Kabir, A. Park, and B. Heisele. Multi-modal face and speaker identification on a handheld device. In *Proceedings of the Workshop on Multimodal User Authentication*, Santa Barbara, CA, December 2003.

243. T. Hazen. Visual model structures and synchrony constraints for audio–visual speech recognition. *IEEE Transactions on Audio, Speech and Language Processing*, 14(3):1082–1089, 2006.
244. E. Weinstein, P. Ho, B. Heisele, T. Poggio, K. Steele, and A. Agarwal. Handheld face identification technology in a pervasive computing environment. In *Short Paper Proceedings, Pervasive 2002*, Zurich, Switzerland, August 2002, pp 48–54.
245. S. McKenna and S. Gong. Recognising moving faces. In H. Wechsler, P. Phillips, V. Bruce, F. Soulie, and T. Huang (Eds.). *Face Recognition: From Theory to Applications*. Berlin Heidelberg New York: Springer, 1998, pp 578–588.
246. T. Hazen, K. Saenko, C. La, and J. Glass. A segment-based audio–visual speech recognizer: data collection, development and initial experiments. In *Proceedings of International Conference on Multimodal Interfaces*, State College, PA, October 2004.
247. C. Bregler and Y. Konig. Eigenlips for robust speech recognition. In *Proceedings of International Conference on Acoustics, Speech, and Signal Processing*, volume 2, Adelaide, Australia, April 1998, pp 669–672.
248. J. Ming and F. Smith. A posterior union model for improved robust speech recognition in nonstationary noise. In *Proceedings of International Conference on Acoustics, Speech, and Signal Processing*, volume 1, Hong Kong, April 2003, pp 420–423.
249. J. Ming, T. Hazen, and J. Glass. Speaker verification over handheld devices with realistic noisy speech data. In *Proceedings of International Conference on Acoustics, Speech, and Signal Processing*, Toulouse, France, May 2006.
250. R. Woo, A. Park, and T. Hazen. The MIT mobile device speaker verification corpus: data collection and preliminary experiments. In *Proceedings of Odyssey, The Speaker & Language Recognition Workshop*, San Juan, PR, June 2006.
251. P.J. Phillips, A. Martin, C.L. Wilson, and M. Przybocki. An introduction to evaluating biometric systems. *IEEE Computer*, 33(2):56–63, 2000.
252. P.J. Phillips, P. Grother, R.J. Micheals, D.M. Blackburn, E. Tabassi, and J.M. Bone. FRVT 2002: evaluation report. Technical Report, March 2003.
253. P.J. Phillips, P.J. Flynn, T. Scruggs, K.W. Bowyer, J. Chang, K. Hoffman, J. Marques, J. Min, and W. Worek. Overview of the face recognition grand challenge. In *Proceedings of IEEE Computer Vision and Pattern Recognition*, San Diego, CA, 20–25 June 2005, pp 947–954.
254. Face recognition vendor test 2006 (http://www.frvt.org/FRVT2006/).
255. X. Lu and A.K. Jain. Deformation modeling for robust 3D face matching. In *Proceedings of IEEE Computer Vision and Pattern Recognition*, New York, NY, 17–22 June 2006, pp 1377–1383.
256. S. Wang, Y. Wang, M. Jin, X. Gu, and D. Samaras. 3D surface matching and recognition using conformal geometry. In *Proceedings of IEEE Computer Vision and Pattern Recognition*, New York, NY, 17–22 June 2006, pp 2453–2460.
257. T. Russ, C. Boehnen, and T. Peters. 3D face recognition using 3D alignment for PCA. In *Proceedings of IEEE Computer Vision and Pattern Recognition*, New York, NY, 17–22 June 2006, pp 1391–1398.
258. W.Y. Lin, K.C. Wong, N. Boston, and Y.H. Hu. Fusion of summation invariants in 3D human face recognition. In *Proceedings of IEEE Computer Vision and Pattern Recognition*, New York, NY, 17–22 June 2006, pp 1369–1376.
259. M. Husken, M. Brauckmann, S. Gehlen, and C. von der Malsburg. Strategies and benefits of fusion of 2D and 3D face recognition. In *Proceedings of IEEE Workshop on Face Recognition Grand Challenge Experiments*, San Diego, CA, 20–25 June 2005, pp 174–181.

260. T. Maurer, D. Guigonis, I. Maslov, B. Pesenti, A. Tsaregorodtsev, D. West, and G. Medioni. Performance of Geometrix ActiveID™ 3D face recognition engine on the FRGC data. In *Proceedings of IEEE Workshop on Face Recognition Grand Challenge Experiments*, San Diego, CA, 20–25 June 2005.
261. A.M. Bronstein, M.M. Bronstein, and R. Kimmel. Three-dimensional face recognition. *International Journal of Computer Vision*, 64(1):5–30, 2005.
262. A.M. Bronstein, M.M. Bronstein, and R. Kimmel. Robust expression-invariant face recognition from partially missing data. In *Proceedings of European Conference on Computer Vision*, Graz, Austria, May 2006, pp 7–13.
263. A. Iannarelli. *Ear Identification.* Forensic Identification Series. Fremont, CA: Paramount, 1989.
264. M. Burge and W. Burger. Ear biometrics for machine vision. In *Proceedings of 21th Workshop Austrian Association for Pattern Recognition*, Hallstatt, May 1997, pp 822–826.
265. K.I. Chang, K.W. Bowyer, S. Sarkar, and B. Victor. Comparison and combination of ear and face images in appearance-based biometrics. *IEEE Transactions on Pattern Analysis and Machine Intelligence*, 25(9):1160–1165, 2003.
266. P. Yan and K.W. Bowyer. Empirical evaluation of advanced ear biometrics. In *Proceedings of IEEE Computer Vision and Pattern Recognition*, San Diego, CA, 2005, pp 41–48.
267. P. Yan and K.W. Bowyer. Multi-biometrics 2D and 3D ear recognition. In *Proceedings of Audio- and Video-Based Biometric Person Authentication*, Rye Brook, NY, 2005, pp 503–512.
268. P. Yan, K.W. Bowyer, and K.I. Chang. ICP-based approaches for 3D ear recognition. In *Proceedings of SPIE Biometric Technology for Human Identification II*, volume 5779, Orlando, FL, 2005, pp 282–291.
269. P. Yan and K.W. Bowyer. Biometric recognition using three-dimensional ear shape. Technical Report 1, University of Notre Dame, 2006.
270. L. Farkas. *Anthropometry of the Head and Face.* New York: Raven, 1994.
271. I.A. Kakadiaris, M. Papadakis, L. Shen, D. Kouri, and D.K. Hoffman. m-HDAF multi-resolution deformable models. In *Proceedings of 14th International Conference on Digital Signal Processing*, Santorini, Greece, 1–3 July 2002, pp 505–508.
272. I.A. Kakadiaris, L. Shen, M. Papadakis, D. Kouri, and D.K. Hoffman. g-HDAF multi-resolution deformable models for shape modeling and reconstruction. In *Proceedings of British Machine Vision Conference*, Cardiff, UK, 2–5 September 2002, pp 303–312.
273. X. Gu, S. Gortler, and H. Hoppe. Geometry images. In *Proceedings of SIGGRAPH*, San Antonio, TX, July 2002, pp 355–361.
274. I.A. Kakadiaris, G. Passalis, T. Theoharis, G. Toderici, I. Konstantinidis, and N. Murtuza. Multimodal face recognition: combination of geometry with physiological information. In *Proceedings of IEEE Computer Vision and Pattern Recognition*, volume 2, San Diego, CA, 20–25 June 2005, pp 1022–1029.
275. A. Johnson. *Spin-Images: A Representation for 3-D Surface Matching.* Ph.D. Thesis, Robotics Institute, Carnegie Mellon University, Pittsburgh, PA, August 1997.
276. P.J. Besl and N.D. McKay. A method for registration of 3-D shapes. *IEEE Transactions on Pattern Analysis and Machine Intelligence*, 14(2):239–256, 1992.
277. G. Turk and M. Levoy. Zippered polygon meshes from range images. In *Proceedings of SIGGRAPH*, Orlando, FL, 1994, pp 311–318.
278. D. Chetverikov, D. Svirko, D. Stepanov, and P. Krsek. The trimmed iterative closest point algorithm. In *Proceedings of International Conference on Pattern Recognition*, volume 3, Quebec, Canada, 2002, pp 545–548.

279. G. Papaioannou, E.A. Karabassi, and T. Theoharis. Reconstruction of three-dimensional objects through matching of their parts. *IEEE Transactions on Pattern Analysis and Machine Intelligence*, 24(1):114–124, 2002.
280. D. Metaxas and I.A. Kakadiaris. Elastically adaptive deformable models. *IEEE Transactions on Pattern Analysis and Machine Intelligence*, 24(10):1310–1321, 2002.
281. C. Mandal, H. Qin, and B. Vemuri. Dynamic smooth subdivision surfaces for data visualization. In *Proceedings of IEEE Visualization*, October 1997, pp 371–377.
282. C. Mandal, H. Qin, and B.C. Vemuri. A novel FEM-based dynamic framework for subdivision surfaces. *Computer-Aided Design*, 32(8–9):479–497, 2000.
283. D. Zorin and P. Schroeder. Subdivision for modeling and animation. In *SIGGRAPH Course Notes*, New Orleans, LA, 2000.
284. J. Warren and H. Weimer. *Subdivision Methods for Geometric Design: A Constructive Approach*. Series in Computer Graphics. Los Altos, CA: Morgan Kaufmann, 2001.
285. R.A. Finkel and J.L. Bentley. Quad trees. A data structure for retrieval of composite keys. *Acta Informatica*, 4(1):1–9, 1974.
286. I.A. Kakadiaris, G. Passalis, G. Toderici, N. Karampatziakis, N. Murtuza, Y. Lu, and T. Theoharis. Evaluation of 3D face recognition in the presence of facial expressions: an annotated deformable model approach. *IEEE Transactions on Pattern Analysis and Machine Intelligence*, 2006 (in press).
287. I.A. Kakadiaris, G. Passalis, G. Toderici, N. Karampatziakis, N. Murtuza, Y. Lu, and T. Theoharis. Expression-invariant multispectral face recognition: you can smile now! In *Proceedings of SPIE Defense and Security Symposium*, Orlando, FL, 17–21 April 2006.
288. J. Portilla and E.P. Simoncelli. A parametric texture model based on joint statistic of complex wavelet coefficients. *International Journal of Computer Vision*, 40:49–71, 2000.
289. E.P. Simoncelli, W.T. Freeman, E.H. Adelson, and D.J. Heeger. Shiftable multi-scale transforms. *IEEE Trans Information Theory*, 38:587–607, 1992.
290. Z. Wang and E.P. Simoncelli. Translation insensitive image similarity in complex wavelet domain. In *Proceedings of IEEE International Conference on Acoustics, Speech and Signal Processing*, volume 2, Philadelphia, PA, March 2005, pp 573–576.
291. Z. Wang, A.C. Bovik, H.R. Sheikh, and E.P. Simoncelli. Image quality assessment: from error visibility to structural similarity. *IEEE Transactions on Image Processing*, 13(4):600–612, 2004.
292. Intel. Open computer vision library (http://sourceforge.net/projects/opencvlibrary/), 2005.
293. X. Chen, T. Faltemier, P.J. Flynn, and K.W. Bowyer. Human face modeling and recognition through multi-view high resolution stereopsis. In *Proceedings of IEEE Computer Society Workshop on Biometrics*, New York, NY, 17–18 June 2006, pp 50–55.
294. H. Chen and B. Bhanu. Shape model-based 3D ear detection from side face range images. In *Proceedings of IEEE Computer Vision and Pattern Recognition*, Washington, DC, 2005, p 122.
295. H. Chen, B. Bhanu, and R. Wang. Performance evaluation and prediction for 3D ear recognition. In *Proceedings of Audio- and Video-Based Biometric Person Authentication*, Rye Brook, NY, 2005, pp 748–757.
296. A. Kale, A. Roy-chowdhury, and R. Chellappa. Fusion of gait and face for human identification. In *Proceedings of Acoustics, Speech, and Signal Processing*, volume 5, 2004, pp 901–904.

297. G. Shakhnarovich and T. Darrell. On probabilistic combination of face and gait cues for identification. In *Proceedings of Automatic Face and Gesture Recognition*, volume 5, 2002, pp 169–174.
298. G. Shakhnarovich, L. Lee, and T. Darrell. Integrated face and gait recognition from multiple views. In *Proceedings of Computer Vision and Pattern Recognition*, volume 1, 2001, pp 439–446.
299. X. Zhou, B. Bhanu, and J. Han. Human recognition at a distance in video by integrating face profile and gait. In *Proceedings of Audio- and Video-Based Biometric Person Authentication*, 2005, pp 533–543.
300. B. Bhanu and X. Zhou. Face recognition from face profile using dynamic time warping. In *Proceedings of International Conference on Pattern Recognition*, volume 4, 2004, pp 499–502.
301. J. Han and B. Bhanu. Individual recognition using gait energy image. *IEEE Transactions Pattern Analysis and Machine Intelligence*, 28(2):316–322, 2006.
302. J. Kittler, M. Hatef, R. Duin, and J. Matas. On combining classifiers. *IEEE Transactions Pattern Analysis and Machine Intelligence*, 20:226–239, 1998.
303. Y. Zuev and S. Ivanon. The voting as a way to increase the decision reliability. *Journal of the Franklin Institute*, 336(2):361–378, 1999.
304. J. Han and B. Bhanu. Performance prediction for individual recognition by gait. *Pattern Recognition*, 2005(5):615–624, 2005.
305. R.Y. Tsai and T.S. Huang. Multiframe image restoration and registration. In *Advances in Computer Vision and Image Processing*. Greenwich, CT: JAI, 1984.
306. P.A. Hewitt and D. Dobberfuhl. The science and art of proportionality. *Science Scope*, 30–31, 2004.
307. M. Irani and S. Peleg. Motion analysis for image enhancement: resolution, occlusion and transparency. *Journal of Visual Communication and Image Representation*, 4:324–335, 1993.
308. L.D. Harmon and W.F. Hunt. Automatic recognition of human face profiles. *Computer Graphics and Image Processing*, 6:135–156, 1977.
309. L.D. Harmon, M.K. Khan, R. Lasch, and P.F. Ramig. Machine identification of human faces. *Pattern Recognition*, 13:97–110, 1981.
310. D. O'Mara. *Automated Facial Metrology*. Ph.D. Thesis, The University of Western Australia.
311. J.C. Campos, A.D. Linney, and J.P. Moss. The analysis of facial profiles using scale space techniques. *Pattern Recognition*, 26:819–824, 1993.
312. B. Dariush, S.B. Kang, and K. Waters. Spatiotemporal analysis of face profiles: detection, segmentation, and registration. In *Proceedings of IEEE Conference on Automatic Face and Gesture Recognition*, 1998, pp 248–253.
313. T. Akimoto, Y. Suenaga, and R.S. Wallace. Automatic creation of 3d facial models. *IEEE Transactions Computer Graphics and Applications*, 13:16–22, 1993.
314. F. Galton. Numeralised profiles for classification and recognition. *Nature*, 83:127–130, 1910.
315. E. Keogh. Exact indexing of dynamic time warping. In *Proceedings of Conference on Very Large Data Bases*, August 2002.
316. S.B. Needleman and C.D. Wunsch. A general method applicable to the search for similarities in the amino acid sequences of two proteins. *Journal of Molecular Biology*, 48:443–453, 1977.
317. J.J. Little and J.E. Boyd. Recognizing people by their gait: the shape of motion. *Videre: Journal of Computer Vision Research*, 1(2):1–32, 1998.

318. A. Sundaresan, A. Roy-Chowdhury, and R. Chellappa. A hidden Markov model based framework for recognition of humans from gait sequences. In *Proceedings of International Conference Image Processing*, volume 2, 2003, pp 93–96.
319. P.S. Huang, C.J. Harris, and M.S. Nixon. Recognizing humans by gait via parameteric canonical space. *Artificial Intelligence in Engineering*, 13:359–366, 1999.
320. S. Sarkar, P.J. Phillips, Z. Liu, I.R. Vega, P. Grother, and K.W. Bowyer. The humanid gait challenge problem: data sets, performance, and analysis. *IEEE Transactions Pattern Analysis and Machine Intelligence*, 27(2):162–177, 2005.
321. R.T. Collins, R. Gross, and J. Shi. Silhouette-based human identification from body shape and gait. In *Proceedings of IEEE Conference on Automatic Face and Gesture Recognition*, 2002, pp 351–356.
322. X. Zhou and B. Bhanu. Feature fusion of face and gait for human recognition at a distance in video. In *Proceedings of IEEE International Conference on Pattern Recognition*, 2006.
323. X. Zhou and B. Bhanu. Integrating face and gait for human recognition. In *Proceedings of Workshop on Biometrics held in conjunction with the IEEE Conference on Computer Vision and Pattern Recognition*, 2006, pp 55–62.
324. R. Bolle, J. Connell, S. Pankanti, N. Ratha, and A. Senior. *Guide to Biometrics*. Berlin Heidelberg New York: Springer, 2003.
325. A.K. Jain, R. Bolle, and S. Pankanti (Eds.). *Biometrics: Personal Identification in Networked Society*. Dordecht: Kluwer, 1999.
326. A.K. Jain, A. Ross, and S. Prabhakar. An introduction to biometric recognition. *IEEE Transactions on Circuits and Systems for Video Technology. Special Issue on Image- and Video-Based Biometrics*, 14(1):4–20, 2004.
327. R.O. Duda, P.E. Hart, and D.G. Stork. *Pattern Classification*. New York: Wiley, 2001.
328. E. Rood and A.K. Jain. Biometric research agenda: report of the NSF workshop. In *Workshop for a Biometric Research Agenda*, Morgantown, WV, July 2003.
329. A.K. Jain, S. Pankanti, S. Prabhakar, L. Hong, and A. Ross. Biometrics: a grand challenge. In *Proceedings of International Conference on Pattern Recognition (ICPR)*, volume 2, Cambridge, UK, August 2004, pp 935–942.
330. M. Golfarelli, D. Maio, and D. Maltoni. On the error-reject tradeoff in biometric verification systems. *IEEE Transactions on Pattern Analysis and Machine Intelligence*, 19(7):786–796, 1997.
331. A. Eriksson and P. Wretling. How flexible is the human voice? A case study of mimicry. In *Proceedings of the European Conference on Speech Technology*, Rhodes, 1997, pp 1043–1046.
332. W.R. Harrison. *Suspect Documents: Their Scientific Examination*. Chicago, IL: Nelson-Hall, 1981.
333. T. Matsumoto, H. Matsumoto, K. Yamada, and S. Hoshino. Impact of artificial gummy fingers on fingerprint systems. In *Optical Security and Counterfeit Deterrence Techniques IV, Proceedings of SPIE*, volume 4677, San Jose, CA, January 2002, pp 275–289.
334. T. Putte and J. Keuning. Biometrical fingerprint recognition: don't get your fingers burned. In *Proceedings of IFIP TC8/WG8.8 Fourth Working Conference on Smart Card Research and Advanced Applications*, 2000, pp 289–303.
335. N.K. Ratha, J.H. Connell, and R.M. Bolle. An analysis of minutiae matching strength. In *Proceedings of 3rd International Conference on Audio- and Video-Based Biometric Person Authentication (AVBPA)*, Halmstad, Sweden, June 2001, pp 223–228.
336. L. Hong, A.K. Jain, and S. Pankanti. Can multibiometrics improve performance? In *Proceedings of IEEE Workshop on Automatic Identification Advanced Technologies (AutoID)*, Summit, NJ, October 1999, pp 59–64.

337. A.K. Jain and A. Ross. Multibiometric systems. *Communications of the ACM. Special Issue on Multimodal Interfaces*, 47(1):34–40, 2004.
338. A. Ross, K. Nandakumar, and A.K. Jain. *Handbook of multibiometrics*, 1st edition. Berlin Heidelberg New York: Springer, 2006.
339. D. Maio, D. Maltoni, R. Cappelli, J.L. Wayman, and A.K. Jain. FVC2004: third fingerprint verification competition. In *Proceedings of International Conference on Biometric Authentication (ICBA)*, Hong Kong, China, July 2004, pp 1–7.
340. C. Wilson, A.R. Hicklin, M. Bone, H. Korves, P. Grother, B. Ulery, R. Micheals, M. Zoepfl, S. Otto, and C. Watson. Fingerprint vendor technology evaluation 2003: summary of results and analysis report. NIST Technical Report NISTIR 7123, National Institute of Standards and Technology, June 2004.
341. P.J. Phillips, P. Grother, R.J. Micheals, D.M. Blackburn, E. Tabassi, and J.M. Bone. FRVT2002: overview and summary. Available at http://www.frvt.org/FRVT2002, March 2003.
342. M. Przybocki and A. Martin. NIST speaker recognition evaluation chronicles. In *Odyssey: The Speaker and Language Recognition Workshop*, Toledo, Spain, May 2004, pp 12–22.
343. International Biometric Group. Independent testing of iris recognition technology: final report. Available at http://www.biometricgroup.com/reports/public/ITIRT.html, May 2005.
344. T.N. Palmer. Predicting uncertainty in forecasts of weather and climate. *Reports on Progress in Physics*, 63:71–116, 2000.
345. G. Vachtsevanos, L. Tang, and J. Reimann. An intelligent approach to coordinated control of multiple unmanned aerial vehicles. In *Proceedings of 60th Annual Forum of the American Helicopter Society*, Baltimore, MD, June 2004.
346. R.S. Blum and Z. Liu (Eds.). *Multi-Sensor Image Fusion and Its Applications*. Boca Raton, FL: CRC/Taylor & Francis Group, 2006.
347. M.A. Abidi and R.C. Gonzalez. *Data Fusion in Robotics and Machine Intelligence*. New York: Academic, 1992.
348. A.H. Gunatilaka and B.A. Baertlein. Feature-level and decision-level fusion of noncoincidentally sampled sensors for land mine detection. *IEEE Transactions on Pattern Analysis and Machine Intelligence*, 23(6):577–589, 2001.
349. D.H. Wolpert. Stacked generalization. Technical Report LA-UR-90-3460, Los Alamos National Laboratory, 1990.
350. H. Drucker, C. Cortes, L.D. Jackel, Y. LeCun, and V. Vapnik. Boosting and other ensemble methods. *Neural Computation*, 6(6):1289–1301, 1994.
351. H. Bunke and A. Kandel (Eds.). *Hybrid Methods in Pattern Recognition*, volume 47 of *Machine Perception and Artificial Intelligence*. Singapore: World Scientific, 2002.
352. M. Tan. Multi-agent reinforcement learning: independent vs. cooperative learning. In M.N. Huhns and M.P. Singh (Eds.). *Readings in Agents*. San Francisco: CA: Morgan Kaufmann, 1997, pp 487–494.
353. K. Woods, K. Bowyer, and W.P. Kegelmeyer. Combination of multiple classifiers using local accuracy estimates. *IEEE Transactions on Pattern Analysis and Machine Intelligence*, 19(4):405–410, 1997.
354. J.A. Benediktisson and P.H. Swain. Consensus theoretic classification methods. *IEEE Transactions on Systems, Man and Cybernetics*, 22(4):688–704, 1992.
355. S.S. Iyengar, L. Prasad, and H. Min. *Advances in Distributed Sensor Technology*. Englewood Cliffs, NJ: Prentice-Hall, 1995.
356. R.A. Jacobs, M.I. Jordan, S.J. Nowlan, and G.E. Hinton. Adaptive mixtures of local experts. *Neural Computation*, 3(1):79–87, 1991.

266   References

357. C. Chiang and H. Fu. A divide-and-conquer methodology for modular supervised neural network design. In *Proceedings of World Congress on Computational Intelligence*, Orlando, FL, June 1994, pp 119–124.
358. K.J. Arrow. *Social Choice and Individual Values*, 2nd edition. New York: Wiley, 1963.
359. L.I. Kuncheva. *Combining Pattern Classifiers – Methods and Algorithms*. New York: Wiley, 2004.
360. J. Kittler, M. Hatef, R.P. Duin, and J.G. Matas. On combining classifiers. *IEEE Transactions on Pattern Analysis and Machine Intelligence*, 20(3):226–239, 1998.
361. L. Xu, A. Krzyzak, and C.Y. Suen. Methods for combining multiple classifiers and their applications to handwriting recognition. *IEEE Transactions on Systems, Man, and Cybernetics*, 22(3):418–435, 1992.
362. J. Ghosh. Multiclassifier systems: back to the future. In *Proceedings of 3rd International Workshop on Multiple Classifier Systems*, Cagliari, Italy, June 2002, pp 1–15.
363. C.C. Chibelushi, F. Deravi, and J.S. Mason. Voice and facial image integration for speaker recognition. In R.I. Damper, W. Hall, and J.W. Richards (Eds.). *Multimedia Technologies and Future Applications*. London: Pentech, 1994, pp 155–161.
364. R. Brunelli and D. Falavigna. Person identification using multiple cues. *IEEE Transactions on Pattern Analysis and Machine Intelligence*, 17(10):955–966, 1995.
365. J. Lee, B. Moghaddam, H. Pfister, and R. Machiraju. Finding optimal views for 3D face shape modeling. In *Proceedings of the IEEE International Conference on Automatic Face and Gesture Recognition (FG)*, Seoul, Korea, May 2004, pp 31–36.
366. A. Kong, J. Heo, B. Abidi, J. Paik, and M. Abidi. Recent advances in visual and infrared face recognition – a review. *Computer Vision and Image Understanding*, 97(1):103–135, 2005.
367. X. Chen, P.J. Flynn, and K.W. Bowyer. IR and visible light face recognition. *Computer Vision and Image Understanding*, 99(3):332–358, 2005.
368. D.A. Socolinsky, A. Selinger, and J.D. Neuheisel. Face recognition with visible and thermal infrared imagery. *Computer Vision and Image Understanding*, 91(1–2):72–114, 2003.
369. R.K. Rowe and K.A. Nixon. Fingerprint enhancement using a multispectral sensor. In *Proceedings of SPIE Conference on Biometric Technology for Human Identification II*, volume 5779, March 2005, pp 81–93.
370. Z. Pan, G. Healey, M. Prasad, and B. Tromberg. Face recognition in hyperspectral images. *IEEE Transactions on Pattern Analysis and Machine Intelligence*, 25(12):1552–1560, 2003.
371. G.L. Marcialis and F. Roli. Fingerprint verification by fusion of optical and capacitive sensors. *Pattern Recognition Letters*, 25(11):1315–1322, 2004.
372. A. Ross, A.K. Jain, and J. Reisman. A hybrid fingerprint matcher. *Pattern Recognition*, 36(7):1661–1673, 2003.
373. X. Lu, Y. Wang, and A.K. Jain. Combining classifiers for face recognition. In *IEEE International Conference on Multimedia and Expo (ICME)*, volume 3, Baltimore, MD, July 2003, pp 13–16.
374. S. Prabhakar and A.K. Jain. Decision-level fusion in fingerprint verification. Technical Report MSU-CSE-00-24, Michigan State University, October 2000.
375. J. Jang, K.R. Park, J. Son, and Y. Lee. Multi-unit iris recognition system by image check algorithm. In *Proceedings of International Conference on Biometric Authentication (ICBA)*, Hong Kong, July 2004, pp 450–457.
376. H. Hill, P.G. Schyns, and S. Akamatsu. Information and viewpoint dependence in face recognition. *Cognition*, 62(2):201–222, 1997.

377. A. O'Toole, H. Bulthoff, N. Troje, and T. Vetter. Face recognition across large viewpoint changes. In *Proceedings of the International Workshop on Automatic Face- and Gesture-Recognition (IWAFGR)*, Zurich, Switzerland, June 1995, pp 326–331.
378. U. Uludag, A. Ross, and A.K. Jain. Biometric template selection and update: a case study in fingerprints. *Pattern Recognition*, 37(7):1533–1542, 2004.
379. C.C. Chibelushi, J.S.D. Mason, and F. Deravi. Feature-level data fusion for bimodal person recognition. In *Proceedings of the 6th International Conference on Image Processing and Its Applications*, volume 1, Dublin, Ireland, July 1997, pp 399–403.
380. E.S. Bigun, J. Bigun, B. Duc, and S. Fischer. Expert conciliation for multimodal person authentication systems using Bayesian statistics. In *Proceedings of 1st International Conference on Audio- and Video-Based Biometric Person Authentication (AVBPA)*, Crans-Montana, Switzerland, March 1997, pp 291–300.
381. K.I. Chang, K.W. Bowyer, and P.J. Flynn. An evaluation of multimodal 2D+3D face biometrics. *IEEE Transactions on Pattern Analysis and Machine Intelligence*, 27(4):619–624, 2005.
382. National Institute of Standards and Technology. NIST biometric scores set. Available at http://www.itl.nist.gov/iad/894.03/biometricscores, 2004.
383. A.T.B. Jin, D.N.C. Ling, and A. Goh. An integrated dual factor authenticator based on the face data and tokenised random number. In *Proceedings of 1st International Conference on Biometric Authentication*, Hong Kong, China, July 2004, pp 117–123.
384. A.K. Jain, K. Nandakumar, X. Lu, and U. Park. Integrating faces, fingerprints and soft biometric traits for user recognition. In *Proceedings of ECCV International Workshop on Biometric Authentication (BioAW)*, volume LNCS 3087, Prague, Czech Republic. Berlin Heidelberg New York: Springer, May 2004, pp 259–269.
385. C. Sanderson and K.K. Paliwal. Information fusion and person verification using speech and face information. Research Paper IDIAP-RR 02-33, IDIAP, September 2002.
386. A.K. Jain and A. Ross. Fingerprint mosaicking. In *IEEE International Conference on Acoustics, Speech, and Signal Processing (ICASSP)*, volume 4, Orlando, FL, May 2002, pp 4064–4067.
387. N.K. Ratha, J.H. Connell, and R.M. Bolle. Image mosaicing for rolled fingerprint construction. In *Proceedings of 14th International Conference on Pattern Recognition (ICPR)*, volume 2, Brisbane, Australia, August 1998, pp 1651–1653.
388. Y.-L. Zhang, J. Yang, and H. Wu. A hybrid swipe fingerprint mosaicing scheme. In *Proceedings of 5th International Conference on Audio- and Video-Based Biometric Person Authentication (AVBPA)*, Rye Brook, NY, July 2005, pp 131–140.
389. K. Choi, H. Choi, and J. Kim. Fingerprint mosaicking by rolling and sliding. In *Proceedings of 5th International Conference on Audio- and Video-Based Biometric Person Authentication (AVBPA)*, Rye Brook, NY, July 2005, pp 260–269.
390. A. Ross, S. Shah, and J. Shah. Image versus feature mosaicing: a case study in fingerprints. In *Proceedings of SPIE Conference on Biometric Technology for Human Identification III*, Orlando, FL, April 2006, pp 620208-1–620208-12.
391. F. Yang, M. Paindavoine, H. Abdi, and A. Monopoli. Development of a fast panoramic face mosaicking and recognition system. *Optical Engineering*, 44(8):087005-1–087005-10, 2005.
392. X. Liu and T. Chen. Geometry-assisted statistical modeling for face mosaicing. In *Proceedings of IEEE International Conference on Image Processing (ICIP)*, volume 2, Barcelona, Spain, September 2003, pp 883–886.
393. X. Liu and T. Chen. Pose-robust face recognition using geometry assisted probabilistic modeling. In *Proceedings of IEEE Computer Society Conference on Computer Vision and Pattern Recognition (CVPR)*, volume 1, San Diego, CA, June 2005, pp 502–509.

394. T. Sim, S. Baker, and M. Bsat. The CMU pose, illumination, and expression database. *IEEE Transactions on Pattern Analysis and Machine Intelligence*, 25(12):1615–1618, 2003.
395. R. Singh, M. Vatsa, A. Ross, and A. Noore. Performance enhancement of 2D face recognition via mosaicing. In *Proceedings of the 4th IEEE Workshop on Automatic Identification Advanced Technologies (AuotID)*, Buffalo, NY, October 2005, pp 63–68.
396. R.-L. Hsu. *Face Detection and Modeling for Recognition*. Ph.D. Thesis, Department of Computer Science and Engineering, Michigan State University, 2002.
397. F.I. Parke and K. Waters. *Computer Facial Animation*. Wellesley, MA: A.K. Peters, 1996.
398. A.K. Jain, R.P.W. Duin, and J. Mao. Statistical pattern recognition: a review. *IEEE Transactions on Pattern Analysis and Machine Intelligence*, 22(1):4–37, 2000.
399. P. Pudil, J. Novovicova, and J. Kittler. Floating search methods in feature selection. *Pattern Recognition Letters*, 15(11):1119–1124, 1994.
400. A.K. Jain and B. Chandrasekaran. Dimensionality and sample size considerations in pattern recognition practice. In P.R. Krishnaiah and L.N. Kanal (Eds.). *Handbook of Statistics*, volume 2. Amsterdam: North-Holland, 1982, pp 835–855.
401. A.K. Jain and D. Zongker. Feature selection: evaluation, application, and small sample performance. *IEEE Transactions on Pattern Analysis and Machine Intelligence*, 19(2):153–158, 1997.
402. A. Ross and R. Govindarajan. Feature level fusion using hand and face biometrics. In *Proceedings of SPIE Conference on Biometric Technology for Human Identification II*, volume 5779, Orlando, FL, March 2005, pp 196–204.
403. A. Kumar and D. Zhang. Biometric recognition using feature selection and combination. In *Proceedings of 5th International Conference on Audio- and Video-Based Biometric Person Authentication (AVBPA)*, Rye Brook, NY, July 2005, pp 813–822.
404. B. Son and Y. Lee. Biometric authentication system using reduced joint feature vector of iris and face. In *Proceedings of 5th International Conference on Audio- and Video-Based Biometric Person Authentication (AVBPA)*, Rye Brook, NY, July 2005, pp 513–522.
405. A. Kumar, D.C.M. Wong, H.C. Shen, and A.K. Jain. Personal verification using palmprint and hand geometry biometric. In *Proceedings of 4th International Conference on Audio- and Video-Based Biometric Person Authentication (AVBPA)*, Guildford, UK, June 2003, pp 668–678.
406. D.W. Scott. *Multivariate Density Estimation: Theory, Practice and Visualization*. Wiley Series in Probability and Statistics. New York: Wiley-Interscience, 1992.
407. S.C. Dass, K. Nandakumar, and A.K. Jain. A principled approach to score level fusion in multimodal biometric systems. In *Proceedings of 5th International Conference on Audio- and Video-Based Biometric Person Authentication (AVBPA)*, Rye Brook, NY, July 2005, pp 1049–1058.
408. R.B. Nelsen. *An Introduction to Copulas*. Berlin Heidelberg New York: Springer, 1999.
409. U. Cherubini, E. Luciano, and W. Vecchiato. *Copula Methods in Finance*. New York: Wiley, 2004.
410. R. Cappelli, D. Maio, and D. Maltoni. Combining fingerprint classifiers. In *Proceedings of 1st International Workshop on Multiple Classifier Systems*, Cagliari, Italy, June 2000, pp 351–361.
411. R. Snelick, U. Uludag, A. Mink, M. Indovina, and A.K. Jain. Large scale evaluation of multimodal biometric authentication using state-of-the-art systems. *IEEE Transactions on Pattern Analysis and Machine Intelligence*, 27(3):450–455, 2005.

412. F.R. Hampel, P.J. Rousseeuw, E.M. Ronchetti, and W.A. Stahel. *Robust Statistics: The Approach Based on Influence Functions*. New York: Wiley, 1986.
413. F. Mosteller and J.W. Tukey. *Data Analysis and Regression: A Second Course in Statistics*. Reading, MA: Addison-Wesley, 1977.
414. P. Verlinde and G. Cholet. Comparing decision fusion paradigms using k-NN based classifiers, decision trees and logistic regression in a multi-modal identity verification application. In *Proceedings of 2nd International Conference on Audio- and Video-Based Biometric Person Authentication (AVBPA)*, Washington, DC, March 1999, pp 188–193.
415. S. Pigeon and L. Vandendrope. M2VTS multimodal face database release 1.00. Available at http://www.tele.ucl.ac.be/PROJECTS/M2VTS/m2fdb.html, 1996.
416. V. Chatzis, A.G. Bors, and I. Pitas. Multimodal decision-level fusion for person authentication. *IEEE Transactions on Systems, Man, and Cybernetics, Part A: Systems and Humans*, 29(6):674–681, 1999.
417. S. Ben-Yacoub, Y. Abdeljaoued, and E. Mayoraz. Fusion of face and speech data for person identity verification. *IEEE Transactions on Neural Networks*, 10(5):1065–1075, 1999.
418. K. Messer, J. Matas, J. Kittler, J. Luettin, and G. Maitre. XM2VTSDB: the extended M2VTS database. In *Proceedings of 2nd International Conference on Audio- and Video-Based Biometric Person Authentication (AVBPA)*, Washington, DC, March 1999, pp 72–77.
419. Y. Wang, T. Tan, and A.K. Jain. Combining face and iris biometrics for identity verification. In *Proceedings of 4th International Conference on Audio- and Video-Based Biometric Person Authentication (AVBPA)*, Guildford, UK, June 2003, pp 805–813.
420. A. Ross and A.K. Jain. Information fusion in biometrics. *Pattern Recognition Letters*, 24(13):2115–2125, 2003.
421. T.K. Ho, J.J. Hull, and S.N. Srihari. Decision combination in multiple classifier systems. *IEEE Transactions on Pattern Analysis and Machine Intelligence*, 16(1):66–75, 1994.
422. A. Agresti. *An Introduction to Categorical Data Analysis*. New York: Wiley, 1996.
423. J. Daugman. Combining multiple biometrics. Available at http://www.cl.cam.ac.uk/users/jgd1000/combine/combine.html, 2000.
424. L. Lam and C.Y. Suen. Application of majority voting to pattern recognition: an analysis of its behavior and performance. *IEEE Transactions on Systems, Man, and Cybernetics, Part A: Systems and Humans*, 27(5):553–568, 1997.
425. Y.S. Huang and C.Y. Suen. Method of combining multiple experts for the recognition of unconstrained handwritten numerals. *IEEE Transactions on Pattern Analysis and Machine Intelligence*, 17(1):90–94, 1995.
426. L.I. Kuncheva, C.J. Whitaker, C.A. Shipp, and R.P.W. Duin. Limits on the majority vote accuracy in classifier fusion. *Pattern Analysis and Applications*, 6(1):22–31, 2003.
427. P. Domingos and M. Pazzani. On the optimality of the simple Bayesian classifier under zero-one loss. *Machine Learning*, 29(2–3):103–130, 1997.
428. G. Rogova. Combining the results of several neural network classifiers. *Neural Networks*, 7(5):777–781, 1994.
429. L.I. Kuncheva, J.C. Bezdek, and R.P.W. Duin. Decision templates for multiple classifier fusion: an experimental comparison. *Pattern Recognition*, 34(2):299–314, 2001.
430. L.I. Kuncheva, C.J. Whitaker, C.A. Shipp, and R.P.W. Duin. Is independence good for combining classifiers? In *Proceedings of International Conference on Pattern Recognition (ICPR)*, volume 2, Barcelona, Spain, 2000, pp 168–171.

431. R. Sharma, V.I. Pavlovic, and T.S. Huang. Toward multimodal human–computer interface. *Proceedings of the IEEE*, 86(5):853–869, 1998.
432. A.K. Jain, R. Bolle, and S. Pankanti (Eds.). *Introduction to Biometrics: Personal Identification in Networked Society*. Dordecht: Kluwer, 1999.
433. J. Daugman. The importance of being random: statistical principles of iris recognition. *Pattern Recognition*, 36(2):279–292, 2003.
434. A.K. Jain, L. Hong, and R. Bolle. On-line fingerprint verification. *IEEE Transactions Pattern Analysis and Machine Intelligence*, 19(4):302–313, 1997.
435. P.N. Belhumeur, J.P. Hespanha, and D.J. Kriegman. Eigenfaces vs fisherfaces: recognition using class specific linear projection. *IEEE Transactions on Pattern Analysis and Machine Intelligence*, 19(7):711–720, 1997.
436. M. Turk and A. Pentland. Eigenfaces for recognition. *Journal of Cognitive Neuroscience*, 3(1):71–86, 1991.
437. S.Z. Li and A.K. Jain (Eds.). *Handbook on Face Recognition*. Berlin Heidelberg New York: Springer, 2005.
438. A.K. Jain, A. Ross, and S. Pankanti. A prototype hand geometry-based verification system. In *International Conference on Audio- and Video-Based Biometric Person Authentication (AVBPA)*, Washington, DC, 1999, pp 166–171.
439. N.A. Schmid and J.A. O'Sullivan. Thresholding method for reduction of dimensionality. *IEEE Transactions on Information Theory*, 47(7):2903–2920, 2001.
440. J. Daugman. High confidence visual recognition of persons by a test of statistical independence. *IEEE Transactions on Pattern Analysis and Machine Intelligence*, 15(11):1148–1161, 1993.
441. W. Shen, M. Surrette, and R. Khanna. Evaluation of automated biometrics-based identification and verification systems. *Proceedings of the IEEE. Special Issue on Automated Biometric Systems*, 85(10):1464–1478, 1997.
442. Special Issue on Automated Biometric Systems. *Proceedings of the IEEE*, 85(9):1341–1516, 1997.
443. R.P. Wildes. Iris recognition: an emerging biometric technology. *Proceedings of the IEEE. Special Issue on Automated Biometric Systems*, 85(9):1347–1363, 1997.
444. A.K. Jain and A. Ross. Learning user-specific parameters in multibiometric system. In *International Conference on Image Processing*, Rochester, NY, September 2002, pp 57–60.
445. A.K. Jain and A. Ross. Multibiometric systems. *Communications of the ACM*, 47(1):34–40, 2004.
446. A.K. Jain, L. Hong, and Y. Kulkarni. A multimodal biometric system using fingerprint, face, and speech. In *International Conference on Audio- and Video-Based Biometric Person Authentication (AVBPA)*, Washington, DC, 1999, pp 182–187.
447. S. Pakanti, S. Prabhakar, and A.K. Jain. On the individuality of fingerprint. *IEEE Transactions on Pattern Analysis and Machine Intelligence*, 24(8):1010–1025, 2002.
448. Biometrics: the future of identification. *IEEE Computer Magazine*, 33(2):46–81, 2000.
449. J.L. Wayman, Error-rate equations for the general biometric system. *IEEE Robotics and Automation Magazine*, 6(1):35–48, 1999.
450. C.C. Leang and D.H. Johnson. On the asymptotic of m-hypothesis Bayesian detection. *IEEE Transactions on Information Theory*, 43(1):280–282, 1997.
451. N.A. Schmid and J.A. O'Sullivan. Performance prediction methodology for biometric systems using a large deviations approach. *IEEE Transactions on Signal Processing, Supplement on Secure Media*, 52(10):3036–3045, 2004.
452. R.M. Gray. *Entropy and Information Theory*. Berlin Heidelberg New York: Springer, 1990.

453. J.A. O'Sullivan and N.A. Schmid. Performance analysis of physical signature authentication. *IEEE Transactions on Information Theory*, 47(7):3034–3039, 2001.
454. J.A. Bucklew. *Large Deviation Techniques in Decision, Simulation, and Estimation.* New York: Wiley, 1990.
455. A. Dembo and O. Zeitouni. *Large Deviations Techniques and Applications.* Berlin Heidelberg New York: Springer, 1998.
456. J.A. O'Sullivan, R.E. Blahut, and D.L. Snyder. Information theoretic image formation. *IEEE Transactions on Information Theory*, 44(6):2094–2123, 1998.
457. J.A. O'Sullivan, R.E. Blahut, and D.L. Snyder. Information theoretic image formation. *IEEE Transactions on Information Theory*, 44(6):2094–2123, 1998.
458. T.M. Cover and J.A. Thomas. *Elements of Information Theory.* New York: Wiley, 1991.
459. R.M. Gray. *Toeplitz and Circulant Matrices: A Review.* Electrical Engineering Department, Stanford University, Stanford, CA 94305, 2000 (http://www-ee.stanford.edu/~gray/toeplitz.pdf).
460. J.R. Hoinville, R.S. Indeck, and M.W. Muller. Spatial noise phenomena of longitudinal magnetic recording media. *IEEE Transactions Magnetics*, 28:3398–3406, 1992.
461. R.S. Indeck and E. Glavinas. Fingerprinting magnetic media. *IEEE Transactions Magnetics*, 29(6):4095–4097, 1993.

Index

AEM, 158
alignment, 144
    ESA, 145
    ICP, 145
    spin images, 145
AND rule, 208
anisotropic diffusion, 96
annotated model
    AM, 144
Area under the curve, 198
automation, 156

Bayes decision theory, 201
Bayes formula, 201
Bayes rule, 210
Bayesian decision fusion, 209
Bayesian framework, 93
Belief function, 210
Biocode, 192
biometric
    fusion, 75, 77, 78, 80–83, 85, 87, 90
biometric fusion, 178
Biweight estimators, 205
black top-hat segmentation, 98
Borda count method, 207
Branch-and-bound search, 198

Capacity of a template, 187
Challenge–response mechanism, 189
challenges, 140
classifier combination, 177
Conditional independence, 210
Confusion matrix, 209
Continuous monitoring, 189
Copula fusion rule, 202
Correlation, 212
Curse of dimensionality, 192, 197
curvature, 172

data
    preprocessing, *see* band-pass filter
Database Indexing, *see* Database Filtering
Decimal scaling, 203
Decision profile, 210
Decision template, 211
Degree of belief, 210
Dempster–Shafer theory of evidence, 210
Dimensionality reduction, 197, 198
Discriminant function, 209, 210
Double sigmoid normalization, 204
DTW, 173
Dual-factor authentication, 192
DWT, 115
dynamic programming matrix, 174
Dynamic Time Warping, 173

ear algorithm, 159
ear issues, 157
ear model, 158
efficiency, 156
eigenspace, 102
enrollment, 142
ESA, 159

Face, 18
face
    appearance, 75, 80, 87–90
    database, 43, 46–49, 84, 86
    quality assessment/measure, 45, 55
    recognition, 44, 50, 56
Face modeling, 196
face profile, 166
face profile recognition, 170
face recognition, biometrics,
    information fusion, 1
Fault tolerance, 190
Feature normalization, 197

Feature selection, 198
Feature transformation, 198
features
    holistic, 78, 80, 85
    local, 78–81, 85
fiducial points, 171
filter
    band-pass, 75, 78–80, 85, 87, 88, 90
Filtering, 193
fitting, 145
FRGC, 140, 150, 164
Fusion
    classifier-based, 206
    decision-level, 208
    density-based, 201
    feature-level, 196
    rank-level, 207
    score-level, 200
    sensor-level, 193
    transformation-based, 202
fusion, 103, 115

Gait Energy Image, 176
gait recognition, 175
GEI, 176
Generalized density, 202
geometry image, 146
glasses, 75, 78, 83–87, 90

Hampel influence function, 205
high-resolution image, 167
Highest rank method, 207
Hybrid systems, 192

IAFIS, 185
ICP, 159
Illumination, 13
image
    deblur, 44, 45, 54–56
    denoise, 54, 55
    enhancement, 44, 55
    set matching, 75, 77, 78, 83, 87
    sharpness measure, 44, 45, 50–55
image fusion, 114
Indexing, 193
indexing-verification scheme, 178
Iterative Closest Point algorithm, 194

Likelihood ratio, 201
Linear Discriminant Analysis, 199
Logistic function, 204
Logistic regression, 207
low-resolution image, 167

M2VTS multimodal database, 206
MAD, *see* Median absolute deviation
magnification
    blur, 44, 49, 53, 54, 56
    high magnification, 43, 44, 47–49, 52, 54
Majority voting, 208
Median absolute deviation, 198, 204
Median normalization, 197, 204
metadata, 146
metrics, 148
    CW-SSIM, 148
    fusion, 148
    haar, 148
Min-max normalization, 197, 202
Mosaicing, 193
    face, 195
Motion, 13
Multi-algorithm systems, 190
Multi-modal systems, 192
Multi-sample systems, 191
Multi-sensor systems, 190
Multi-unit systems, *see* Multi-instance systems
Multiple classifier systems, 188
multispectral face recognition, 92

Naive Bayes rule, 210
Noise, 189
normalization, 176, 178

observation distance, 43, 44, 46–48
    long range, 43, 44
occlusion
    glasses, *see* glasses
OR rule, 208

Pattern recognition system, 185
preprocessing, 143
    hole filling, 143
    median cut, 143
    smoothing, 143
    subsampling, 143
Principal Component Analysis, 102, 199

Product rule, 202
pyramid, 148

Recognition, 18
recognition, 142
recognition metric, 179
results
    expressions, 152, 155
    multisensor, 152
    transforms, 152

Score normalization, 202
Sequential backward floating search, 198
Sequential backward selection, 198
Sequential forward floating search, 198
Sequential forward selection, 198
set
    matching, *see* image set matching
spectrum
    infrared, *see* thermal spectrum
    thermal, 75, 76, 84, 86–90
    visual, 75, 84, 86–90
Spoof attacks, 189
storage space, 156
Sum rule, 202

superficial blood vessels, 96
surveillance, 43–45, 48

Tanh normalization, 204
thermal infrared, 91
Thermal Minutia Points, 99
transforms
    haar, 147
    pyramid, 147
Two-quadrics, 204

UH, 150, 160, 161, 164
UND, 160, 161, 164

vascular network, 92
verification, 85, 88–90
Video, 18
voting, 114

wavelet, 114
Weighted majority voting, 209

XM2VTS multimodal database, 206

z-score normalization, 203